ALMOST LIKE A WHALE

THE ORIGIN OF SPECIES UPDATED

STEVE JONES

Doubleday

LONDON · NEW YORK · TORONTO · SYDNEY · AUCKLAND

TRANSWORLD PUBLISHERS LTD
61–63 Uxbridge Road, London W5 5SA

TRANSWORLD PUBLISHERS
c/o Random House Australia Pty Ltd
20 Alfred Street, Milsons Point, NSW 2061, Australia

TRANSWORLD PUBLISHERS
c/o Random House New Zealand
18 Poland Road, Glenfield, Auckland, New Zealand

TRANSWORLD PUBLISHERS
c/o Random House Pty Ltd
Endulini, 5a Jubilee Road, Parktown 2193, South Africa

Published 1999 by Doubleday
a division of Transworld Publishers Ltd

A catalogue record for this book is available from the British Library.
ISBN 0385 409850

Typeset in 12/13½pt Granjon by Falcon Oast Graphic Art

Printed in Great Britain
by Clays Ltd, St Ives plc

To Alex and Anna Trench

CONTENTS

THE ORIGIN OF SPECIES: FACSIMILE TITLE PAGE AND LIST OF CONTENTS............ix

AN HISTORICAL SKETCH OF THE PROGRESS OF OPINION
ON THE ORIGIN OF SPECIES...xvii

INTRODUCTION..1

I   VARIATION UNDER DOMESTICATION...23
II   VARIATION UNDER NATURE...43
III   STRUGGLE FOR EXISTENCE...59
IV   NATURAL SELECTION..74
V   LAWS OF VARIATION...110
VI   DIFFICULTIES ON THEORY..128
VII INSTINCT..156
VIII HYBRIDISM..183
IX   ON THE IMPERFECTION OF THE GEOLOGICAL RECORD..........................206
X   ON THE GEOLOGICAL SUCCESSION OF ORGANIC BEINGS.......................231
XI   GEOGRAPHICAL DISTRIBUTION..254
XII GEOGRAPHICAL DISTRIBUTION – CONTINUED.....................................278
XIII MUTUAL AFFINITIES OF ORGANIC BEINGS; MORPHOLOGY;
       EMBRYOLOGY; RUDIMENTARY ORGANS...297
INTERLUDE: ALMOST LIKE A WHALE?...333
XIV RECAPITULATION AND CONCLUSION...356

FURTHER READING...380

INDEX...389

ON

THE ORIGIN OF SPECIES

BY MEANS OF NATURAL SELECTION,

OR THE

PRESERVATION OF FAVOURED RACES IN THE STRUGGLE
FOR LIFE.

By CHARLES DARWIN, M.A.,

FELLOW OF THE ROYAL, GEOLOGICAL, LINNÆAN, ETC., SOCIETIES;
AUTHOR OF ' JOURNAL OF RESEARCHES DURING H. M. S. BEAGLE'S VOYAGE
ROUND THE WORLD.'

LONDON:

JOHN MURRAY, ALBEMARLE STREET.

1859.

CONTENTS.

INTRODUCTION .. .. Page 1

CHAPTER I.

VARIATION UNDER DOMESTICATION.

Causes of Variability — Effects of Habit — Correlation of Growth — Inheritance — Character of Domestic Varieties — Difficulty of distinguishing between Varieties and Species—Origin of Domestic Varieties from one or more Species — Domestic Pigeons, their Differences and Origin — Principle of Selection anciently followed, its Effects — Methodical and Unconscious Selection — Unknown Origin of our Domestic Productions — Circumstances favourable to Man's power of Selection 7–43

CHAPTER II.

VARIATION UNDER NATURE.

Variability — Individual differences — Doubtful species — Wide ranging, much diffused, and common species vary most—Species of the larger genera in any country vary more than the species of the smaller genera—Many of the species of the larger genera resemble varieties in being very closely, but unequally, related to each other, and in having restricted ranges 44–59

CONTENTS.

CHAPTER III.

STRUGGLE FOR EXISTENCE.

Bears on natural selection—The term used in a wide sense—Geometrical powers of increase — Rapid increase of naturalised animals and plants—Nature of the checks to increase—Competition universal — Effects of climate — Protection from the number of individuals—Complex relations of all animals and plants throughout nature—Struggle for life most severe between individuals and varieties of the same species; often severe between species of the same genus—The relation of organism to organism the most important of all relations .. Page 60-79

CHAPTER IV.

NATURAL SELECTION.

Natural Selection — its power compared with man's selection — its power on characters of trifling importance — its power at all ages and on both sexes — Sexual Selection — On the generality of intercrosses between individuals of the same species — Circumstances favourable and unfavourable to Natural Selection, namely, intercrossing, isolation, number of individuals — Slow action — Extinction caused by Natural Selection — Divergence of Character, related to the diversity of inhabitants of any small area, and to naturalisation — Action of Natural Selection, through Divergence of Character and Extinction, on the descendants from a common parent — Explains the Grouping of all organic beings 80-130

CHAPTER V.

LAWS OF VARIATION.

Effects of external conditions — Use and disuse, combined with natural selection; organs of flight and of vision — Acclimatisation — Correlation of growth — Compensation and economy of growth — False correlations — Multiple, rudimentary, and lowly organised structures variable — Parts developed in an unusual manner are highly variable: specific characters more variable than generic: secondary sexual characters variable — Species of the same genus vary in an analogous manner — Reversions to long-lost characters — Summary 131-170

CONTENTS.

CHAPTER VI.

DIFFICULTIES ON THEORY.

Difficulties on the theory of descent with modification—Transitions—
Absence or rarity of transitional varieties—Transitions in habits
of life—Diversified habits in the same species—Species with
habits widely different from those of their allies—Organs of
extreme perfection—Means of transition—Cases of difficulty—
Natura non facit saltum—Organs of small importance—Organs
not in all cases absolutely perfect—The law of Unity of Type
and of the Conditions of Existence embraced by the theory of
Natural Selection   ..   ..   Page 171–206

CHAPTER VII.

INSTINCT.

Instincts comparable with habits, but different in their origin —
Instincts graduated — Aphides and ants — Instincts variable —
Domestic instincts, their origin — Natural instincts of the cuckoo,
ostrich, and parasitic bees — Slave-making ants — Hive-bee, its
cell-making instinct — Difficulties on the theory of the Natural
Selection of instincts — Neuter or sterile insects — Summary
207–244

CHAPTER VIII.

HYBRIDISM.

Distinction between the sterility of first crosses and of hybrids —
Sterility various in degree, not universal, affected by close inter-
breeding, removed by domestication—Laws governing the sterility
of hybrids — Sterility not a special endowment, but incidental
on other differences — Causes of the sterility of first crosses and
of hybrids — Parallelism between the effects of changed conditions of life and crossing — Fertility of varieties when crossed
and of their mongrel offspring not universal — Hybrids and
mongrels compared independently of their fertility — Summary
245–278

CONTENTS.

CHAPTER IX.

ON THE IMPERFECTION OF THE GEOLOGICAL RECORD.

On the absence of intermediate varieties at the present day — On the nature of extinct intermediate varieties ; on their number — On the vast lapse of time, as inferred from the rate of deposition and of denudation — On the poorness of our palæontological collections — On the intermittence of geological formations — On the absence of intermediate varieties in any one formation — On the sudden appearance of groups of species — On their sudden appearance in the lowest known fossiliferous strata

Page 279–311

CHAPTER X.

ON THE GEOLOGICAL SUCCESSION OF ORGANIC BEINGS.

On the slow and successive appearance of new species — On their different rates of change — Species once lost do not reappear — Groups of species follow the same general rules in their appearance and disappearance as do single species — On Extinction — On simultaneous changes in the forms of life throughout the world — On the affinities of extinct species to each other and to living species — On the state of development of ancient forms — On the succession of the same types within the same areas — Summary of preceding and present chapters   312–345

CHAPTER XI.

GEOGRAPHICAL DISTRIBUTION.

Present distribution cannot be accounted for by differences in physical conditions — Importance of barriers — Affinity of the productions of the same continent — Centres of creation — Means of dispersal, by changes of climate and of the level of the land, and by occasional means — Dispersal during the Glacial period co-extensive with the world 346–382

CONTENTS.

CHAPTER XII.

GEOGRAPHICAL DISTRIBUTION—*continued*.

Distribution of fresh-water productions — On the inhabitants of oceanic islands — Absence of Batrachians and of terrestrial Mammals — On the relation of the inhabitants of islands to those of the nearest mainland — On colonisation from the nearest source with subsequent modification — Summary of the last and present chapters .. .. Page 383–410

CHAPTER XIII.

MUTUAL AFFINITIES OF ORGANIC BEINGS: MORPHOLOGY: EMBRYOLOGY: RUDIMENTARY ORGANS.

CLASSIFICATION, groups subordinate to groups — Natural system — Rules and difficulties in classification, explained on the theory of descent with modification — Classification of varieties — Descent always used in classification — Analogical or adaptive characters — Affinities, general, complex and radiating — Extinction separates and defines groups — MORPHOLOGY, between members of the same class, between parts of the same individual — EMBRYOLOGY, laws of, explained by variations not supervening at an early age, and being inherited at a corresponding age — RUDIMENTARY ORGANS; their origin explained — Summary
411–458

CHAPTER XIV.

RECAPITULATION AND CONCLUSION.

Recapitulation of the difficulties on the theory of Natural Selection — Recapitulation of the general and special circumstances in its favour — Causes of the general belief in the immutability of species — How far the theory of natural selection may be extended — Effects of its adoption on the study of Natural history — Concluding remarks 459–490

INDEX 491

'How extremely stupid not to have thought of that!'

T. H. HUXLEY, on reading *The Origin of Species*

'If a man will begin with certainties, he shall end in doubts; but if he will be content to begin with doubts, he shall end in certainties.'

BACON: *The Advancement of Learning*

AN HISTORICAL SKETCH OF THE PROGRESS OF OPINION ON THE ORIGIN OF SPECIES

Two of the worst of all lines of English poetry, written in 1799 by John Hookham Frere:

> 'The feather'd race with pinions skim the air –
> Not so the mackerel, and still less the bear!'

However poor that verse, it has a moral. The lines come from Hookham Frere's somewhat neglected poem 'The Progress of Man; Poetry of the Anti-Jacobin'. Birds, bears and fish carry a political message. Things are as they are and it is folly to change them. The French Revolution disturbed the God-given order: to proclaim the rights of man was as absurd as to suggest that mackerel – or even bears – might fly.

A pair of quotations from *The Origin of Species*, published sixty years later:

'It is conceivable that flying-fish, which now glide far through the air, slightly turning and rising by the aid of their fluttering fins, might have been modified into perfectly winged animals. If this had been effected, who would have ever imagined that in an early transitional state they had been inhabitants of the open ocean, and had used their incipient organs of flight exclusively, as far as we know, to escape being devoured by other fish?'

'In North America the black bear was seen by Hearne swimming for hours with widely open mouth, thus catching, almost like a whale, insects in the water ... we might expect, on my theory, that such individuals would occasionally have given rise to new species, having anomalous habits, and with their structure either slightly or considerably modified from that of their proper type.'

Thus Charles Darwin on what evolution might do. Although it had not yet made a whale-bear from a land animal with a taste for aquatic insects, it could. After all, evolution had already produced such improbable creatures as a fish that skimmed the air. The idea of life as fixed in a divine mould was dead. Instead, all was change.

Before Darwin, the great majority of naturalists believed that species were immutable productions, and had been separately created. Today, his theory that they undergo modification and are the descendants of pre-existing forms is accepted by everyone (or by everyone not determined to disbelieve it). Most people would, if asked, find it hard to explain why. We all know that men and chimps are relatives and that plants, animals and everything else descend from a common ancestor. The struggle for existence, the survival of the fittest and the origin of species are wisdom of the most conventional kind. Evolution happened: and, in 1996, even the Pope agreed (although he would admit only that 'new knowledge leads us to recognize in the theory of evolution more than a hypothesis').

All this is rather like Galileo and the earth's journey around the sun. We know he was right, but what was his evidence? Why, in his dispute with the Vatican, did he mutter *'eppur si muove'* – 'but still it moves'? In the absence of high technology his proof was subtle. It involved the movement of 'wandering stars' – planets – against a background of fixed bodies deep in space. The pattern made sense only if the Earth itself was a planet and not a stationary object around which the sky turned. Such evidence, persuasive though it might be, is unknown to most of those who believe his ideas.

Evolution is much the same. Although the notion is as simple as that of the solar system, it is not the obvious explanation of how the world works. Common sense tells us that life – like the Sun – revolves around ourselves. The idea has but one fault: it is wrong.

Satellites make it impossible to deny the structure of the universe. In the same way, genes are a triumphant proof of the fact of evolution. Darwin's theory of common descent does for biology what Galileo did for the planets. It was laid out in a book written for the general reader, the only bestseller to change man's conception of himself. An idea put forward in 1859 is still the cement that binds the marvellous discoveries of today. *The Origin of Species* is, without doubt, the book of the millennium.

Nowhere else is the case for evolution better put. Darwin called his work 'one long argument'. Modern biologists can but agree. To read his four hundred pages is to be amazed by how well their reasoning accommodates each new breakthrough as it appears. *The Origin*'s logic is as powerful today as it was when it was written.

Its facts, however, are those of a century and a half ago and leave many gaps before its case can be considered proven. All – or nearly all – have now been filled. My own book brings Darwin up to date. It is, as far as is possible, an attempt to rewrite *The Origin of Species*. I use its plan, developing as it does from farms to fossils, from bee-hives to islands, as a framework, but my own Grand Facts (another phrase beloved of Darwin) are set firmly in the late twentieth century. *Almost like a Whale* tries to read Charles Darwin's mind with the benefit of scientific hindsight and to show how the theory of evolution unites biology as his millennium draws to an end.

Evolution has become more than a science, as its ideas are used, wittingly or otherwise, in economics, politics, history, art and more. No educated person can afford to ignore them; and they have no excuse for doing so, as parts of the Darwinian story are so well told in the many excellent books that describe aspects of his theory for a general audience. No book, though, tells it all; there is no modern and non-technical treatment of evolutionary biology, in all its depth and vast diversity. To rewrite *The Origin of Species* is more than most biologists would dare, but daunting (and in some ways hope-less) though that task might be, I attempt it in this book. If an apology is called for, I make it here.

The main difficulty in achieving my goal has been to know what to leave out. *The Origin* marks the foundation of modern biology, and of large parts of geology and psychology as well. I have attempted to reflect its breadth and have, no doubt, failed. My own volume tries to use Darwin's logic to illuminate the discoveries of today. It is not a history of evolutionary ideas, or of life; nor a biography of Darwin, or of animals and plants, but an argument that will, I trust, persuade my readers of the truth of evolution. To keep it within bounds (and it is the same length as the original, with much the same division among chapters) I have had to pick and choose, and to omit certain topics in order to allow others a chance.

My standard for inclusion is a mention in Darwin's great work. I allow a marginal exception. That volume contains but one substantial sentence on humans ('light will be thrown on the origin of man and his history'), but so much is now known about our past that I devote my final chapter to the subject. Darwin himself rules out several topics. His book has nothing to do with the origin of the primary mental powers, any more than with life itself. The beginnings of life and of consciousness are also absent from my own pages; and I avoid all reference to moral aspects of the theory, so far as they exist.

It would be presumptuous to present this essay as more than a shadow of its original, in content or in form. *The Origin* is the high point of the literature of fact. Darwin wrote well because he read well. In a single summer, his diary records, he enjoyed *Hamlet*, *Othello*, *Mansfield Park*, *Sense and Sensibility*, Boswell's *Tour of the Hebrides*, the *Arabian Nights* and *Robinson Crusoe*. His own prose is like a Victorian country house. It radiates confidence from whatever direction it is viewed, as literature, as autobiography or as brilliant science.

Compare Darwin's account of his first sight of the Galapagos with that of Herman Melville, whose *The Encantadas* (another name for the islands) was published in 1854. Darwin is vivid and direct: 'In the morning (17th) we landed on Chatham Island, which, like the others, rises with a tame and rounded outline, broken here and there by scattered hillocks, the remains of former craters. Nothing could be less inviting than the first appearance. A broken field of black basaltic lava, thrown into the most rugged waves, and crossed by great fissures, is everywhere covered by stunted, sun-burnt brushwood, which shows little signs of life. The dry and parched surface, being heated by the noon-day sun, gave to the air a close and sultry feeling, like that from a stove: we fancied even that the bushes smelt unpleasantly'.

Melville is, in contrast, feeble: 'But the special curse, as one may call it, of the Encantadas, that which exalts them in desolation above Idumea and the Pole, is that to them change never comes; neither the change of seasons nor of sorrows. Cut by the Equator, they know not autumn, and they know not spring; while, already reduced to the lees of fire, ruin itself can work little more upon them. The showers refresh the deserts, but in these isles rain never falls. Like split Syrian gourds left withering in the sun, they are cracked by an everlasting drought beneath a torrid sky'.

Nobody could copy Darwin's language. I have not tried to do so (although I have filched a few of his sentences in the hope of improving the tone of my own work). As a hint at what is to be gained from reading *The Origin* in the original, I include (where they exist, from his fourth chapter onwards) his summaries, his own list of chapter contents and *The Origin*'s final 'Recapitulation and Conclusion'.

It would be possible – like the Borges character who rewrote *Don Quixote* word for word – to read each sentence of *The Origin of Species* unchanged in a modern context in the hope of revealing meanings unknown to the nineteenth century. I do not propose to do so. Although the structure of its argument is intact and the order of chapters the same, I use Darwin's masterpiece as a scaffold rather than a straitjacket. This is a post-modernist treatment of evolution, with the strengths and weaknesses so implied. Its architecture is of an earlier age, but it is constructed with today's materials. *The Origin* is also, it must be said, a work of high Victorian seriousness, with no concession to any desire to be entertained. In these more flippant times I yield to the temptation to leaven a scientific narrative with tales from the curious history of evolution and those who study it.

Students of evolution face another danger. Boris Vian's mystical novel *Froth on the Daydream* has a character who dedicates his existence to the petrified vomit of Jean-Paul Sartre. Biologists (like Marxists) share that inelegant habit. What, they ask, did the patriarch mean? Could he be wrong, or is it forbidden even to suggest such a heretical idea? To interpret sacred texts has a fascination of its own. I have tried to avoid it.

Darwin is the best biographized of all scientists, and among that dull and sometimes arrogant race stands out as attractive and modest. Even his family has an allure. His great-grandfather

Erasmus published a theory of evolution in heroic couplets and appears in *Frankenstein*: 'The event on which this fiction is founded has been supposed, by Dr Darwin, and some of the physiological writers of Germany, as not of impossible occurrence'. He was a Lunatick, a member of a group that included Joseph Priestley and Josiah Wedgwood. When not engaged in the discovery of oxygen or the introduction of industry into England they designed a machine 'capable of pronouncing the Lord's Prayer, the Creed and the Ten Commandments in the Vulgar Tongue'. Robert Darwin, Charles's father, was a physician who mixed with (and lent money to) the aristocracy and Charles's own grandson Bernard played golf for England.

This book is about Darwin's science, the heart of biology. Its roots are in the past, but it is the key to the present. Its subjects include the AIDS virus and the human race, dog shows and the garbage that floats in the Pacific. Milton, some say, was the last man to know everything (or to know enough about most things to discuss them with authority). Darwin was the last biologist who could claim that. His mind was, he said, 'a machine for grinding general laws out of a large collection of facts'. The hundreds of books and papers referred to in the manuscript of which *The Origin* was to be a 'sketch' include *The Cottage Gardener and Country Gentleman's Companion*, the *India Sporting Review* and the *Philosophical Transactions of the Royal Society of London*. Charles Darwin wrote to scores of people, expert and amateur, in search of information, and wove their lore into his case.

Nobody could do that now. So great is today's knowledge that there are no Miltons even of biology, no one who has sufficient command of the field to debate it with any colleague, from whatever sphere. To understand evolution involves interests so disparate that it is impossible to embrace them all. That is the joy – and the tragedy – of modern science.

Because we now know so much about how life works, evolution has become the science of the exceptions. It would be tedious to consider the feuds about details that consume the subject (although not to do so guarantees that every biologist will find something to dislike in this book). Even so, and bitter as the disputes are, no scientist denies the central truth of *The Origin*, the idea of descent with modification.

Darwin did not have that comfort. He had to persuade an audience unused even to the notion that life could change that it all shared a pedigree. Against much opposition he succeeded. The *Daily Telegraph*, no less, urged its readers to vote against an election candidate who had given a favourable review to *The Origins* and John Ruskin, who had decried his 'filthy heraldries', was forced to say of geology: 'Those dreadful Hammers! I hear the clink of them at the end of every cadence of the Bible verses'. There had been ideas about evolution before Darwin's time, but he was the first to provide not just a mechanism but the proof that it worked.

In spite of his twenty-year search for evidence, Darwin was so conscious of the gaps in his thesis that he might never have made it public. His book is full of apologies: 'To treat this subject at all properly, a long catalogue of dry facts should be given; but these I shall reserve for my future work ... It is hopeless to attempt to convince any one of the truth of this proposition without giving the long array of facts which I have collected, and which cannot possibly be here introduced ... I must here treat the subject with extreme brevity, though I have the materials prepared for an ample discussion'. Today's readers may feel a certain relief that the promised book never appeared. By happy chance, Darwin was stung into publication of a summary of his ideas by an unexpected letter from Alfred Russel Wallace containing the same notion.

Evolution is inevitable. It depends on mistakes in reproduction. Descent always involves modification, because any copy, be it of a picture or a gene, must be less than exact. Information cannot be transmitted without loss, and a duplicate of a copy is, in its turn, less perfect than what went before. To reproduce in succession an original again and again is to make – to evolve – something new. What went in emerges transformed by errors of descent, the raw material of biological change.

That ore is turned into precious metal by natural selection, the furnace within which diversity is tried. Life is a struggle. As more individuals are born than can possibly survive, a grain in the balance will determine which individual shall live and which shall die. The slightest advantage in any one being, at any age or during any season, over those with which it comes into competition, will turn the balance. Natural selection is simple. It picks up inherited differences in the capacity to reproduce. If one version multiplies itself better than others it will take over and, in the end, a new form of life – a new species – will emerge.

Errors of descent are the stuff of evolution. Variation in the ability to copy them – natural selection – gives it a direction. Nature does not favour beauty, or strength, or ferocity; all it can do is to advance those best able to multiply themselves. Although its products include the most beautiful and most repulsive of beings there is no mystery to Darwin's machine: it is no more than genetics plus time.

Darwin's own great weakness was his failure to understand what now seems simple. In 1859, ignorance about inheritance was as general as it had been for a thousand years. In 1726, Mary Toft, of Godalming, saw a rabbit while she was pregnant. Then, she said, she gave birth to one rabbit after another. After the first dozen, George I sent his court anatomist to examine her. In his *Short Narrative of an Extraordinary Delivery of Rabbets* he attested the truth of the tale, and suggested that the animals had jumped down her Fallopian tubes. Mary Toft was soon exposed as a fraud (and became the subject of a ballad by Pope and a sketch by Hogarth). In spite of a noble attempt to transcend such tales, Darwin remained confused. In 1866, Gregor Mendel at last got things right; and his work, in its clarity and elegance, even gained a mention in the *Encyclopaedia Britannica*. Darwin never saw it.

Its importance was not appreciated until the first years of the twentieth century. Darwin had a messy scheme of his own, based in part on the mixing of substances present in the blood. He soon saw that it was flawed. A heredity based on dilution leads any useful character to be thinned as the generations succeed. It would blur any emerging division among species and evolution would stop. Although Darwin agonized about the problem he had no reason to be anxious. Inheritance is based not on liquids but on particles – genes – that can be recovered unchanged at any time. DNA speaks a digital rather than an analog language and even a slight advantage can be summed over the years. Genetics saved *The Origin*, and with its help the study of evolution has been transformed. Its central question – how varieties emerge into species – can now be answered in Mendelian terms.

To modern biologists, species are republics of genes, separated from their neighbours by sexual barriers. Any change in the DNA that is good for all its members – an ability to manage on half the food, or to have twice the number of offspring – will spread to fill the state, but will never get into another one. To define species by genes does not always work, because some are caught in the act of promotion towards an identity of their own and because distance can put a stop to sex even among members of the same one. Even so, to interpret the origin of species in genetic terms is the keystone of the arch that now bridges the ancient gap between the study of inheritance and of evolution.

Darwin's thesis was that the world's variety arose, not from forgotten disasters, but through processes visible today. For him, the present was the key to the past. It led him to believe that evolution is driven by the simple, slow and potent mechanism known as natural selection. As this acts solely by accumulating slight, successive, favourable variations, it can produce no great or sudden modification; it can act only by very short and slow steps. What limit, he asked, can be put to this power, acting during long ages and rigidly scrutinizing the whole constitution, structure, and habits of each creature – favouring the good and rejecting the bad? With such a machine at its disposal, nature had no reason to make leaps.

Geology persuaded Darwin that there was no need to call on ancient cataclysms, be they Biblical floods or gigantic earthquakes, to shape the earth. A tiny stream, given long enough, could carve a giant valley and a shallow sea make, as it dried, a plain a thousand miles across. Could not life be formed in the same way? If landscapes could be transfigured by slow change, so, surely, could flesh. The idea of a universe preserved since the Creation was dead.

That belief made biology into a system of knowledge rather than a set of random facts. Any theory with such ambitions was bound to attract criticism. It did, and, ever since 1859, it has continued to do so.

The Origin is two things: a bold statement of the idea of evolution, and a work of advocacy as to how it took place. It contains a mass of

evidence that makes a compelling case for both the fact of evolutionary change and for one view of how it happened. In his old age, faced
with a wave of inconvenient discoveries, Darwin began to complicate
his ideas. There is an irony in my title. The phrase 'almost like a
whale' comes from the Sixth Edition of *The Origin*, published in
1872. In 1859, Darwin was more confident. His Leviathan was unrestrained: 'I can see no difficulty in a race of bears being rendered,
by natural selection, more aquatic in their structure and habits, with
larger and larger mouths, till a creature was produced as monstrous
as a whale'. I base my own book on the clarity of his First Edition,
but that apologetic 'almost' is at the heart of biology.

The fact of evolution has survived. It is a dance to the music of
time that unites all who join in. Its theme is more elaborate than once
it seemed. Darwinism is often rendered as a dignified waltz to the
melody of natural selection. That is less than fair to its author.
Darwin himself saw that descent with modification could happen in
many ways. Accident might dictate what plants or animals arrive on
an empty island, and some variation might remain as fluctuating
elements, to increase or to be lost by chance. He was, however,
convinced that natural selection has been the main means of
modification. Species, to him, were but varieties writ large, a single
step in the process of slow change that unites biology.

Darwin's great idea – of life as a series of successful mistakes – is
simple; so simple, indeed, that it seems almost impossible that it
could make such complicated things. Its enemies still maintain –
without cease – that evolution is so blind that it could never build
an eye and that to understand the mystery of life (or, at least,
of humankind) must demand mysterious forces from outside the
world of science. Such claims are easy to dismiss, but they cast
their venerable shadow wide. Although all biologists accept the
truth of evolution, some have almost a compulsion to complicate its
ideas.

Darwin's theory has been much revised, but rarely to its advantage. The idea of natural selection as evolution's only agent and of all
change as gradual are each, perhaps, too simple. Faced with the facts
of the 1850s, Darwin complained about 'difficulties so grave that to
this day I can never reflect on them without being staggered'. He has

more to be concerned about today. Charles Darwin's feet, like those of any idol, have been much inspected for signs of clay and a few traces have been found. However, those who hope to replace his ideas often fail to notice quite how well his icon has lasted in the face of a century's explosion of knowledge.

Every evolutionist has, by definition, the benefit of hindsight. Some of Darwin's problems have been solved, others restated in modern terms. To read today's biology is, quite often, to relive the argument of *The Origin* in modified language. There have, of course, been many shifts in opinion, and an evolutionary 'theory of everything' of the kind proclaimed by physics is still far away. Even so, and in spite of the many wonderful facts that illuminate the science, there have been rather few new ideas since his time. As a result, this book emerges (somewhat to my surprise) as a work of refreshing conventionality. Darwin's thesis is perfectly able to support a century and more of scientific advance.

I have never met a biology undergraduate who has read *The Origin of Species*. Even scientists, familiar as they are with its contents (or what they believe them to be), honour it in the breach rather than the observance. It is, though, much studied by students of the humanities as an element of a Philosophy or an English Literature course. There is nothing wrong with that. It was, after all, quick to enter the literary canon: Anna Karenina's last thought before she decides on suicide is, 'Yes, the struggle for existence and hatred is all that holds men together'. Unfortunately, the subject's bible is often presented for what it is not, a work of metaphysics rather than of science. Darwin and Melville both say a lot about whales, but to rely on *The Origin* in a philosophy class is like using *Moby Dick* as a zoology text. One is fact, the other metaphor. The Arts Faculty often finds it hard to tell the difference.

Evolution is to the social sciences as statues are to birds: a convenient platform upon which to deposit badly digested ideas. Humans are, of course, constrained by biological history, as pigs are limited in the ability to fly by their ancestors' lack of wings. We are all branches on a common tree and share descent with primates and, for that matter, with pigs. Biology tells us that we evolved, but when it comes to what makes us human is largely beside the point. There might be inborn drives for rape or for greed, but *Homo sapiens*, unique among animals, need not defer to them. This has not stopped those unable to explain society by other means from pressing evolution into service. Darwinism has been debased since it began by those who use it to support their own creed. It was not the first (and will probably not be the last) science to be abused for political ends.

Bishop Berkeley saw sociology as a branch of physics. In 1713 – soon after Newton's *Principia* – he came up with the view of society as a parallel case of the universe. It was ruled by gravity, by a Law of Moral Force, a 'principle of attraction in the Spirits or Minds of men' that draws them into 'families, friendships, and all the various species of society'. Civilization, like the planets, finds distant objects less attractive than those close at hand. If men are governed by the Earth's attraction (and to jump off a cliff shows that they are), why should society not be so? The Founding Fathers themselves saw in the United States Constitution 'how by the attraction of gravitation the various parts are held in their orbits, and represent in Congress, the judiciary and the President a sort of imitation of the solar system'. Other thinkers of the time preferred what they called a political anatomy, derived from William Harvey (who discovered the circulation of the blood). Two legislative houses were needed, a stronger and a weaker, because the heart had two ventricles, different in size.

Anyone who came up with a planetary or cardiac philosophy of life would today be laughed out of court. Evolution has joined those sciences in the dock. Walter Bagehot, in his *Thoughts on the Application of the Principles of Natural Selection and Inheritance to Political Society*, proposed that 'what was put forward for mere animal history may, with a change of form, but an identical essence, be applied to human history'. Most of what his many successors claim about the same thing does no more than restate the obvious in biological terms. The rest – whatever it might be – is not science.

The Olympian vagueness of their notions is illustrated by the writings of Teilhard de Chardin. He linked biology to the Spirit of Christmas in a gaseous envelope called the noösphere: 'Life physically culminates in man, just as energy culminates in life ... The phenomenon of Man was essentially pre-ordained from the beginning'. *The Origin* does not have much sarcasm, but the 'Historical

Sketch' that begins its later editions mentions a Dr Freke who had, in a paper of wonderful obscurity, claimed precedence for its ideas. As Darwin says: 'As Dr Freke has now (1861) published his Essay . . . the difficult attempt to give any idea of his views would be super-fluous on my part'. That does for Teilhard and his heirs too.

I once spent thirty years studying the evolutionary genetics of snails. Although my research decorated the margins of the subject (and is so incidental to this book as scarcely to appear in it), I still have no real idea what makes them tick. Society is, it seems, easier to explain. Charles Darwin saw where the importance of his theory lay ('species are not – it is like confessing a murder – immutable') and was opposed to the naïve use of his ideas in human affairs. This book has nothing on the various attempts, more or less infantile, to apply Darwinism to civilization.

Darwin and Wallace presented their ideas to the Linnean Society of London in 1858. They had little impact. Thomas Bell, a dentist with an interest in reptiles and then President of the Society, claimed in his review of the year that it had not 'been marked by any of those striking discoveries which at once revolutionise, so to speak, the department of science on which they bear; it is only at remote intervals that we can reasonably expect any sudden and brilliant innovation which shall produce a marked and permanent impression on the character of any brand of knowledge, or confer a lasting and important service on mankind'.

Bell's lack of judgement is reflected in his own book, *Kalygonomia, or the Laws of Female Beauty* (with plates bound as a separate volume to allow them to be locked away from inquisitive eyes). It listed 'defects in the Intellectual system of Women (4); Defects in the Mechanical system of Women (17) and Defects in the Vital system of Women (9)'. The reptilian dentist had failed to notice the alibi that his Society's paper was to provide for students of society. He dis-missed the talk that was to change biology for ever, and went on to note that 'A Bacon or a Newton, a Davy or a Daguerre, is an occasional phenomenon, whose existence and career seem to be specially appointed by Providence'. Darwin and Wallace did not count.

The sad truth is that the idea of descent with modification did not need Providence, or Darwin, at all. Sooner or later, like any dis-covery, it would have appeared in another guise. In science, revolutions are bound to happen. Nowadays, no biologist could work without Darwin's theory. Evolution is the grammar of their science. It accepts his painfully recognized fact that life, like the English language, works to rules that, even if filled with exceptions, make sense.

In spite of the Thomas Bells of this world, science (unlike the arts) can be detached from those who do it. For that reason, I refer to no living scientist by name. I owe a debt of gratitude to the many friends and colleagues who have commented on (and often disagreed with) my manuscript. They include Douda Ben-Sasson, Sam Berry, John Brookfield, Bryan Clarke, Michael Coates, Jerry Coyne, Andrew Leigh-Brown, Adrian Lister, James Mallet, Ursula Mackenzie, Ruth Mace, John McCririck, Michael Morgan, David Parkin, Norma Percy, Mark Ridley and Kay Taylor. All have been gracious about my use of their time and, too often, my reluctance to accept their criticisms.

Darwin's ability to rule over his disciples from beyond the grave is such that I hope that he will be forgiven an occasional appearance in this volume. His spirit is on every page.

ON

THE ORIGIN OF SPECIES

INTRODUCTION

ACCORDING TO A 1991 opinion poll, a hundred million Americans believe that 'God created man pretty much in his present form at one time during the last ten thousand years'. A large majority saw no reason to oppose the teaching of creationism in schools. They followed in a long tradition. A text of 1923, *Hell and the High Schools*, claimed that: 'The Germans who poisoned the wells and springs of northern France and Belgium and fed little children poisoned candy were angels compared to the text-book writers and publishers who are poisoning the books used in our schools . . . Next to the fall of Adam and Eve, Evolution and the teaching of Evolution in tax-supported schools is the greatest curse that ever fell upon this earth'.

Fifty pieces of legislation tried to put a stop to the subject. All failed. Undeterred, Alabama called for a note to be pasted into text-books: 'This book may discuss evolution, a controversial theory some scientists give as a scientific explanation for the origin of living things, such as plants, animals and humans . . . No one was present when life first appeared on earth. Therefore, any statement about life's origins should be considered as theory, not fact'. In 1998, the State of Washington tried the same trick, and others will no doubt follow.

Such intolerance is new. At the end of the last century few clerics opposed the idea of evolution. In spite of polemic against a 'gene-alogical table which begins in the mud, has a monkey in the middle and an infidel at the tail' most were ready to accept a compromise between *The Origin* and the Bible. A Day of Creation might be millions of years long, or might represent six real days that marked the origin of a spiritual Man after the long ages it took all else to evolve. Real bigotry had to wait for modern times.

The creationist movement is part of a triumphal New Ignorance that rules in many places, the United States more than most. In fact, the majority of those determined to tell lies to children believe in Darwin's theory and understand how it works, without noticing. Evolution is embedded in the American consciousness for a simple and terrible reason. For the past two decades the nation has lived through an episode that has, with extraordinary speed, laid bare the argument of *The Origin of Species*. The organism involved was unknown in the nineteenth century, but is now familiar. It is the AIDS virus.

Creationists find it easy to accept the science of AIDS. Its arrival so close to the millennium and the Last Judgement is a useful illustration of God's wrath. Homosexuals, they claim, have declared war on nature, and nature has exacted an awful retribution. Fundamentalists admit the evolution of a virus as nature's revenge but will not concede that the same process acts upon life as a whole.

Even to anti-evolutionists, AIDS is proof of descent with modification because they can see it happening. Its agent has changed in its brief history and has adapted to overcome the many challenges with which it is faced. As death approaches, a patient may be the home of creatures – descendants of those that infected him – as different as are humans and apes. Every continent, with its own sexual habits, has its own exquisitely adjusted set of viruses; and AIDS has its relatives in animals quite different from ourselves. Darwin would have been delighted to see the workings of his machine so starkly exposed.

Science makes patterns from ideas. If AIDS can evolve, so can anything else. *The Origin* uses freshwater bears and flying-fish in evidence to make a case that applies to all forms of life. For its opponents, in contrast, what is true for viruses cannot be true of birds or fish, let alone a man. The existence of an animal as unlikely as a whale is, for them, proof that evolution does not work.

The other view of the origin of whales, men or viruses is simple. As many more individuals of each species are born than can possibly survive and as, consequently, there is a frequently recurring struggle for existence, it follows that any being, if it vary however slightly in any manner profitable to itself, under the complex and sometimes varying conditions of life, will have a better chance of surviving, and thus be naturally selected. From the strong principle of inheritance, any selected variety will tend to propagate its new and modified form.

Every part of Darwin's thesis is open to test. The clues – from fossils, genes or geography – vary in each case, but from all of them comes the conclusion that the whole of life is kin. That is no mere assertion, but a chain of deduction with every link complete. The biography of the AIDS virus, Nature's newest and tiniest product, is almost complete and that of whales – the largest animals ever seen – is fragmentary, but they are cousins under the skin. The AIDS virus is change seen under the microscope, and the whale the same process viewed, in glimpses and over long ages, through a biological telescope. Evolution at the extremes of size is an apt prelude to the great drama that is Darwinism.

Creationists often deny the possibility of an intermediate between two species. Take whales and land animals. What use are flippers on solid ground, or feet in the sea? 'There are simply no transitional forms in the fossil record between the marine mammals and their supposed land mammal ancestors . . . It is quite entertaining to attempt to visualize what the intermediates may have looked like. Starting with a cow, one could even imagine one line of descent which prematurely became extinct, due to what might be called an udder failure'. The complaint (and the leaden humour) is not new. A London newspaper of 1859 said of Darwin's 'whale' passage that: 'With such a range and plasticity . . . we know not where to stop – centaurs, dryads and hamadryads and (perhaps) mermaids once filled our seas'.

Nobody has ever seen a dinosaur, let alone a mermaid. Evolution is, most of the time, an attempt to reconstruct a history with a pace far slower than those who study it. AIDS is unique because genes and time come together on a human scale. Darwin himself saw disease as a model of change. Almost the first recorded hint of his theory is in a note made on the *Beagle*. He was told by the surgeon on a whaling ship that lice from Sandwich Islanders will not survive on Europeans. How, he asked, could this be – unless each had diverged from the same ancestor? Why should a Creator, if parasites were needed, not make a universal louse for all mankind?

AIDS came to notice in 1981 with a report of a sudden increase in a certain form of pneumonia. As the sober language of the 'Morbidity and Mortality Weekly Report' of the United States Center for Disease Control put it: 'The fact that these patients were all homosexuals suggests an association between some aspect of homosexual life-style or disease acquired through sexual contact and *Pneumocystis* pneumonia in this population'. The illness became notorious with the death of the actor Rock Hudson in 1985. By then, more than twelve thousand Americans were dead or dying. Within a decade, half a million had perished. Nobody guessed that such a rare disease could become a pandemic. Camus, in *The Plague*, has it that: 'A pestilence isn't a thing made to man's measure; therefore we tell ourselves that pestilence is a mere bogey of the mind, a bad dream that will pass away. But it doesn't always pass away, and from one bad dream to another it is men who pass away'. They did and, more and more, they will.

AIDS, like the Great Pox of the fifteenth century, is spread by sex. The ground was well prepared before its seeds were planted. In the 1970s, five thousand gay men moved to San Francisco each year. By 1980, venereal disease was widespread – and four out of every five of the patients were homosexual men. A typical AIDS victim admitted to sex with eleven hundred people in his lifetime, while some claimed as many as twenty thousand partners. Most of the city's homosexual males had the viral illness known as hepatitis B, and many suffered from Gay Bowel Syndrome, multiple gut infections acquired from the curious sexual habits of part of their community. Casual sex in bath-houses – the Cornhole, the Boom Boom Room, the Toilet Bowl – helped the diseases to spread. AIDS, though, was new.

It was greeted with hysteria. Some claimed that the virus had been placed in Tutankhamen's tomb to punish those who defiled his grave and had come to America with an exhibition of his treasures. An analyst studied what he called its psycho-incubation. AIDS victims, he said, had suffered an emotional emergency in childhood that made them feel abandoned and later led to illness. The editor of *Burke's Peerage* went further. To preserve the purity of the human race his publication would not list any family in which a member was known to have the disease: 'We are worried that AIDS may not be a simple infection, even if conveyed in an unusual way, but an indication of a genetic defect'.

Although some dissenters tried to associate its symptoms with the use of capsules of amyl nitrite to enhance erotic pleasure, the real cause was soon found. The culprit is a virus, the Human Immunodeficiency Virus, or HIV.

Like a whale, the virus is built on an inherited plan coded by genes, each one liable to accident every time they are copied. HIV is unusual even among viruses. As a retrovirus, its genes are based not on DNA, but on its relative RNA (a molecule used in most creatures to translate, rather than to transmit, the genetic message). All retroviruses – and they come in many forms – contain about ten thousand RNA units, or 'bases'. The AIDS virus subverts its host's cells. It forces them to make replicas of itself with an enzyme whose job is to copy information from the invader's RNA into human DNA. Each new particle hides itself in a cloak of cell membrane into which it inserts a protein. This is the key to the infection as it fits into matched molecules on the surface of blood cells and opens the door to their interior.

The lock that turns to its enemy's key is most abundant on certain cells of the immune system. These multiply in response to infection, but cannot cope with the challenge. Billions of new particles are made each day, and although most are at once destroyed, they soon prevail. Soon after the virus arrives, the number of protective cells falls, only to rise as the fightback begins. Then, the immune system begins to collapse. The first sign of illness is a malaise no worse than influenza. This clears up, but HIV stays at work. As the body's defenders are driven back, other diseases gain a hold. For most people, the transition from infection to overt illness takes from six to ten years.

As AIDS advances there may be pneumonia, fungal infections, diarrhoea, weight loss and a viral form of blindness. A cancer called

Kaposi's sarcoma, otherwise found among aged Jewish men, quite often appears. Its first sign is purple marks on the skin, but as it progresses it kills. Kaposi's sarcoma is caused by a herpes-like virus common in the homosexual community. It gains entry – as do the other infections – because the body's defences have been undermined. If the patient does not first surrender to a fungus, bacterium or cancer, he wastes away.

The history of AIDS, over days, years and centuries, is simple. It involves descent, accompanied by modification. Each virus divides once a day. Mutation may be followed by natural selection that allows the invader to adapt to the body's defences, to the drugs used to treat it and to the sexual habits of the society in which it lives. Other changes are, it seems, unheeded by selection and build up at random as the generations pass. In time (and it does not take long) new forms of virus emerge.

The genes tell the whole story. They link a patient with the person who infected him, with others long dead, and with the viruses of apes, cats and whales. Except in its details, and the trivial matter of size, the evolution of the AIDS virus is that of every other being. Its nucleic acids reconstruct its past.

The disease is not very contagious. The chance of infection for a woman each time she has sex with an HIV-positive man is one in five hundred; and a stab with a contaminated needle is more likely to pass on hepatitis than to transmit the AIDS virus. Nevertheless, by 1992 it was the main cause of death for young American men. By 1998, one in a hundred of the world's sexually active population carried it, most as yet unaware of their plight. Ignorance played a part: many Tanzanians, for instance, believed that insecticide sprays protected against contagion. The epidemic has moved on from the bath-houses. Most cases are in the tropics, with a new outbreak in the states of the former Soviet Union. There, drug injection is rife and the incidence of syphilis (a sign of an AIDS outbreak on the way) has shot up by a hundred times.

The long incubation period means that no Third World country has yet faced the full truth of what is to come. In London, one person in six died in the last great epidemic, the Great Plague of the 1660s, which then faded away. By 1998 half the adult population of some

African cities was HIV-positive. AIDS is already Africa's greatest killer and the continent's death rate may increase by five times in the next decade. In Botswana, children born today have a life expectancy of twenty years. Without AIDS, the figure would be nearer seventy. In much of the tropical world, the virus has entered the public domain. It will be almost impossible to drive it out.

AIDS, like plague or lung cancer, is an illness of social change; of travel and of promiscuity. In the United States most infections are passed on by homosexuals, by addicts who use dirty needles and by those unfortunate enough to receive a transfusion of contaminated blood. In Africa, India and Asia almost all cases come from sex between men and women or by transfer from mothers to babies. Like all diseases – or flowers or songbirds – the virus is delicately adjusted to the challenges it must face.

HIV is bad at making exact copies of itself, which is one reason why it does so well. The error rate of its copying enzyme gives it a mutation rate a million times higher than that of its host. The virus is, like Mr Micawber, always waiting for something to turn up, and – most of the time – it does. Each new particle has, on average, a single change in its RNA. As a result, any patient soon contains a vast diversity of intruders. Most are defective and do not survive, let alone infect anyone else. Others withstand all that the body, or human ingenuity, can do.

A patient is an island to which its invaders must adapt or die. Natural selection is hard at work from the moment they arrive. The immune system is good at its job and billions of HIV particles succumb to it. However, a few emerge unharmed because genetic accidents alter their signals of identity. They are missed by the body's defences, multiply and prevail. Selection also explains why drugs failed for so long. There has been no lack of research. Fifty times as much has been spent per AIDS death in the United States as on each American who dies from stroke. In spite of great progress in treatment, the main conclusion from years of work and millions of dollars is that evolution is a relentless foe.

The first drugs attacked the copying enzyme in the hope that if the virus could not translate its information it would fail. Others use chemicals that look enough like components of DNA to fit into the chain as it grows and to stop it dead. Some treatments attack HIV proteins, or use 'anti-sense' nucleic acids, a mirror image of genetic information that can bind to viral genes and block their action. Much has been spent, to little effect, on the search for a vaccine.

As years went by with no success, desperate patients turned to quacks, who stimulated their immune systems with ozone blown into the rectum or with expensive potions made of ground-up tortoise-shells. Now, at last, there is hope. The survival of Americans with the disease has doubled in the past decade (in the main through improved treatment of AIDS-related illnesses) and in Europe the death rate is a fifth of what it was. By the late 1990s doctors dared to ask whether the virus itself could be wiped out. In some people a combination of drugs reduced it to undetectable levels and – perhaps – led to a cure. The optimism did not last. The viruses hide in the blood cells and reappear when the medication is stopped.

Now, for those able to withstand (and, at ten thousand dollars a year, afford) the treatment, the illness can at least be kept under control. In 1997 the gay newspaper *The Bay Area Reporter* was, for the first time in two decades, able to use the headline: 'No Obituaries'. The therapy takes dozens of pills a day, and to stop even for a short time allows the virus to rebound. Many abandon the treatment as it is so exacting, but the days when diagnosis meant death are over.

Every AIDS patient is a monument to the theory of evolution. Natural selection alters the identity of the virus as the disease progresses. Drugs, too, lead to evolutionary change, each drug generating its own response. Alterations in a mere five crucial sites in its tiny genome may allow HIV to escape from the best that medicine can do.

The best evidence for a theory comes when two experiments give an identical result. The drug Ritonavir was first used in the mid-1990s. At first, there was success, with – in most patients – the number of viruses reduced to a hundredth of its former abundance. Within months, all suffered an ominous slow climb back to the original levels and, as the treatment lost its power, the disease continued its course. Every resistant virus, from London to San Francisco, has an identical mix of four mutations. What is more, and although they occur at random, evolution utilizes them in the same order each time. The first change to be seized upon happens in a site to which Ritonavir attaches. It causes a slight improvement in the virus's ability to cope, but is at once picked up by selection. That single small advance is followed, in the same sequence, by natural selection's choice of three further mutations in different viral genes. In combination they increase resistance twentyfold.

Each point in the viral RNA has an error rate of around one in ten thousand. The chance of the four changes happening at once is that figure, multiplied by itself four times – one in ten million billion, a total greater than the number of particles made in the entire course of an illness. It could never be reached by the accidents of mutation in a single individual, let alone within the hundreds who have evolved resistance. Evolution triumphs because it turns to natural selection, the plodding accumulation of error. It gives the virus the ability to generate the same improbable result each time it is challenged.

AIDS is Darwinism unadorned: a faulty copying machine that alters faster than its opponents. The shadow of evolution also lies over the epidemic as a whole. For HIV, in its dawn, with San Francisco in a new Summer of Love, it was bliss to be alive. Any agent of infection was guaranteed a welcome. The pressure was on, and a gene that helped a virus to attack a new victim was favoured, whatever harm it did to its host. From the viral point of view, the death of a patient (and his HIV particles) meant little, as long as he had already infected dozens more.

In those early days, virulence was all. AIDS became more lethal as the epidemic grew. Americans infected in the plague's first few months survived for longer than those who caught it later. By then, selection had brought to the fore the variants quickest to copy themselves, even if they killed their carriers, because they spread faster than the others. Soon, homosexuals became more cautious. For the virus, the change in behaviour was bad news because it made it more difficult to reach a new host. The new erotic environment forced the disease to be kinder. Variants slow to kill were favoured over those that led to a rapid demise because their bearers lasted long enough to pass the virus on.

Where did AIDS come from? The records reveal a fragment of its history, but the genes say much more. For viruses (as for everything else) they reveal a shared past of descent, accompanied by modification. The early epidemic in Europe infected not just homosexuals but African immigrants who belonged to quite a different sexual community. Why should it attack these disparate groups?

Genes gave the answer. The 'signature site', a length of three hundred or so RNA bases, changes at great speed. It was a hint that AIDS is more than one disease. When any group of organisms — viruses, mice, butterflies or whales — is studied in enough detail, what seems at first sight an entity often turns out to be a set of distinct but related organisms; in other words, different species. HIV is no exception.

Some humour to relieve a grim story. A man goes into a Szechuan restaurant in Aberystwyth (a doggedly Celtic town) and is served by a Chinese waiter who speaks perfect Welsh. Beckoning the boss, the customer asks where he found this prodigy. The answer: 'Keep your voice down, he thinks he's learned English!'

In other words, from a Chinese speaker's point of view, Welsh and English are mere dialects of each other, each just as easy or hard to understand. They are members of the Indo-European group of languages, descended from a common ancestor. Although anglophones find the Celtic tongue impenetrable, the only way to test how distinct it might be is to put it into context, with Chinese as a separate group with which Welsh and English can be compared. The difference between the two is then seen, in global context, to be tiny.

There is a hidden structure in the languages of the world. The various degrees of difference of those from the same stock can be expressed by groups subordinate to groups: and the proper or even only arrangement would be genealogical. This would be strictly natural as it would connect together all languages, extinct and modern, and would give the filiation and origin of each tongue. Cockney and Geordie are closer than English and French; and those two Indo-European dialects are more distant relatives of Greek. Greek, in its turn, is not much like Urdu; but each retains enough of its past to hint at a common ancestry. The ancestral language can itself be reconstructed from the hints held in its much diverged descendants.

Information is transferred by genes in much the same way as it is by words. For each, evolution is inevitable. Literary fragments – fossil speech – and some audacious assumptions about the rate at which words alter show their family ties. French and English split not long ago (to Dumas, after all, English was just French badly pronounced); and, strange though it sounds, there is also nothing special about Welsh. It broke away from its sisters long after Chinese separated from the ancestor of them all. Language is the key to history. If we all spoke the same one – Chinese, Welsh, or English – and there were no records of the past, it would be impossible to tell where our shared tongue came from. Of course, we do not. Instead, descent with modification uncovers the biography hidden in every sentence and reveals the history of those who speak them.

For the human immunodeficiency virus, changes within a patient – a mere shift in genetic accent – or in the particles that pass from one victim to the next track its evolution over months or years. If (and the assumption is a bold one) the great limbs of the AIDS tree follow the same rules as those of its twigs, the epidemics of today, it should be possible to work out the past growth of any branch, however ancient. To compare the human immunodeficiency virus with its distant relatives – the Chinese-speakers of the viral world – hints at the origin of the plague itself.

Genes reveal two separate human immunodeficiency viruses. They differ in almost half the sequence of letters in the signature site. The virus involved in the homosexual outbreak is known as HIV-1, that mainly responsible for the illness in African heterosexuals as HIV-2. Each type is divided into secondary clusters. HIV-1 (like the Indo-European languages, with their eastern and western offshoots) has two great divisions. The first is universal, while the other, much rarer, has its home in Cameroon, Gabon and Equatorial Guinea. A third distinct type was found in 1998 in just two people in Cameroon. The main cluster contains ten or so subtypes (HIV-1A, -1B and so on), each of which diverges by up to a quarter in the genetic autograph. Minor variants in signature also disclose the presence of distinct strains of each subtype.

The virus has a structure of relatedness that traces its roots further and further into the past. Its hint at a grand natural system proclaims quite plainly that the innumerable variety of viruses have all descended, each within its own class or group, from common parents, and have all been modified in the course of descent. The modified descendants proceeding from a single progenitor become broken up into groups within groups. The rate of growth of recent shoots of the tree of illness can be used to reveal its distant past.

A species can originate but once. Few biologists are lucky enough to see it happen, even at the giddy pace of evolution among viruses. How is it possible to trace the origin of a disease most of whose victims are dead? Nucleic acids tell the story. They reveal a recent eruption of diversity. In a family tree of the agent of AIDS drawn for the United States, differences explode, like a firework, from a common source – the mark of the disease's sudden arrival and rapid spread. The pattern of global relatedness, too, looks more like a shrub than an oak. For HIV-1, all the main branches join in a common node. The HIV-2 pedigree looks much the same. For each, the point where the shoots all meet marks the origin of an epidemic.

A clock is hidden within all words, and in each length of nucleic acid. It depends on a simple and regular mechanism, the build-up of errors with time. If the moment when its hands were set is known and the mechanism ticks smoothly, its rate of movement can be measured. This molecular clock, as it is known, can be used to estimate when members of the HIV family last shared an ancestor. One site in several hundred of the virus's genes changes per year. Just to tally up the changes among lines misses a lot, because any letter in the sequence may alter back and forth several times. The timer's rate also varies because patients who die soon after infection leave no time for it to tick, while those who struggle on for years undergo more evolution in the RNA. Even worse, it seems to move at different speeds in different subtypes. Nevertheless, the clock in the genes makes sense, and HIV-1 samples collected in the 1990s have moved on from those taken ten years earlier. To turn back the hands to zero puts the start of the global outbreak at some time in the 1940s. HIV-2, the genes show, began its international career at much the same moment.

The earliest known specimen of the human immunodeficiency virus was found long after the death of its victim. It came from a fossil, the preserved remains of an anonymous African inhabitant of the Congo city of Leopoldville (now called Kinshasa and the capital of a Democratic Republic). More than a thousand blood samples left over from the first years of Congolese independence were tested, but although twenty or so were HIV-positive, just one, from a patient who died in 1959, retained any viral genes. The order of the few RNA bases to survive shows it to have been quite similar to the common ancestor of today's HIV-1 viruses as reconstructed from the genes of their descendants. AIDS, it seems, started within an African (a discovery greeted by the *Ghanaian Times* as 'a shameful and vulgar attempt to push this latest white man's burden onto the door of the black man'). The similarity of the inferred and the actual ancestor suggests that the global epidemic started soon after the Second World War, perhaps from a single virus particle. Startling as that seems, it is no more than what happens every few years as new waves of Asian flu sweep across the world.

The virus did not take long to escape from its native land. Only one route out of Africa has been explored. In 1976, a young Norwegian sailor died of a mysterious illness. A sample of his tissue was, many years later, found to contain traces of the Cameroon branch of HIV-1. He was infected on a trip to West Africa, soon after his fifteenth birthday, in 1960. The voyage from Norway to Senegal, Ghana and Cameroon gave him plenty of chances for sex. His records show that, somewhere on the journey, he caught gonorrhoea. The doctor missed his other, fatal, illness.

Fourteen years after the Norwegian brought home his viral cargo, a bisexual German musician who liked to hire prostitutes to take part in orgies became the first European to be diagnosed with AIDS. The sailor was by then a lorrydriver, who often visited Cologne, the orgiast's home town. The next victim was a French barmaid from Reims. Reims, too, was on his regular route. The history of a dead mariner, preserved in a bottle, uncovers one viral pathway into Europe. Most of the others will never be found.

The genes show that the invasion force involved just a few in-trepid voyagers. All the subtypes of HIV-1 are found in its native continent, but other places have only a few. Thus, the earliest samples taken in North America and Europe all belong to subtype 1B, proof that a small number of travellers brought the virus from Africa to the developed world.

Once in a new-found land the migrants must evolve to cope. HIV-1, the killer of the Western world, has remained an exclusive beast. Each of the eight or so modern subtypes sticks to its own community and has adapted to fit its sexual tastes. In Thailand, HIV-1B was the main form at the start of the 1990s. It had been brought from the West by a drug-user and became something of a specialist at travel by the anal route. The subtype is, in its new nation of sex tourists, in retreat before HIV-1E, a virus that prefers conventional sex. In Russia, drug-users have four subtypes of HIV-1, while female prostitutes and homosexuals each lay claim to their own.

AIDS' ability to cope with human vice is helped by its own sex-life. Sex, in all its guises, is no more than a way to mix up genes. If someone is infected with two HIV subtypes, the viral enzymes can reshuffle their RNA to give a generation of viruses with new combinations of genetic material. About one HIV-1 subtype in ten is a blend of the RNA of other, older, forms. HIV-1E – today's Thai form – is a hybrid between the African HIV-1A (which has not itself reached Asia) and an undiscovered donor of the genes for Thai coat protein. Some strains are a patchwork of genes from four or more sources.

One definition of what species are (and it can be hard to tell) turns on sex. If two individuals – virus or whale – can blend their genes to make young with elements from each parent, they belong to the same species. If they cannot, they are distinct. People who pick up both HIV-1 and HIV-2 (and such unfortunates do exist) never produce hybrids derived from each type. In contrast, those inflicted with, say, HIV-1A and HIV-1B often generate new mixtures. In the world of traditional biology HIV-1 and HIV-2 would be defined as distinct species, 1A and 1B as varieties of the same one. A mutation for drug resistance within HIV-1 might soon spread to every subtype of that virus, but would never enter HIV-2. Among the agents of AIDS, descent with modification has, it seems, gone so far that what was once an entity has been divided into two: and a virus has completed its journey along the Darwinian road to an identity of its own.

Much of the story of AIDS is evolution on a human scale, the tale of an opportunist in the modern world. The genes also hint at its distant ancestors.

Primate genes show that chimps, humans and gorillas are close kin, while orangs are more distant and monkeys further still from the almost-human trio. The monkeys themselves have a deep split between the Old World and the New. However, the family trees of host and the AIDS-like viruses found in most African primates are not at all alike. The disease and its victims have, it seems, trodden different paths, and AIDS must have entered humans from another animal.

An accidental experiment shows how a change of scene can lead to disaster. A harmless African monkey virus was passed in error to Asian monkeys. They all died within a few months. When blood from the first Asian monkey was then injected into a second victim and from him in rapid series to a third, fourth or fifth, the virus became more lethal at each step. Later generations of monkeys died in weeks. As in the gay men of San Francisco, rapid transfer favoured the viruses fastest to copy themselves, whatever harm they did. In one zoo, macaques picked up an AIDS-like illness from the talapoin monkey. In the talapoin the virus is harmless, but in macaques it kills – and moves from animal to animal by simple contact, with no need for sex. Humans have, so far, been spared such a fate.

Genes show that HIV-1 and HIV-2 are not much related and come from different sources. HIV-1 resembles a virus found in West African populations of chimpanzees while HIV-2 is closer to another from the sooty mangabey. They entered humans more than once. The three main groups of HIV-1 each resemble different lineages of chimp virus, and HIV-2 has half a dozen distinct types. People who handle primates for one reason or another may also be infected by AIDS-like viruses from mandrills and macaques, although these cause no symptoms in humans. The agent of AIDS is, they show, just one of a great family of viruses, most of which do little harm to those who bear them.

Why should a primate virus attack humans? Plenty of hunters have been bitten by monkeys, and Victorian travellers ate them with more or less enjoyment (although one found 'something extremely disgusting in the idea of eating what appears, when skinned and dressed, so like a child'). The great French hunter Paul du Chaillu recorded that in 1861 the Fang people of Cameroon gave up the purchase of corpses as food in favour of gorillas. Those animals are now protected, but the Fang still talk of gorilla tongue as a delicacy; and, with today's opening of the forests by timber companies, there is a new call for 'bush-meat' (apes included) to feed their workers. Ape-meat has also become a popular restaurant dish among the emerging middle class in parts of Africa.

Many – perhaps most – diseases come from animals, rabies from dogs, anthrax from cattle and Lyme Disease from deer. To have a pet is a good way to become ill. Parrots give psittacosis to their owners, puppies a roundworm that can lead to blindness and pet snakes often pass on salmonella. Pharaonic tombs have images of African green monkeys kept at home; and the god Thoth, the inventor of speech, was worshipped in the form of a baboon. In the Pithecussae, the Monkey Cities, these animals, it was said, waited at table and no doubt bit their customers from time to time too. Pets, though, may not be to blame. Some claim that AIDS is an epidemic sparked off by scientists who used monkeys to develop vaccines, or by doctors who used unclean needles in the fury of vaccination in Africa after the War.

However it began, the human immunodeficiency virus must be far older than any of its extant strains. The fossil virus from Kinshasa is more similar to modern HIV-1 than to any chimp virus, so that the transfer to humans happened long before the young Zairean met his fate. Perhaps, over the centuries, there were local outbreaks of AIDS in Africa, each the result of a separate infection from an ape. They did not spread in an earlier age without travel, transfusion or frantic promiscuity. Not until the 1950s could the illness expand to fill the world.

Genes also hint at a prehistory for AIDS. Many animals have viral diseases of the immune system and a virus quite similar to those of primates causes a similar affliction in cats. In the 1980s such an epidemic led to the death of thousands of Mediterranean dolphins. Its cause was a distant relative of the agent of AIDS.

The human immunodeficiency virus contains in its brief history the entire argument of *The Origin of Species*: variation, a struggle for existence, and natural selection. Geography tells part of its story, as do fossils, and its genes are a link to distant relatives with which it shared an ancestor long ago. They reveal a hierarchy of order, of groups nested within groups, as evidence of descent from a common source pushed further and further into the past.

Our lives are too short to understand the evolution of other beings in such detail. Take the aquatic bear for which Darwin suffered such mockery. Could it ever have made its way towards the sea? What does it take to become a whale, to live at the other end of the scale from a virus? Now, we know.

Whales, like all mammals, breathe air and give milk. When did they take to the water and what were they before they made the move? Their new life involved more than a change of medium. They grew, to a hundred and fifty tons in the case of the blue whale (which is to humans as we are to mice). The skull and neck became shorter and the nose moved backwards. The ear closed and sound now passes through a layer of fat. Legs evolved into fins, with extra bones in the back to match. Beneath the skin were other changes. The deepest diver can make it to four hundred feet without artificial aids and holds his breath for a few minutes to do so. The sperm whale dives to a mile and more, and can stay under for two hours. The change is in its chemistry rather than its lungs. Whale muscles contain large amounts of myoglobin, a protein that pulls oxygen from the blood. Their oxygen is kept not as a gas, but as a chemical compound. This in turn allows the lungs to collapse at depth as a defence against the bends (nitrogen bubbles in the blood) as the animal comes to the surface.

The remains of extinct creatures mark each step in the move from land to sea. The fossil evidence is confirmed by the record of the genes. If whales survive, their history will soon be as well understood as is that of the AIDS virus. The technology is much the same, and the evolutionary logic is identical.

Most of the fossils suggest that the distant ancestors of whales were hyena-like beasts called mesonychids, scavengers for carrion and hunters of fish. They underwent a radical change of habit. The Simla Hills of Northern India, with their mountain climate, were a holiday haven for the British rulers of the Raj. There, several thousand feet above sea-level, was found a fifty-three-million-year-old jawbone from *Himalayacetus*, the first known ancestor of today's whales. The fossil came from a beast that seems to have spent time both in fresh water and in a long-lost sea.

A fifty-million-year-old skull discovered in the Kala Chitta Hills of Pakistan came from an animal further on the way to whaledom. *Pakicetus*, as it was called, lacks the fatty earplugs of its descendants. Its days were passed between land and water, with an inner ear midway between those of whales and of land animals, allowing it to hear both in the air and beneath the surface. Those oldest whales lived in a vanished ocean, the Tethys Sea, now replaced by India and Pakistan, its floor thrust into the skies to build the highest peaks in the world. Only later did their descendants escape to fill the seven seas.

A younger version, found in Egypt, was christened *Ambulocetus natans* (the swimming walking-whale) after its large back legs, with seven-inch toes. *Ambulocetus* was about the size of a sea-lion, with a long tail quite different from a whale's flukes. Another relative, from three million years later, has its limbs reduced by a third. Yet another version, *Basilosaurus* (whose name reflects its first designation as a 'king lizard', regal indeed at seventy feet) lived about forty-five million years ago. It had small but perfectly formed rear limbs projecting a few inches from its body. With these fossils, almost all the steps from land animal to leviathan have been found.

The mammals of today hint at how the first whales leapt into the waves. Many can, with more or less embarrassment, make their way in water. Dogs paddle, humans do the breaststroke, seals swim better still. Modern whales do the job so well that they cannot walk on land. The feet of *Ambulocetus* put it between wind and water, in the otter league, with back feet bigger than those in front. It swam better than any dog and may have been as good as a sea-otter, as it moved not just with kicks of its rear legs but by flexing its spine. This was a large step towards the whales, which do the same with the help of an evolutionary novelty, a pair of giant flukes.

Although old bones are quite persuasive about what is needed to make a whale they are, like all ancient remnants, above all rare. Nobody will ever reconstruct the biography of whales from the fragments of their ancestors. Fortunately, whales – as much as viruses – are living fossils. The vestiges of limb bones show that, once upon a time, they had legs (and now and again a modern whale is born with small hind legs of its own) but the genes say more about whence they came.

The chronicle of the DNA is most obvious when it goes wrong. One whale suffered from '. . . a peculiar snow-white wrinkled fore-head . . . The rest of his body was so streaked, and spotted, and marbled with the same shrouded hue that, in the end, he had gained his distinctive appellation of the white Whale'.

Moby Dick was not quite a fiction, for a white sperm whale (Mocha Dick, named after the island off the coast of Chile where he made his onslaughts) attacked whale-boats in the 1830s. A snow-white example was once caught by the Japanese. The beluga whale's name in Old Norse means 'corpse whale' because of its resemblance to a drowned body. It shares (with white mice and albino people) an inherited defect in the ability to make the dark pigment called melanin. Just one whale gene is on the genetic map. The human disease Chediak-Higashi Syndrome arises from a mutation that causes silvery hair. In mice the same gene is called, from its effects on coat colour, the *beige* mutation. Patients (or mice) who inherit it get cancer and die young, because they lack a class of white blood cell whose job is to destroy tumours. Chediak-Higashi is connected to AIDS, because the HIV virus does some of its damage when it attacks such cells. The altered gene has also been found in mink, cattle – and a single killer whale.

Genes can track down the Moby Dicks of this world as well as they can its viruses. One whale sampled in the North Atlantic in 1964 found its way three decades later to the meat-counter of a department store in Osaka after having been captured, supposedly as part of a scientific survey. Whole populations can be traced in this way. Sperm whales in the southern hemisphere share a small gap in the DNA not found in those of the north, as evidence of an independent past. The genes of other whales, too, reflect their history. The humpbacks of Hawaii almost all carry the same pattern of DNA, with virtually no variation among them. The Hawaiians had no names for whales and the old whalers never found them around the islands. Perhaps, like

AIDS on its first forays into Europe, they are recent immigrants, descendants of a few founders and a limited pool of genes.

DNA links whales to the other mammals. They are not, alas, related to bears (who sit firmly among the dogs). Instead their molecular heritage shows them to be close to the hoofed mammals, the ungulates. Within that group (which includes deer, horses, pigs, giraffes, hippos and elephants) whales are nearer to those with an even number of toes – pigs, deer or hippos – than to ungulates with one, three or five toes, such as horses and rhinoceri.

The real key to their past lies, by chance, among some relatives of the AIDS virus. They have been hidden in the DNA of those great beasts since long before the first whale took to the oceans.

For some viruses, the war between host and parasite gives way to truce. They become integrated into the genes of their carriers and are copied each time the proprietor's cell divides. Because such hangers-on do no harm, they are transmitted for millions of generations, dormant in the same place in the DNA as that of an ancient invader. Much of our own genetic material is made of such decayed retro-viruses, some shared with other primates, others with more distant beasts. Their arrival may have been marked by an epidemic rather like AIDS. If it was, all hint of bad blood between vehicle and passenger has long disappeared. Perhaps, in the distant future, the sole evidence of AIDS itself will be a few silent sections of DNA scattered among the genes of our remote descendants.

The biochemical hitchhikers can be used to search for kinship; not of the viruses, but of those who carry them. They reveal the history of animals billions of times their size as, if two species have the same retrovirus inserted into the same position in their genes, they are likely to descend from a common ancestor.

The retroviruses of whales hint at an unexpected past. Whales, hippos, deer and giraffes have three shared elements of this kind, each in the same place in the DNA. The crucial trio of passengers is absent from all other even-toed ungulates, pigs and elephants included. They show that a deer is more related to a whale than it is to a pig. Had Darwin speculated about an aquatic and open-mouthed stag (or even giraffe) he would not have been far from the truth.

Other parts of the DNA narrow the search for the whales' immediate kin. Whales do not, it must be said, look much like any other animal. Fossils may tie them to the ungulates, but they seem quite distinct from any extant member of the group. The genes tell a different story. The DNA responsible for the proteins in milk and in blood-clots shows whales to be closely related to hippopotami.

The molecular tree of hoofed mammals sprang from a common root sixty-five million (rather than sixty) years ago. Like that of immunodeficiency viruses it is bushy, which makes it hard to sort out where the branches leading to horses, rhinos and the others split off. Nevertheless, the marriage of hippopotamus with whale is clear. The molecular clock, its rate set by the bones that fill the rocks, points at an ancestor common to whales and hippos some fifty-five million years ago, just before *Himalayacetus*, the earliest known fossil whale.

Whales and hippos may not much resemble each other nowadays, but retain some hints of kinship. Hippos spend their time in the water and on land; and some early whales did the same. Young hippos swim before they can walk, and hippos and whales each nurse their young underwater as they squeak and squeal through river or sea. Both are hairless, neither can sweat and their males each keep their testicles inside the body rather than in a convenient bag.

Although genes hint that the animal most like a whale is a hippopotamus, there is no manifest link between that ponderous river-dweller and those swift predators, the mesonychids. New fossils, though, show the mesonychids to be descendants of an ancestor shared with whales, rather than on the direct whale line. Although – as for AIDS viruses – much remains to be learned about the ancient history of leviathans, the case for their evolution is impossible to deny.

The virus and the whale each tell a story of how descent with modification leads, in time, to new forms of life. Each, in its disproportionate way, affirms the truth of evolution. Although much is obscure, it is impossible to entertain any doubt, after a century of the most deliberate study and dispassionate judgement, that the view which most naturalists once entertained – namely that each species has been independently created – is erroneous.

To deny the truth on grounds of faith alone debases both science and religion. The point was made by Galileo himself. Summoned to explain his views and their conflict with Scripture, he argued that the Church had no choice but to agree with the discoveries of science. It would, he said, be 'a terrible detriment for the souls if people found themselves convinced by proof of something that it was made a sin to believe'. Creationists have not yet faced that fact.

No biologist can work without the theory of evolution. Like Galileo's notion of a solar system with the sun at its centre, Darwin's long argument makes sense of their subject. Ideas of origin were once, like *Moby Dick*, allegories. They helped to comprehend not the structure but the meaning of the Universe. Some still hope to find symbolic significance in Darwinism. They will not: but his work turned the study of life into a science rather than a collection of un-related anecdotes.

CHAPTER I

VARIATION UNDER DOMESTICATION

Character of Domestic Varieties — Relation between Man and his Domestics — Origin of Domestic Varieties from one or more Species — Principles of Selection, anciently followed — Domestic dogs, their Differences and Origin — Methodical and Unconscious Selection — Breed and identity — Evolution on the Farm — Zoological Gardens: the Call of the Tame — Loss of Variety under Domestication — The Wolf beneath the Skin – Difficulty of distinguishing between Varieties and Species

MAN HAS A strange relationship with his domestic animals. The Victorian explorer William Burchell found himself unable to eat zebra when he was near starvation in Africa, because of its resemblance to his favourite mare. The French government, alarmed by the waste of good protein, had managed in the 1860s to persuade its citizens to feed on horse; but in London, the Society for the Propagation of Horse Flesh as an Article of Food failed in the endeavour, in spite of a launch banquet of Salmon with Racehorse Sauce, Filet of Pegasus, and, to follow, a *Gâteau Vétérinaire*. Even so, at about that time, the *Live Stock Journal and Fancier's Gazette* complained that 'in some parts of England cats are not wholly despised as an article of diet' and that a notorious gang of cat-eaters in West Bromwich meant that fanciers 'cannot keep a favourite a week'.

Animals, as they become domestic, enter an uncertain domain between the real and the artificial. They persuade man to accept the living world as part of himself, promoted from food to member of the family. In the Middle Ages pigs were tried and hanged for murder, and only forty years ago a female rhinoceros was elected, by a large majority, to the São Paulo City Council. In an equivalent confusion today, a third of all dog-owners are happy to identify their pet as closer to their heart than is anyone else in the household.

As the wilderness creeps into the home, boundaries that were once distinct become blurred. Greeks, Egyptians and Icelanders each had sacred dogs – Cerberus, Anubis and Garm – to guard the entrance to the next world. In them, and in William Burchell's mare, the wild undergoes a spiritual transformation beyond the reach of science.

It is just one step further than a change that affirms the central truth of evolution: that variation within existing forms can, with human help, bring forth new kinds of creatures quite different from their ancestors. Evolution on the farm is a small-scale version of that in Nature. A wild beast does not at once become tame, or a new breed arise in an instant. Their passage from one way of life to another obeys the rules that apply to the real world. Much can be learned about the course of the great stream of evolution from domestication, its minor tributary. It shows that species are not set in stone, but are always in flux.

Many people turn to *The Origin* in a search for a philosophy of life. Most are disappointed to find that the first chapter is mainly about pigeons. It shows how breeders have produced birds as distinct as the pouter, the runt, the barb, the turbit and the laugher from that drab source, the rock pigeon, as proof of how animals can change. Darwin himself joined the Philoperisteron, the smartest club of the London Fancy. He visited breeders in their gin palaces in the Borough and became a considerable expert on the birds. His own became a pleasure: 'the greatest treat ... that can be offered to any human being'. Darwin did his job too well. The publisher's reader who commented on his manuscript suggested that it should be turned into a handbook on how to breed the birds: 'Everyone is interested in pigeons ... The book would be reviewed in every journal in the kingdom, and would soon be on every library table'.

Variation under domestication – in fields, in zoos and in living rooms – is still powerful evidence for his theory. When we look to the individuals of the same variety or sub-variety of our older cultivated plants and animals, one of the first points which strikes us is that they generally differ much more from each other than do the individuals of any one species or variety in a state of nature. Pigeons alone have dozens of breeds. The transformation of what was once a plain and unambiguous bird – and its equivalents in the garden or around the fireside – tells an evolutionary story.

Pigeons have been succeeded in Cockney affections by dogs (as is manifest to anyone who walks in London's streets or parks). The city's dogs are beasts very different from a timber wolf, its municipal roses quite distinct from their wild relatives. All are a product of selection. Some wolves or roses were, for some reason, better able to cope with man and, unlike their savage kin, survived and multiplied. Their progeny in turn were winnowed and, in time, the wild became domestic. Those who chose servitude changed to do so. Lord Somerville, quoted in *The Origin*, had it of breeders that: 'It would seem as if they had chalked out upon a wall a form perfect in itself, and then had given it existence'. As a result, the best place to see evolution is on the farm.

For most people, humans are at the top of a pyramid whose foundations were laid some four billion years ago. Mankind is, it seems, the master of the fate of all other beings, most of all those used for food or work.

But who is at the centre of the worldwide web of exploitation? The answer is not humans, but pigeons, dogs or roses, for their lives depend on man's readiness to be used. More than nine-tenths of all the flesh in Britain – except that of man himself – is of domestic animals. Mice, deer and all these islands' aboriginal inhabitants are left far behind. Although the servants repay their debt with milk or bread, affection or tennis courts, their victory is clear. Fifty million dogs live in the United States, while ten thousand wolves remain. Wolves are hunted down because they dare to kill a few of man's billion sheep, themselves the relatives of the last mouflons. To be born free was, for wolves or wild sheep, a dead end.

The grand enslavement began in the Middle East, with rye, thirteen thousand years ago. Those who grew it were hunters, and long remained so. The hunter-gatherers of the Yangtze cultivated rice twelve thousand years ago, and the first squash was grown in clearings in South American jungle a thousand years later. Until the modern age there were no domestic plants with an origin in Australia or Southern Africa. It is not that these countries, so rich in species, do not by strange chance possess the aboriginal stocks of any useful plants, but that the native plants have not been improved by continued selection up to a standard of perfection comparable with that given to the plants in countries anciently civilized. This is itself evidence for evolution, as not until variation was noticed by man did he create new animals and plants. Cows or wheat are special only because they were exposed to human choice.

Wheat descends from three grasses that hybridized on the Anatolian steppes. One, einkorn, still grows there and is in a small way cultivated. The DNA of wild and cultivated einkorn shows that just the plants from the Karaçadag Mountains, between the Tigris and the Euphrates, gave genes to wheat. There the modern world was born, as those hills, with the lowlands to the south, were also the birthplace of grapes, olives, chickpeas and bitter vetch. These crops were, in time, enriched by plants from other places – rice from China, maize, potatoes and more from South America – and by the domestic animals of today (among whom pigs came first, followed by goats, sheep and, much later, cattle). Their descendants are evidence of how nature has been moulded to human ends.

On the Breeds of the Domestic Dog. Nowhere has domestication gone further than in the household. Dogs bear witness to its power. The first dog show in England was held in Newcastle in 1859, the year of publication of *The Origin.* It featured familiar breeds such as pointers, setters, and spaniels. The Americans took up the sport a few years later with a display of Queen Victoria's own deerhounds, a dog called Nellie who walked on her hind legs because the front pair was missing, and the prize of a pearl-handled revolver for the overall winner.

Now, the United States has eleven thousand dog shows a year and seven billion dollars are spent on veterinary fees alone. The forms generated in the brief search for perfection are remarkable. The champion of the 1998 Westminster Kennel Club Show, America's premier event, was Fairwood Frolic, a Norwich terrier. She was selected from the best dogs in groups that included a malamute, a toy poodle, a longhaired dachshund and an old English sheepdog. Previous winners have come from breeds as extreme as the pug (whose eyes have sometimes furtively to be pressed back into their sockets by their owners). To survey the arena at Madison Square Garden or at the National Exhibition Centre in Birmingham (where Cruft's, the English championship, is held) is to see how plastic flesh can be. A dog show is evolution chalked out for all to behold.

All these breeds descend from some wild ancestor. There are thirty-five named species of wild dog, which gives plenty of candidates for domestication to choose from. Some are solitary like the fox, others social like the jackal; some as small as the crab-eating zorro of South America, a few as large as the Arctic wolf. The bones of an even bigger animal, the dire wolf, are preserved in the La Brea Tar Pits in Los Angeles. It died out ten thousand years ago.

Just one dog (given the Latin name *Canis familiaris* by the great classifier Linnaeus) has been domesticated. Linnaeus defined that familiar beast as distinct from all others on the basis of its upturned tail, found in none of its kin. Other characteristic features were that domestic dogs lick their wounds, are often infected by gonorrhoeae, and urinate when they hear certain musical notes.

The variation within the single canine to place its future in the hands of man transcends that in all its relatives put together. Linnaeus commented on the domestic dog's diversity and mentions the Naked Dog, the Fat Alco and the Techichi (a barkless New World dog with a wild and melancholy air) as variants of his *Canis familiaris*. Chihuahuas stand six inches high at the shoulder, Irish wolfhounds four feet. A St Bernard weighs fifty times as much as a Pomeranian, and the range of colour, shape and temper of pet dogs is almost beyond imagination. Certainly, if they were not our familiar domestics, at least a score of dog breeds might be chosen, which, if shown to a mammalologist, would be ranked by him as well-defined species. If the history of dogs as a product of man's whim were not so familiar, the numbers given distinct labels by science would rocket. Only because we see them as mere breeds are they confined within a single Latin name.

The sole alternative to admitting the variation of dogs as evidence for evolution is to believe that each breed descends from a separate wild ancestor, now extinct with no token of its passing. The doctrine of the origin of domestic races from several aboriginal stocks was once carried to absurd extremes. As Nature was thought to be static and its products unable to change, breeders claimed that there must have existed at least a score of species of wild cattle, as many sheep, and several goats in Europe alone, and several even within Great Britain.

Even now, the idea of separate ancestry for dog breeds is not dead. It is, however, wrong. But what was the cur whose descendants burst into such variety? So different are the domestic kinds that it seems impossible that they could descend with modification from a single source. Now, the genes make it plain. The dog's sole ancestor is one celebrated animal, the wolf.

Wolf bones are found near those of humans as long ago as the time of Boxgrove Man in Surrey, four hundred thousand years before the present. Nobody knows the nature of the relationship. Perhaps men and wolves fed in the same place, or wolves were pests who raided the hunters' camps. The people of Boxgrove may sometimes have picked up a cub to be tamed, fattened and eaten later. Whether neighbour, camp-follower or convenient snack, the wolves changed little as their owners were transformed. A mere ten thousand years since, when humans were more or less what they are today, the skulls of wolves recovered from a drowned camp of the nomads who crossed from Siberia to Alaska were almost the same as those of their Boxgrove ancestors. They had shorter faces and their teeth were more crowded, but in all appearance they were wild animals.

At about the same time, hunters began to use arrows rather than spears. Wolves – dogs, as they soon became – became more useful as they could chase and pull down wounded prey. Such a creature took at once a large step towards the fireside. They deserved honour, and Darwin noted with disapproval how the people of Tierra del Fuego would devour their old women rather than their dogs in times of shortage. At Ein Mallah in Israel, in a grave of the earliest farmers, is the skeleton of a puppy buried next to a child. A wild animal had become a member of the family. Soon, its muzzle shrank, its teeth became smaller, its eye grew large and round, and the modern dog had arrived.

The singing dog of New Guinea is a relic of that primitive beast. The dingo, too, is not much changed since those days. Dogs got to most of the world (except Africa, which was dog-free until the Iron Age of AD 500) as soon as humans did but, worldwide, changed little for thousands of years.

One set of genes, in dog or man, is special. Mitochondrial DNA is found in certain cell organelles whose job is to generate energy. It passes (like the mirror image of a human surname) through females alone; from mothers to daughters and sons, but daughters alone transmit it to the next generation. It contains within itself the history of bitches. Like any set of genes or surnames it accumulates changes as the generations pass. How distinct such DNA might be in any pair of lineages hints at how long it has been since they last shared a mother.

A survey of dozens of breeds of dogs, and of wolves, coyotes and foxes, shows that wolves and dogs are indeed the closest kin. Dog 'surnames' fall into two distinct groups, evidence that the animals were tamed twice in different places. As their ancestors were in contact with humans for thousands of years, a mere two domestications suggests that to become a servant is not as easy as it seems.

Although the bones suggest a split between wolf and dog just ten thousand years ago, mitochondrial DNA hints at an earlier divergence. Fossils show that wolves and coyotes separated about a million years ago. If (as with the origin of AIDS, and with no more evidence in favour of the idea) the genes have changed at a steady rate since the split, the molecular clock of divergence of dogs from wolves can be set in the context of the change in coyote genes. It puts the division between wild and domestic at more than a hundred thousand years before the present. This is far earlier than the first skull to look like a modern dog, and is quite near the time when humans of modern form emerged. Why an animal so much like a wolf should sever itself from its kin, stay close to humans for so long, and yet remain unchanged, is not clear. The clock within dog genes is filled with assumptions. Fossils are less ambiguous as evidence of the past; and a canine history of ten millennia makes more sense than one ten times as long.

Whenever it happened, what did it take for the dog to abandon the wild? Biology and culture worked together. Nobody claims ownership of a wolf, but a dog without a home is a pariah. Its existence depends on its relationship with man. Dogs choose their masters well, with a close fit between family income and dog-ownership. Sometimes, though, they are ejected. Italy has a million feral dogs, driven back some way towards Nature. They live, like their ancestors, in packs of a dozen or so. Their existence is miserable by comparison with that of their domestic cousins. For food they roam from rubbish tip to rubbish tip. They have a parasitic rather than an affectionate relationship with ourselves. Servitude has even destroyed their native ability to raise young. As a result, only one feral pup in twenty lives through its first year, and were it not for a steady influx of other outcasts, the packs would soon die out. For our favourite pet, to become housebroken led to a dramatic increase in the quality of life. It was, however, a journey down a one-way street.

Dogs have paid a price for easy living. To become domestic stifles the world of the senses. Wolves are fierce, fearful and filled with stress; dogs calm, docile and, for most of the time, carefree. Pets are by their nature a parody of a wild animal. What made the wolf an emblem of dread has been much diluted. Its ears, once pricked, are floppy and the sounds of the world are dulled. Its sharp eyes are blurred by a fringe of hair and can no longer stare an opponent into submission. The lupine tail, an expression of rage or delight, is in many breeds so curled as to bear no message at all. Most pets cannot even raise their hackles in anger as their hair is too long. All this comes from an unconscious preference by man for an animal that knows its place.

What was once done without thought has been echoed by science. In 1950s Russia, silver foxes were farmed for fur. They were savage, suspicious and liable to die from anxiety. On a certain collective, in an attempt to improve matters, only those willing to accept human company were chosen as parents. Within twenty years and a mere ten thousand foxes, the farmers saw a great shift in their charges. The ranch was filled with well-behaved animals that looked more like dogs than foxes, with a lowered tail and drooping ears. Many had piebald coats, quite unlike their unrestrained kin, and the females reproduced – like dogs – twice rather than once each year.

To breed for tameness was enough to make the change. The other characters followed.

Many of the qualities of today's dogs arise from the simple human taste for young over adults. For most of the time, a pet's job is to be a surrogate child. Any animal that looked like a pup – a short muzzle with small teeth, round and attractive eyes – was bred from at the expense of his more brutish sibs. An amiable disposition also helped. Familiarity alone prevents our seeing how universally and largely the minds of our domestic animals have been modified by domestication.

The end result of evolution through man's desire was to produce a sexually mature wolf pup. A wolf-sized dog, a Labrador, say, has a brain a fifth smaller than that of its wild relative; the size, indeed, of the brain of a three-month-old wolf. It is scarcely possible to doubt that the love of man has become instinctive in the dog. Any pet behaves like a juvenile version of its ancestor. It sits around in the hope of a meal and licks the hand that feeds it, as wolf cubs lick their mother's face to persuade her to disgorge meat. Most wild dogs co-operate to raise their young, either in packs or (as in jackals) with each pair helped by a 'maiden aunt'. The success of their tamed heir turns on its ability to persuade men to act as aunts.

How far any dog is allowed to progress into maturity depends on what its owners need: mere affection, or much more. The key is man's power of accumulative selection: nature gives successive variations; man adds them up in certain directions useful to him.

The hounds that protect sheep in the mountains of southern Europe – the Pyrenean mountain dog or the Anatolian shepherd – look fierce, but are in fact enormous infants. They are brought up with the flock and come to regard them as members of their family. An Anatolian dog is quite unable to hunt. Even if starved it will not eat a carcass unless it has already been cut open – a job done by its mother in the years before domestication. If a wild cat attacks the flock, the guardian scares it off not with ferocity, but with clumsy friendship. It alarms the nervous predator with hospitable barks as it wags its tail and runs up in welcome. Such canine Peter Pans even have low levels of the brain chemicals used in nerve transmission.

Herd dogs are granted an adolescence. A good collie enters into the spirit of the hunt, up to a point. It fixes a sheep with its eye and stalks it, but does no more. Under no circumstance will it bite its prey. Corgis – a favourite of the Royal Family – take the next step. Because cattle, their usual target, are so large, to move them they must eye, stalk and nip at the animal's heels. Catahoula leopard cowhog dogs go still further towards the adult wolf. They form packs to harry cattle across the range until the herd runs to where it should.

Distinct breeds have been around for a long time. An image of a pharaoh hound is on the tomb of Tutankhamen. The Pekinese appeared in second-century China as a pet bred to look like the spirit-lion of Buddha. So sacred were they that, when the British occupied Peking, the Empress ordered the execution of her favourites to prevent their capture by the white devils.

A new form, with its own appearance and personality, can arise with great speed. Although the Inuit used dogs to haul sledges across the Arctic long before the arrival of the white man, the tandem harness, with its many animals linked together, was introduced by Europeans. Today's sled dogs are, over more than ten miles, the fastest mammals on land. Their marathon, the Iditarod, extends over a thousand miles of ice and snow from Anchorage to Nome. The winners do it in ten days. Teams at the turn of the century were a mixed and disreputable bunch but, within a few years the victors began to converge on a common type. Soon, they looked like the familiar beasts of today, and now all the champions share the same thick coat, long lope and ability to work as a team. Even their feet have changed, with a specialized pad that does not pick up snow.

Some of the diversity in size or in temperament of various breeds descends from the ancestral wolves. Some – like that in 'sporting plants'; by which term gardeners mean a single bud or offset, which suddenly assumes a new and sometimes very different character – has arisen since the animals became part of our household arrangements. Such changes happen because the genetic material of dogs, like that of viruses, is not copied to the letter. Any mammal, with some seventy thousand genes, suffers from a constant influx of mutations. Most are harmless or are hidden by unaltered versions of the same gene. Some, though, cause changes that can be selected by an alert breeder.

However the variation arises, it is used by fanciers to make new forms that, sooner or later, attract labels of their own. More than three hundred distinct dog breeds are recognized. Quite what that means is a matter of taste, because a breed does not exist until it has been named. The American Kennel Club has as its mission 'the maintenance of the purity of thoroughbred dogs'. Their self-imposed task is impossible. Each stock, pure though it may seem, is always evolving as it responds to genetic change and to its master's desires.

Some of those are open, but most are not. There is a kind of selection which may be called unconscious, and which results from every one trying to possess and breed from the best individual animals, with no wish or expectation of permanently altering the breed. Slow and insensible changes of this kind could never be recognized unless actual measurements or careful drawings of the breeds in question had been made long ago, which might serve for comparison.

For a few dogs, the job has been done. The bulldog was much painted in the days when bulls were baited. It was a savage animal, 'unequalled for high courage and stoutness of heart', whose speciality was to fly at the face of its quarry and to use its massive jaws to bite and hold on to the animal's nose. Its own set-back nostrils helped the dog to breathe as it did so. The sport was outlawed in 1835. By 1900 the bulldog had – with no conscious attempt to change it – become 'a ladies' dog as its kindliness of disposition admirably fits it'. The purists were far from happy. Compared to its fierce original, 'the disgusting abortions exhibited at the shows are deformities from foot to muzzle . . . and totally incapable of coping with a veteran bull'.

All breeds have changed, on purpose or by accident. A 1570 book recognized just seventeen: Terar, Harier, Bloudhound, Gasehunde, Tumber, Stealer, Setter, Wappe, Turnspit, and others. By 1850, one writer recognized forty breeds, and the Cruft's show of 1890 had two hundred and twenty classes (although, admittedly, the last included 'stuffed dogs, or dogs made of wood, china etc.').

One lineage reveals how new forms appear. The King Charles spaniel, much incorporated in portraits of Charles II, was in the seventeenth century a black-and-tan animal. It has been much modified since his time. Now, it has four varieties: the original, another with white marks, yet another reddish-brown, and the last reddish-brown with white spots. The decision as to what to accept as a distinct class is quite arbitrary. The American Kennel Club is conservative and recognizes a mere hundred and fifty or so breeds (of which the King Charles Spaniel is – or was until 1997 – just one). A certain nationalism takes hold. Canada, for instance, insists on a separate name for the Nova Scotia duck-tolling retriever, accepted nowhere else. The Kennel Club of England is happy to subdivide its pets. The brown variety of the royal spaniel has its own identity as the Blenheim King Charles spaniel, and a form which looks like the original but has a long face is graced with the title of the cavalier King Charles. In 1997, a new American dog was born. The cavalier King Charles was at last acknowledged by the American Kennel Club and gained its own category in their shows.

Try the patience of owners as it might, such uncertainty is testament to the fact of evolution. A breed, like the dialect of a language, can hardly be said to have had a definite origin. It is hard to delimit any new kind of dog because when a system of its nature in flux is forced into fixed categories, boundary disputes are inevitable.

Whether accepted by purists or not, breeds must, because they look distinct, vary in their genes. Such differences, from duck-tolling retriever to cavalier King Charles, are quite small. They involve only the differences selected by man. He cannot select any deviation of structure excepting such as is externally visible; and indeed he rarely cares for what is internal. As a result, variants safe from human scrutiny have changed less than have size, shape or behaviour. Much of the hidden diversity of the camp-followers remains, silent and unaltered, within the stocks of today. Distinct as a terrier and a greyhound might appear, they are alike under the skin. Breeds are pure only in the eyes of their owners. So much variability remains within each canine stock that some cities plan to use DNA fingerprints on faeces on the street to identify serial offenders.

As fanciers select their preferred types, the stream of ancient genes for behaviour, shape and size splits into a series of rivulets that narrow as selection goes on. Once a lineage is well established, such patricians of the canine world have their mates chosen from among their relatives. As a result, they become inbred; they share a heritage because of descent from a common ancestor. A noble dog bonds to noble people and Irish wolfhounds have, their owners say, an inborn ability to recognize those who bear the blood of Irish kings.

Some canine genes are (like many of those for human disease) harmful when present in double dose. The deformations that emerge from such genetic mishaps are seized on by fanciers as fuel for their explorations of the wilder shores of taste. In the kennel, after all, we see adaptation, not to the animal's own good, but to man's use or fancy. The malign effects suffered by inbred animals show how evolution can exploit hidden diversity.

Time magazine claimed in 1994 that a quarter of all pedigree dogs suffer from a genetic disorder. The estimate may be too high, but without question many have been damaged through the efforts of their owners. If man goes on selecting, and thus augmenting, any peculiarity, he will almost certainly unconsciously modify other parts of the structure. Some canine difficulties arise because a change bred for in a particular character alters the development of a different one. Almost all stocks have their own problems. Chows are almost blind because their turned-in eyelids give them a quizzical look, while bloodhounds, selected for a droop in the lower eyelid and a mournful face, suffer from inflamed eyes. In the same way, Dobermann pinschers suck their flanks and there are problems of paranoia among basset hounds. The three hundred and fifty canine diseases known to be inherited include some – cataracts, retinal cancers, epilepsy – close to those of humans. The dog may repay part of its debt to its owners as it helps them to understand their illnesses.

Other domestics have also been subdivided by man in his search for the excellent. There, the problem is the loss of breeds rather than their constant emergence. Some three thousand named types of ass, cow, goat, horse, pig, sheep and water buffalo are known, a third of them at risk of extinction. Their history – and their raw material – is in their DNA. The genes show that horses (like dogs) were tamed more than once, from Europe's huge herds. Cattle, too, were domesticated twice. Those from Africa and Europe are on their own branch, those from India on the other. Africa has two varieties, a small form found in the west, and zebu, large animals with humps, common throughout the continent. Zebu are susceptible to a fatal blood parasite and can be helped with genes for resistance bred in from their kin.

Farm animals show how selection can soon promote new forms. Cattle, sheep and goats never forget a face (even from photographs) and recognize not just individuals but whole breeds by the way they look. Given the choice, a cow prefers to associate, and to mate, with one of its own kind, be it a Friesian or a Hereford. What is more, a male tends to fall for a female who reminds him of his mother. After all, she brought him up and belongs – by definition – to his own kind. The Oedipal effect is strong. Young male lambs raised by goats, or young male kids raised by sheep, are interested only in sex with animals like their foster-mother – male sheep with female goats, and male goats with female sheep. A male's preference for sex with a partner like his mother keeps a breed to itself – which is the first step in any promotion from variety to the elevated status of species.

Selection. Farmers have much simplified their plants and animals. As they select from the finest, they lose the second-best – and their genes. Like dog-fanciers, they have begun to realize that diversity, their raw material, has been eroded at a fearful rate by the efficiency of selective breeding. They try, by careful crossing, to limit the damage. Even so, their domestics have become a small sample of the wild. The entire United States soybean crop – sixty million tons of it – descends from a dozen strains collected in North-Eastern China. In the 1970s most of the country's maize harvest was lost as all the billions of identical plants, superior as they were, succumbed to the same fungus. Animals also suffer from man's obsession with the excellent. Sunny Boy, a Dutch bull of superb quality, died in 1997 after his two millionth donation of semen. Half a dozen other bulls have each bestowed more than a million ejaculates upon the world's cows, which means that the genes of untold others have been lost.

Nature is now pillaged for what remains. Animals are preserved as frozen embryos or as sperm freeze-dried to a powder and brought back to life. The world has seven hundred seed banks, each a reservoir of diversity. Botanists search for genes in the ancestors of today's crops, and for new sources to exploit. The anti-cancer drug Taxol came, after all, from an American yew, and a contraceptive is found in a Brazilian tree. In the siege of Leningrad, people starved rather than eat the seeds of wild grains collected by the great Soviet geneticist Nikolai Vavilov (who died on Stalin's orders) and the Third World today complains about collectors who hunt for wealth among the local plants.

The first attempt to rescue Nature's diversity began in 1828. The goal of the new Zoological Society of London, as set forth in its Prospectus, was to introduce new kinds of animal into England 'for domestication or for stocking our farm-yards, woods, pleasure-grounds and wastes'. The society had not a single success. It triumphed in a different way, as a display cabinet of curious animals for Victorian London.

Now, London Zoo, like all others, has changed its image. Its aim is no longer to pillage the wild for man's use, but to protect what remains against domestication. Zookeepers see themselves as the last hope for many of the products of evolution, but – like the first farmers and the members of the American Kennel Club – they are testimony to its inexorable force.

To some extent zoos have succeeded. Without them we would have no Arabian oryx or European bison at all. They face two prob-lems. First, a zoo can sample just a tiny part of the diversity present in the wild. An animal that once ranged over thousands of miles is forced to migrate to a tiny island. As in European AIDS viruses or the humpback whales of Hawaii, just a small proportion of the ancestral genes can make it to the new home. Keepers must breed from what little they possess – which itself means evolution. They must also, perforce, choose as parents those animals best able to cope with their new environment. As a result, their charges begin to change, as cows and dogs have changed.

A zoological garden bears the unwelcome message that, because of man's inadvertent selection, any animal taken from the wild becomes domestic, a travesty of its natural self. Evolution is as hard at work on caged animals as on those born free. In time they will emerge as beings quite different from what they were. Those who conserve animals in the hope of returning their descendants to Nature may be disappointed by what they let loose. Their failure shows how descent with modification is impossible to avoid.

In an attempt to save the wild, more and more has been tamed. Pessimists claim that by the middle of the next century the world's zoos

– an area no larger than that of Glasgow – will be the last stronghold of all the hundred and sixty kinds of primate, sixty out of seventy-two wild cats, and most of the world's wild dogs. Two thousand vertebrates have their future behind bars. Already, many are reduced to a wild population of less than a thousand. The prospect for most of them is bleak.

The first zoo was founded by King Shulgia of Mesopotamia in 2000 BC. He kept lions in his park near the divine city of Nippur. He was followed by the Chinese Emperor Wen Wang with his 'Garden of Intelligence' and by Henry VI of England, who caged a lion in the Tower of London (and gave free entry to those who brought a dog or cat for its lunch). Caged beasts testified to wealth; and Aristotle, Charlemagne and Theodore Roosevelt all had zoos of their own. The liberation of the inhabitants of the Paris Zoo at the time of the Siege in 1870 had practical as well as symbolic significance. Most of the emancipated creatures were eaten.

At first, the inhabitants of such cabinets of curiosity died soon after arrival, because nobody knew how to look after them. In one, the gorillas were given sausages and a pint of beer for breakfast, followed later by cheese sandwiches, boiled potatoes and mutton, and more beer. Few survived (the visitors did not help; when Philadelphia Zoo opened in 1874, its sloth was poked to death by umbrellas within a week) and those that did were reluctant to breed in public.

Akbar, the great sixteenth-century Mogul of India, is mentioned in *The Origin* as the proprietor of twenty thousand pigeons. He crossed his stocks, 'which method was never practised before', and 'improved them astonishingly'. He also owned a thousand cheetahs, but these undomestic beasts were contrary when it came to sex. The whole menagerie produced just a single litter. The next cheetah to be born in captivity was in Philadelphia in 1956. Nothing is more easy than to tame an animal, and few things more difficult than to get it to breed freely under confinement (in the cheetah, the female's need for a sexual chase by several males before she can ovulate is part of the problem). That may be why so few animals have been domesticated and why dogs, horses and cows were housebroken just once or twice.

For wild animals, from wolves to cheetahs, to accept human hospitality reveals variation usually hidden and provides selection with new and often unwelcome opportunities. Zoos must cope with a legacy of damaged genes. Although of no apparent value in themselves, they hint at how much diversity is concealed beneath the face of Nature. If faulty genes are rare and the animal common, then few will be unlucky in the biological lottery, as most of the defective copies are masked by a normal version. Sometimes, though, two damaged copies get together and the carrier pays the price. This happens most often when relatives mate, as – by definition – each descends from a shared ancestor who might himself have carried a single faulty replica of the gene. When brother and sister white-footed mice are mated in the comfort of a cage, their progeny released into the wild die at four times the rate of those of unrelated parents.

For zoos, evolution is a problem harder to solve than is the diet of the gorilla. A limited pool of parents means a subsequent genetic change. White tigers have been known since 1834 and were once much pursued by hunters. Now hundreds are in captivity. Each descends from a single male, Mohan, captured by the Maharaja of Rewa in the 1950s. More than a hundred have been bred in Cincinnati Zoo alone. Albino cats are popular, and the Philadelphia white lion brought in a million dollars a year in tickets. They are a symptom of domestication (helped in this case by careful family planning, as its father is also its grandfather and great-grandfather). Animals forced into cages hence reveal the genetic secrets of the wild.

Zookeepers suffer from other kinds of unplanned selection. As wild animals are hard to breed in captivity, their guardians tend to choose those best able to cope with their new circumstances. That means change – towards tameness and, as in the foxes kept for fur, towards other things as well. All Przewalski's horses – reduced to a core of thirty-two animals, worldwide, in the 1940s – descend from a dozen wild ancestors caught in Mongolia, and from a single female Mongolian pony. In their early years, inadvertent preference for the animals most ready to accept captivity increased the contribution of their tame progenitor until some herds traced a fifth of their ancestry to that single pony. Although keepers have now set out to purge the stock of their ancestor's insistent genes, the instant response to selection for good behaviour has a lesson for all other inhabitants of zoos. Although the horse's saviours hoped to keep a wild animal safe, what they have done is to evolve a new one.

Today's conservationists are well aware of such dangers. Their philosophy is quite opposed to that of farmers (who hope to select the best as hard as they can). Hundreds of 'species survival plans' have been made in the hope of combating the power of natural selection. They involve the exchange of animals among zoos to keep up diversity and reduce local adaptation and, at the same time, strenuous attempts to ensure that all families have the same number of young in order to minimize the chance of inherited differences in reproductive success. All this is an attempt to bend the Darwinian rules and will delay rather than solve the problem of evolution behind bars.

Not all prisoners change their nature. No matter how much an unnaturally tame goose – or a white tiger – alters as its ancestors retreat into the past, each retains much of its history in its genes. Arabian oryx returned to Oman after eight generations in captivity could still navigate across the desert to find water a hundred miles away. No inmate of a zoo has been so domesticated as to change its scientific name, to lose its natural identity in favour of one imposed by man. Nobody enters a lion's cage unescorted and, in the United States, elephant-keepers have the most dangerous job of all – more so than the police, with one keeper in six hundred killed each year.

Dogs, too, retain a past much older than their alliance with man. Each is a barely evolved wolf, a descendant of a dangerous animal. That is part of their attraction. Their owners glory in their pet's vicarious tie to Nature. Jack London, in his 1903 novel *The Call of the Wild*, tells the tale of a pampered domestic animal, a cross between a St Bernard and a Border collie, stolen and taken to the Klondike. The animal learns the truth: 'Kill or be killed, eat or be eaten was the law; and this mandate, down out of the depths of Time, he obeyed'. The book sold two million copies. London, a social Darwinist of virulent stamp, had a political agenda but his story has a scientific message. After a series of bloody adventures, Buck, the canine hero, emerges in Nietzschean style as the leader of a pack of pure-bred wolves. The 'dominant primordial beast' – the wolf within – had at last come to the fore. Domesticity was a mere stage in his history, reversed when the wild called with more insistence than did the fireside.

By owning a dog, any dog, men welcome into the home a beast that preserves much of its primordial self. Overgrown juveniles though they are, evolution by human choice has not removed the instincts of their ancestors. In the United States, a hundred times more people are murdered each year than are killed by dogs. Even so, their pets slaughter about twenty people a year (although the wolf itself has killed no one within the past century). Like wolves, dogs attack the weak, be they children, old, or drunk. Packs of feral animals have pulled infants from bicycles and eaten them, and a mere half-dozen beagles, dachshunds and terriers once devoured an eighty-year-old woman. The homicidal packs relive their past. All domestics retain some of their ancestry and every child licked – or savaged – by its pet is fresh proof of the fact of change.

Now, the dog has disappeared, victim of evolution. Two centuries after it gained its scientific name, the International Commission on Zoological Nomenclature has stripped it of its identity. Their action accepts the transition between breed and species.

As anyone who walks in a park knows, any dog is ready to discuss its genes with any other, as long as it is on heat. Even so, there are barriers to full sexual agreement among breeds. No owner of a pedigree bitch is happy to see it mate with a mongrel, and breeders ensure that most of their aristocrats never come across an animal not of their own kind. If an owner's guard slips and male chihuahua meets female great Dane, mechanical constraints to copulation come into play, or (if roles are reversed) the birth of a giant pup to a mother so much smaller than the father causes problems. Even when large forms like great Dane and St Bernard are mated, the young are defective as they inherit so many genes for abnormal growth.

Linnaeus – a classifier of what he saw as a fixed biological universe – named the domestic dog as a separate species, *Canis familiaris*. For him, it had the same status as the elephant or the tiger: an animal so removed from its relatives as to demand a label of its own. The International Commission on Zoological Nomenclature defines its species more stringently. Dogs as an entity were doomed, because once out of its owner's sight, any dog is happy to have sex with any other – and, even, given the chance, with a wolf. The DNA traffic was in both directions. Dog mitochondria hint at an influx of Russian wolf genes and, because of crosses between wild and tame, most European wolves have accepted genes from their descendants.

Sex was, for the lexicographers of life, enough to destroy canine independence. In 1993 the Smithsonian Institute's *Mammal Species of the World*, the *Who's Who* of mammals, admitted the domestic dog only as a subspecies of the wolf *Canis lupus*. As *Canis lupus familiaris* it joins the American wolf *Canis lupus occidentalis* and the European wolf *Canis lupus lupus*; each so much alike as to be ranked as mere varieties of the same animal. The Smithsonian's decision is at the centre of the theory of evolution. It shows how arbitrary is the distinction between breed and subspecies, and how a subspecies may, through sexual choice, gain a personality of its own. The accumulative action of selection, whether applied methodically and more quickly, or un-consciously and more slowly, will always cause life to change. Dogs are still wolves beneath the skin; and for every creature, domestic or wild, once evolution is at work a change of identity must follow.

CHAPTER II

VARIATION UNDER NATURE

Variability — Doubtful species — Essence versus individual difference — Hidden or cryptic species — Species resemble varieties in being closely, but unequally, related — Defining life's frontiers — The undiscovered universe and the overcrowded Ark — Races of mice, in Laboratory and Nature — Hybridism and Identity — The law meets evolution — Genetics and the bird-watcher's dilemma

BIRDWATCHERS AND ORNITHOLOGISTS are not at all the same. To the latter, everything about birds is of interest – how they migrate, where they breed, or what they eat. Birdwatchers have a single concern, which is to see as many kinds as they can. Once seen, as soon forgotten, or, at least, ticked off and added to the Life List that is the basis of their self-esteem.

I went through the same phase. After the usual interest in stamps and an eccentric deviation into cheese labels, I was given a pair of binoculars. At the age of twelve I eavesdropped on a group of excited amateurs (twitchers, as they call themselves nowadays) as they peered at some gulls bobbing, on a dim winter day, on the then filthy waters of the River Mersey. All agreed: one of the birds was not an ordinary herring gull, but the much scarcer glaucous gull, seldom seen so far south. My problem was that I could not see any difference. A member of the flock was a rarity, but which was it? Did it count? Could I check the box in my bird book? It was my introduction to the ethics of science. I admit it: I made the tick, but then I rubbed it out.

Twitchers, like scientists, belong to a fellowship of faith. They play cards against Nature. It is possible to win every time by faking one's hand, but to do so removes the point of the game. That is the strength of science, and its greatest weakness. Without collective trust it could not work. Instead there would be the dismal apparatus of mutual suspicion familiar to every accountant.

Birding is refreshingly free from fraud. It has had its scandals, such as the notorious case of the Hastings Rarities (a set of bizarre sightings on the South Coast in Edwardian times), the dubious claim of a Dalmatian pelican in Colchester in the 1960s, and more. Even so – and whatever the rivalry amongst the twitching fanatics – most of those involved in the sport play by the rules.

One problem baffles the most ethical birdwatcher. Stamps (or cheese labels) are easy. The 1853 One Shilling Cape of Good Hope Triangular (or the 1954 Vache qui Rit) is, or is not, genuine. Fakery apart, there is no reason to question the object itself. But what if it is ambiguous? Then opinion, the enemy of science, creeps in. Is one kind of bird really unlike another? How different does it have to be to count as distinct? What, indeed, is a 'species' in the first place? Does it have a scientific definition, or is it all in the eye of the beholder?

Most people can tell gulls from terns. Many can separate the herring gull from the lesser black-backed (look at the back, which is pale in the first and dark in the other). My friend the glaucous gull is more subtle – as large as a greater black-backed, but as pale as a herring gull, without its black wingtips. A real expert can even sort out the Iceland gull (like a small glaucous, with longer wings). Birders still argue about the existence of the yellow-legged gull as a distinct entity, but its 'bold, confident look' is said to be diagnostic.

But how to deal with variation within each bird? Herring gulls from Estonian bogs in the 1950s had, some say, yellow legs, but these have now disappeared in favour of the pink legs found everywhere else. Those from the Atlantic islands are on occasion blown to Britain. Their backs are almost as black as those of lesser black-backeds. A 'generally stouter bill' might help, but what use is a generalization when an individual must be sorted out? If the despondent twitcher were to travel to Iceland he would be frustrated by hybrids between the Iceland gull and its commonplace relatives. Where does he put his tick?

Species and nations have a lot in common. What, for example, is a German? The tribe has a shared and guttural means of communication that interrupts intercourse of most kinds, but the character is equivocal, for Austrians speak the same language. Since 1913, the country has defined its own citizens by descent, by German blood (whatever that might be). It includes within the realm the remnants of the Saxon diaspora (many of whom – Romanians included – cannot speak German at all), but cuts out children born in Berlin of Turkish parents. One badly behaved teenager was deported to Turkey even though he was born in Germany; while at the same time a hundred thousand Romanian-speakers of approved blood were allowed in. Until the fall of the Berlin Wall, indeed, a geographical barrier made many German citizens more alien to each other than Westerners or Easterners were to the French or the Poles. A century ago German identity itself meant little, as there were only Prussians, Bavarians and Rhinelanders, political entities of their own, each now reduced to variants within some greater Teutonic whole.

The problem of how to define Germans, or any nation, arises because the question is ambiguous: does it turn on shared appearance and behaviour, on geography, or on descent? Is a country an historical entity, or should it be identified only on criteria that apply today? How much can frontiers be allowed to leak before a nation loses its essence? When will Germans be seen as Europeans, as Prussians have become German?

Such problems of identity turn on natural variation, the raw material of evolutionary change. Like a politician, the twitcher has to deal not just with differences among individuals but with the subtle distinctions that separate each kind. The difficulty of how to define domestic breeds has been magnified and transferred to the world as a whole. Twitchers are asking a question older than the theory of evolution. How should they deal with forms that possess in some considerable degree the nature of species but are not classified as such?

Taxonomy, the science of ordering life, has to worry about that problem. Needless to say, many animals and plants are easy to tell apart.

If they were not, birdwatching and natural history museums would each go out of business. One tribe in New Guinea recognizes a hundred and thirty-six kinds of bird, just one fewer than that accepted by the experts. Experts and tribesmen have the same philosophy. Each needs an archetype, a gold standard, to allow their specimens to be put in the correct cabinet.

Once, all taxonomists worked in the same way. An animal was killed and its remains stuffed, pinned or bottled. Then, it was described in the scientific literature. The cadaver was the 'type' against which others could be checked. In 1868, in China, the French missionary Père Armand David saw the skin of a black-legged white bear. It resembled animals shown in ancient works of art and until then assumed to be polar bears brought back from the north by hunters. The first specimen of the mysterious beast was collected in 1929 by Theodore and Kermit Roosevelt, the sons of the President. They shot a giant panda asleep under a tree. Its body gave the animal entry to the pantheon of mammals as *Ailuropoda melanoleuca*. It joined the world of science as had all its relatives, as a corpse.

Now, fewer than a thousand pandas are left. In China, to kill one means the death penalty. Taxonomists, too, are more careful with their material than once they were. The essence of a species can now (or so museum-keepers believe) be preserved not in its bones but in its genes. The Bulo Burti boubou shrike of Somalia was recognized in 1991 on the basis of the DNA sequence in a feather shed by a captive bird. The type specimen – the very substance – of this new form is a set of dark bands on a photographic plate. The rest of the bird was released.

Not all pandas – or Bulo Burti boubou shrikes – are alike. They may look the same but are, like whales, dogs, or viruses, full of diversity. Classifiers hence face a fatal temptation: to split their animals into too many groups. As in the Kennel Club or the United Nations, quarrels break out between those who like to subdivide the world and those who hope to unify it.

A rich nineteenth-century collector, Isaac Leigh, was interested in the freshwater mussels of North America. He named more than a thousand kinds on the basis of tiny variations in shell-shape and size.

Now the number has been reduced by two-thirds. A hundred and two of his types are classified as one. Isaac Leigh was too enthusiastic about his varieties. His cherished diversity was no more than that between people with brown or blue eyes or between the pink- and yellow-legged herring gulls that once infested Estonian marshes. He had, nevertheless, put his finger on a problem that still plagues museums. How should they fix the frontiers between supposed entities when each is filled with variation?

Genetics, the science of differences, has not made their job any easier. Before it began it had often been asserted – but the assertion was quite incapable of proof – that the amount of variation under nature is strictly limited in quantity. Now, the claim can be tested, and it fails.

Most members of most species do not look much different one from the next. Any fruit-fly is much like another, and even their best friends find it hard to tell mice apart. In spite of some exceptions – the colourful snails or butterflies that come in dozens of forms and are still studied by a few outmoded naturalists – to share a Latin name imposes, almost by definition, a certain uniformity upon those who bear it. That comforts both creationists and experts on taxonomy. They like to see existence as a set of neat ideals, each filled with some pure Platonic essence. However, a great deal is hidden within even the most uniform creature. Genetics shows that no one – not even the glorified chemists which most biologists have become – can any longer suppose that all the individuals of the same species are cast in the very same mould.

Systematists are far from pleased at finding variability in important characters, and that there are not many men who will laboriously examine internal and important organs, and compare them in many specimens of the same species. When they look for diversity – with microscopes or with DNA-sequencing machines – they usually find it. A new science of classification has emerged, based not on characters visible to the eye, but on the genes themselves. Individuality is everywhere. Part of a plant or an animal's variability in shape or size comes from the conditions of life. Nevertheless, to move the eggs of small birds to the nests of large, or to plant the same seeds in different places shows that about half the differences in such characters reflect genetic variation.

Much more variety lies beneath the surface. Most animals, even the simplest, have inherited cues of identity on their cells; precursors of the blood groups and the other genes of the immune system found in more complicated creatures. Chromosomes vary in shape, size and arrangement; and the proteins themselves are filled with difference. Most molecules move when placed in an electrical field. They slow down when passed through a jelly that filters them by shape, size and electrical charge, to a degree that is altered by slight changes in their structure. The technique reveals a mass of variation in certain proteins, the products of genes. Now, it is much applied to DNA itself, either to measure the lengths of the pieces that emerge when the molecule is cut in particular places, or to help put its individual units into order. The method was the key to a universe of variability and to what became known as the 'find them and grind them' era of evolution.

For a time it seemed that all Darwinian problems could be solved by a glance at differences in protein structure, cheap, simple and tedious as the work was. All conceivable plants and animals (and some scarcely so) were subject to the art. Variation goes from zero in a few plants and in sea-elephants to situations in which half the genes tested in a particular species may be present as alternative forms. Species can, in the new world of the molecules, no longer be seen as absolutes. They are not units, but groups of individuals, each with a biological personality of its own. That poses a question about their very nature. The DNA of two unrelated mice or deer is separated by a million or more differences. How can taxonomy work if its subjects can no longer claim an unblemished identity?

In fact, genetics gives the science a status it once lacked. An animal's place need no longer depend on the opinion of an expert. Instead, it can be put into context, its inborn uniqueness in fact shared with neighbours, with other populations of its own kind and with forms more or less distinct. The genes reveal a hierarchy of difference, from population to variety to species and on to life's higher divisions. Sometimes the gradation is clear, and sometimes less so, but order nested within order is all around. The pedigree of life can now be drawn in nucleic acids.

At first sight, some animals seem detached from such biological reality. The differences between two populations of the same snail in adjacent Pyrenean valleys are greater than those between man and chimp. Nevertheless, the overall picture is clear. Thousands of plants and animals reveal a close fit between the divisions revealed by genes and the groups long used by classifiers. From seaweeds to mammals, separate populations of the same species share about nine-tenths of their protein diversity, while the variation common to any pair of related species ranges from a third to threequarters (although birds, it must be said, do not fit into this neat picture as the differences among them are small). When mammals are compared with birds, or flies with snails, descent with modification is undisguised because it has gone on for so long.

Genes also show the actual numbers of distinct kinds of plant or animal to be far higher than experts once thought; in part because so little of the world has been explored, but, quite often, because DNA shows that what was at one time classified as a single form of life is in fact several.

Nobody knows how many different species there might be, even in a taxonomy based on external appearance. It takes a long time to sort out even simple groups. The first list of British butterflies was made in the year of the Great Fire, and by 1710 almost half of those known today had been discovered. Not until the late nineteenth century, the era of the great amateur naturalists, was the list complete.

The situation is worse for less familiar beings. A million and a half kinds of plant and animal have their own names, but even among well-studied groups, such as insects or worms, three times as many may remain to be described. To gas the inhabitants of a tropical tree can reveal a thousand new kinds of beetle. About seventy thousand fungi have been given a label, but the experts agree that a million are still undiscovered. The sea is an unknown – and enormous – land; if its waters were evenly spread, the whole globe would be drowned under a mile of water. The oceans are filled with mystery and may hold from half a million sorts of animal to twenty times as many. A handful of marine mud can hide a hundred different kinds of nematode worm. In 1995 a whole new phylum (the largest

inclusive group of creatures), one of just thirty-five in the macro-
scopic world, was found on the sea bottom, its members attached to
the lips of lobsters. So much is unexplored – the deep sea, the rainforest, even the soil in suburban gardens – that the world may
contain a hundred million different kinds of plant and animal.

Some are hidden not beneath the oceans, but nearer at hand. As
the Victorians surveyed the globe, five hundred new mammals such
as the mountain gorilla and the terrible mouse (the size of a
fox-terrier) were brought to light each year. The number has
dropped to an annual hundred or so, but splendid discoveries are still
made. In 1992 strange horns were seen in the homes of Laotian
hunters. Two years later, their owners were tracked down, scattered
over a range of wooded mountains. They are the saola, a bridge
between oxen and antelopes. Eleven of the eighty known kinds of
whales and dolphins were discovered in the present century. In the
last few years remote places have revealed new deer and antelope, ten
new primates and a whole host of rodents. The world's list of
mammals is about five thousand long, but three thousand more may
be waiting in the wings. Plants, too, hide a multitude of unknown
forms. Almost half the palms of Madagascar have been discovered in
the past decade. In Mexico, a new relative of maize was found, a
plant of great potential whose presence had been missed by generations of collectors.

Molecular variation under nature reveals divisions invisible to the
most skilled taxonomist. To journey into the rainforest or the garden
will uncover new kinds of being, but to sequence the DNA of those
we know may reveal many more.

The pipistrelle is Europe's commonest bat. What seemed a single
animal is now known to be two, distinct in their genes and with
squeaks of different pitch. When it comes to the leopard frog of North
America, what was once classified as a species is now thought to be at
least twenty-seven, different in their DNA and quite unable to cross.

Genes ask old questions about species in a new way, often to the
discomfiture of those who classify the world. Most African elephants
roam in herds through the savannah, but others live a more solitary
existence in forests. They meet at salt licks and live on leaves and
fruit. The molecules show the forest elephants to be distinct from

their familiar kin. Whether each deserves a scientific name of its own is so far undecided. In determining whether a form should be ranked as a species or a variety, the opinion of naturalists having sound judgement and wide experience once seemed the only guide to follow. Today, evidence of separation also lies in the genes: in whether two animals can mate and have offspring. Nobody has yet dared to try to test the sexual desires of the forest and the savannah elephant, and in most new-found species the chance to do so never arises. As a result, biochemists have taken on the naturalists' role and the genes have become the touchstone of identity.

The issue of who belongs where in the natural world can sometimes be side-stepped with 'varieties', 'races' or 'subspecies'. What these are often depends on who studies them. If one compares several floras – lists of the plants in a particular place – drawn up by different botanists, it is possible to see what a surprising number of forms have been ranked by one as good species, and by another as mere varieties. The same is true of animals. Seventeen and a half thousand species of butterfly have been described – but they are divided into a hundred thousand subspecies. All this points at the quandary faced by those who make lists. Where do the boundaries lie?

Noah, the world's second taxonomist (after Adam) had to decide who to allow on to his Ark. He took on board seven pairs of each of the biblical clean animals (ruminants, those who chew the cud and have cloven feet) and a pair of each of a selection of the unclean beasts (including, it seems, all the insects – those that 'walk on many feet'). His Ark is estimated by Biblical scholars to have been four hundred and fifty feet long. Nowadays it would have to be a lot bigger. As well as three hundred and thirty thousand kinds of beetle, the Ark might – depending on Noah's views on classification – have to accept the half-dozen named varieties of tiger and twice as many leopards. Chevrotains – a group of foot-high deer found in Asia and Africa – have a hundred and twenty subspecies (or so some experts claim), each of which would have to argue itself on board.

There is a rift in the fraternity of museum-keepers. They need to classify their specimens in order to preserve them. But what is to be done about those awkward cases in which a bird or a mosquito cannot be checked off in its own box? The essence of any collection of stamps or of teapots must be that each specimen belongs in a distinct category. If they do not, the whole system breaks down. In biology, the urge for order has to defer to the reality of change. For plants and animals – unlike cheese labels – even to discuss whether a particular form is a species or a variety is, quite often, vainly to beat the air.

Familiarity breeds varieties. It is within the best-known countries that we find the greatest number of forms of doubtful value, because only in such places is enough known to blur the boundaries between what might appear at first to be distinct. Seven forms of European wild cat have been named – and their ranges coincide exactly with national boundaries. In the same way, many subspecies and races of voles, butterflies and more have been described from the islands around the British coast. Most attain that status only because Britain's plants and animals have been so much studied.

If any animal or plant in a state of nature be highly useful to man, or from any cause closely attract his attention, varieties of it will almost universally be found recorded. The mouse has always been in the public eye, although its image has changed from pest to scientific specimen. It reveals, as does no other animal, how much varieties and species have the same general character and how hard it can be to tell one from the other.

Mouse bones are found in association with man six thousand years ago, at Catal Hüyük in Turkey. There they were useful, as the skeletons suggest that they were skinned (perhaps to make clothes). Apollo Smintheus, God of Mice, was worshipped in Homeric times and, some say, in Crete until the Middle Ages. The ancients knew of their diversity. Pliny speaks of white mice used in fortune-telling, and, in China, government records list thirty such animals caught between AD 307 and 1641. Japan had mouse-fanciers two millennia ago and Mendel himself bred grey and white specimens in his monastery room (but had to keep his results secret as pets were forbidden). Not until the present century did the mouse come into its own.

Just before the First World War, Miss Abbie Lathrop, a Massachusetts schoolteacher and failed poultry farmer, set up a business for the sale of pet mice. Her first attempt with 'waltzers', mice with an inner-ear defect that caused them to dance, did not much amuse the public. Her business took off when she began to sell animals to laboratories.

From her stocks came many of the mice now used by their millions in research. As brothers were mated with sisters, forced incest caused the animals within each inbred line to become, in effect, identical twins, all with the same genes. Those refined beasts were a sample of the diversity of their wild ancestors. Variation no longer circulated through the population as a whole, but was parcelled out among lines, each a repository of a part of the total. For mice in Nature, individuals are different but populations are much the same. In the laboratory the opposite is true. How much the lines differ is, as a result, a measure of variation in the wild. Each stock has become, in effect, a new and artificial race.

Mouse lines are distinct in many ways. Some are fat, some thin, some active, some passive. Some prefer sweet foods, others salty. Some, given the choice between water and a beverage the strength of gin go, to a mouse, for the latter, while others are strict teetotallers. A hundred thousand mice are used each year to test what substances might cause cancer. A certain chemical increases the incidence of cancer in some strains, but decreases it in others. Size, weight, colour, heartbeat, sexual activity – all vary among lines and all are testimony to how easily a wild beast can be subdivided.

To read the catalogue of strains kept by the Jackson Laboratory in the United States (the mouse's Library of Congress) is to retrace their history. The C57Bl stock, the most used of all, originated with Miss Lathrop in 1921. It is now divided into substrains, kept separate in the lab, which since those days have diverged in how much they rattle their tails, the rate at which they age and in their liability to cancer.

Wild races of mice, too, can be defined by genetic differences that seem too great to be accommodated in a single scientific name. Most come from isolated places: not laboratory cages, but islands and remote valleys. There, they can evolve undisturbed. More than forty races have been named in Europe alone. The mice of the Welsh island of Skokholm are a quarter again as large as those on the

mainland, while those on the Isle of May are docile. Some populations even have different numbers of chromosomes. Most mice
have twenty pairs, but in many places certain chromosomes become
fused together. More than a hundred fusions are known, to give local
races in the Alps, in Greece and on the Orkneys. The Val Tellavina
in the Italian Alps contains, within ten miles, five distinct chromo-
somal types. Because such animals find it difficult to cross with
others, each race can be defined not just by how it looks, but by how
easy it finds it to have sex with the neighbours.

The biography of mice is much the same as that of gulls, or
mussels, or any other animal studied in enough detail. At first, there
seemed to be just a single kind. Pliny introduced the term *musculus*
or 'little mouse' to separate mice from rats. It was borrowed by
Linnaeus as the scientific name for his familiar Swedish animal, *Mus
musculus*. By 1940, there were almost one hundred and fifty species
of mice, each with its own Latin name, but then, in a dramatic move
towards simplicity, they were all lumped into a worldwide mouse
under the Linnaean label (although a dozen or so subspecies were
allowed, to satisfy the splitters).

That was too bold. At least seven genetically separate units are
hidden within the supposed entity and each now has its own
Linnaean name. Some are widespread like *Mus domesticus*, common
in Western Europe, the Americas and Australia, while some are
confined (as in Spain and Portugal, which have a mouse of their
own). Some have odd habits – the steppe mouse, for instance, makes
hillocks in which it stores grain and spends the winter. Although a
few of the various kinds can be crossed in the lab, most never mate
in the wild. Their DNA is quite distinct (which means that wild
mice are a valuable source of new genes). Genetic tests on inbred
strains show some to be of hybrid origin. Indeed, part of the heritage
of C57Bl comes from quite another species, as a hint at a chequered
past under Miss Lathrop's care.

The mouse's hierarchy of change is not unbroken. Where the true
Mus musculus (an animal of Scandinavia and North-East Europe)
meets its western and southern relative *Mus domesticus*, as it does in
a great curve from Denmark through Germany and across the
Balkans to the Black Sea, they hybridize. Different as they are, their

males and females can mate; but the two do not merge. Just a few miles away from the zone, their DNA patterns are distinct. Within it, the offspring of mixed pairs are unhealthy and full of parasites, and their sons are sterile. This is enough to save European mice from fusion.

Even so, the frontiers are blurred and, over a territory thousands of miles long, but no more than ten across, have sprung a leak. That reflects a history of divergence as varieties make the painful transition towards an identity of their own. The centre of genetic diversity of the world's mice is in Pakistan and India. *Mus musculus* and *Mus domesticus* began their journey there, in a common homeland and with a shared pool of genes. Their ancestors travelled in man's wake, in separate waves north and south around the icy Alps as farmers moved west. As they went, they evolved, until, at last, when the circle was closed and the mice met, each had changed enough to render them incompatible. As a result, each kind has been promoted to the status of a species rather than remaining as mere varieties of the same one.

Some places – the Pyrenees, or the Alps – are full of such hybrid animals and plants, from newts and toads to hedgehogs and oaks. At the time of the ice ages their inhabitants were driven into fastnesses in South-West and South-East Europe. There they were transformed, and descent with modification created forms that, when they met, were less able to exchange genes than before. Each, like the mice, is an evolutionary adolescent, poised on the edge of independence, but not yet ready to cut itself off from its relatives.

The inescapable conclusion from all this is that species are strongly marked and permanent varieties; their permanence assured by a block to the exchange of genes. Hybrid zones show how hard it is to draw a distinction between the two. Mice, dogs, newts, oaks and more are alive and are always in flux. Because of evolution, all of them must change as the years or the miles roll by. As descent always involves modification, resemblance decreases as a shared ancestor recedes into the past. Just as in families or nations, the decision as to who to let in is often a matter of taste. What seem like solid frontiers may soon melt away.

The courts have long been involved in demarcation disputes. The conservation movement means that they have no choice but to be

dragged into evolution as well. The United States Endangered
Species Act is stout in its defence of those at risk of extinction: it can
protect 'any distinct population segment of any species of vertebrate
fish or wildlife which interbreeds when mature' or is 'an important
component in the evolutionary legacy of the species'. What, if anything, can that mean and who is to decide on which population needs
protection?

The dusky seaside sparrow – an animal defined by its unusually
dark feathers – once lived in a patch of scrub around Cape
Canaveral. In the 1960s it went into decline. The US Fish and
Wildlife Service spent five million dollars on its last scrap of habitat,
but still the numbers went down. Five of the last six birds (all males)
were captured. The final hope for their precious genes was to cross
them with another kind of sparrow from nearby. The courts quashed
the plans on the grounds that the progeny would be hybrids, with no
rights under the law. The bird was declared extinct in 1987.

Conservationists were outraged. How could this piece of
American heritage be allowed to die? Although the government
refused to become involved, such is the majesty of a Latin name that
the fight goes on. The attempt to save some version of the bird continues, at Disney World.

The precious sparrow was defined, as are many birds, by a slight
change in the way it looks. It takes an expert to separate the dusky
version from its vulgar and unprotected neighbours. The tale has a
biological twist. Dusky seaside DNA is, as far as can be seen,
identical to that of other, commonplace, sparrows nearby. The
famous bird is not a separate entity at all. The genes do reveal a deep
split between the seaside sparrows of the Atlantic coast of Florida
(the dusky form included) and those – to the naked eye almost the
same – on the Gulf of Mexico. Where does the boundary lie? Which
is the species and which the variety?

In 1990, the Fish and Wildlife Service gave up the anti-Darwinian
struggle. It no longer placed hybrids outside the law. Texas cougars
were released into Florida in the hope that they would breed with
and sustain the last Florida panthers (already inbred, with most of
the males plagued by undescended testicles). The Federal
Government is now reduced to the preservation of 'evolutionarily

significant units', whatever those might be. The dignity of the law
has met the volatility of Nature and has been forced to retreat.

Whatever species may be (and they are not what birders or govern-
ments might hope), they are not fixed. Instead, their boundaries
change before our eyes. What is a mere variety to some is granted its
own identity by others. Quite often, animals that are similar on the
surface differ in their genes. All this is grist for evolution, for the
transition between variation within a single kind and the origin of a
new one.

The biggest difficulty about species is to decide what they are. The
problem is time; to describe in the two dimensions of today something that evolved in three. For classifiers, what matters most is the
future. All plants and animals are, with hindsight, the same because
they all descend from an ancestor three billion years old. At the
present day they may hybridize or stay apart; and biologists can
spend useful lives in deciding where lines should be drawn. To
taxonomy, though, their essence lies in years to come. Museums
assume (and it seems fair) that cats and dogs are separate because
there will never be an animal that traces a shared descent from dogs
and cats. That assumption is impossible to test.

When it comes to the nature of species, a certain pragmatism
helps. The word comes from the Latin *specere*, to look at; but simple
appearance is not enough. No one definition has yet satisfied all
naturalists; yet every naturalist knows vaguely what he means
when he speaks of a species. The amount of difference considered
necessary to give to two forms the rank of species is quite indefinite.
One text on evolution reviews seven statements of what the mythic
word might mean and – as does every attempt to impose order on the
chaos of life – fails. Of course, to an evolutionist, it should.

Most biologists have a guilty secret: they started as birdwatchers.
Lord Rutherford claimed that all science is either physics or stamp-
collecting. In some ways he was right, but biologists have the excuse that
they were the first to see that life is not just an album of specimens.
Instead, it is fluid, and in its inconstancy reflects the fact of change.

Birdwatching is a more refined pastime than it was in the days
when Liverpool was a port and a twelve-year-old could tick off gulls

and at the same time check funnels for shipping-line symbols. Now the twitchers have their own association. The UK400 Club is so named because to see four hundred different kinds of bird in Britain is the mark of a lifetime's dedication. Twitching is a competitive business. Three of the top five enthusiasts for the sport are called Steve (and three of the top hundred are women). All have more than five hundred feathers in their caps. For them, the glaucous gull is dull, given the choice of great black-headed, Mediterranean, laughing, Franklin's, little, Sabine's, Bonaparte's, black-headed, slender-billed, ring-billed, common, herring, Thayers', yellow-legged, Iceland, lesser black-backed, greater black-backed, Ross's and ivory gulls to be checked off on the British List.

Even for the heroes of the birding world, evolution raises its vexatious head. Take the yellow-legged gull, admitted as a 'tick' after much vacillation about its status (confirmed when it spread to Northern Europe in the 1970s and lived alongside its cousins but did not breed with them). The UK400 guide has it that: 'This species is distinct from Herring Gull. In fact, it is more related to the Lesser Black-backed Gull, with which it sometimes hybridises. The races *atlantis*, *michahellis*, *cachinnans*, *barabensis* and *mongolicus* are included within this complex, whilst Armenian Gull *armenicus* is considered by some to be a further species. This isolated form breeds on the Armenian Lakes, Turkey and Iran and winters in northern Israel. The Arctic races *heuglini* and *taimyrensis* are best treated as Siberian races of Lesser Black-backed Gull, whilst the race *vegae* is best lumped with Herring Gull or treated as a separate species'.

That statement, in its petulant tone, contains within itself the theory of evolution: that differences blend into each other in an insensible series; and a series impresses the mind with the idea of an actual passage. For twitchers to treat life as a glorified stamp album is entirely to miss its point.

CHAPTER III

STRUGGLE FOR EXISTENCE

Geometrical powers of increase in Man and other creatures — Rapid increase of
naturalized animals and plants, from cows to waterweeds — Competition
universal: cycles, stalemate or disaster may each result from a Struggle for
Existence — The term used in a wide sense — Nature of the checks to
increase — Effects of climate, and the risks of the marketplace — Shortage
of food — Complex relations of all animals and plants throughout Nature —
The battle for sex — The fisherman and the struggle — The relation
between organisms and the failure of conservation

THE UNITED STATES National Park Service is expert at letting in
daylight on magic. Ellis Island was, for forty years after it closed, a
magical place. Because it could not be visited it could be invented. As
the gateway to seventeen million immigrants, it was a metaphor for
America itself.

Now, the island is just another New York tourist attraction, run
with federal efficiency for the descendants of those who passed
through it and no more mystical than Disney World. Instead it is
instructive. A helpful bar chart in the reception hall shows the size of
the United States population from its foundation, projected into the
future. The picture is of continuous growth to the middle of the next
century, when the display stops with the promise of four hundred
million Americans.

The nation now holds two hundred and seventy million
people, most of them descendants of European immigrants.
Its history has a message for evolution: that the existence of any

creature is a constant struggle against relentless forces.

America is full of failures. The Irish arrived and died out, as did the Welsh. The Vikings were around even earlier but abandoned the country in 1013. Some stopped on the way, in Greenland. Their numbers reached six thousand, but by 1361, when a priest came to see what had happened to his flock, all had gone. They had ignored the sea in order to farm. The record of the ice shows that a series of bad summers in the 1350s had starved them. The colonists even killed and ate their dogs.

Later attempts by the English to find a toehold in the Americas also collapsed or fell into terrible difficulty. The 'Lost Colony' of Roanoke was founded on an island off the coast of the present North Carolina by a hundred and fourteen people, who landed under the direction of Sir Walter Raleigh in 1587. Although the first New World child of English parents was born there, three years later neither she nor anyone else was left. Another venture in Virginia received seven thousand immigrants in the two decades after 1606. In that time, six thousand died, their struggle over. Tree rings show that the worst drought for five hundred years had struck just before their arrival, and another almost as fierce followed a few years later. Crop failure and famine killed the settlers.

Most migrants share a history of repeated disaster by many and dramatic success by a few. Not many attempts to reintroduce threatened animals into the wild have worked. Millions of dollars have been spent to airlift the eggs of Kemp's Ridley sea-turtle to Texas. Not a solitary 'head-started' animal made it back to the beach. Most of the claimed successes are mitigated failures. The California condor was declared extinct in 1987. Since then, some have been released from their Los Angeles 'condorminium' but the few who survive must be trained to come to food. They, too, have come to grief in the struggle for life.

All colonists have a natural desire to make their new land more like home. In the nineteenth century, all over the world, acclimation societies were founded to do the job. The rabbits let loose in South Australia in 1859 succeeded, but they followed an earlier Antipodean attempt that did not. In New Zealand, a hundred and thirty kinds of bird were set free. Just twenty lasted to the 1980s. What mattered was how many were released. Eight out of ten species with more than a hundred set loose prospered, compared to a quarter of those with fewer birds. Four hundred yellowhammers were released in New Zealand, where the bird flourishes, thirty into Australia, where it sank from sight. Any small group of animals gambles with limited capital in the great casino of Nature. As a visitor to Monte Carlo soon finds out, that means inevitable ruin. Sooner or later, the money has gone and it is impossible to continue.

In the end, America – and the Americans – made good. Two decades after its foundation, Virginia was ten thousand strong. Throughout the eighteenth century its population grew, mainly through its own sexual efforts rather than by immigration. Patrick Henry ('Give me liberty or give me death'), first Governor of an independent Virginia, himself had eighteen brothers and sisters. Now the state has seven million people. A few years after the Revolution, an elderly Rhode Island woman claimed five hundred descendants, half of them alive at the time of her death. The population of England and Wales increased by a quarter in the first half of the eighteenth century, while that of North America multiplied six times. Benjamin Franklin, in his 1751 essay *Observations concerning the Increase of Mankind, peopling of Countries, etc.*, concluded that 'our people must at least be doubled every twenty years . . . in another century . . . the greatest number of Englishmen will be on this side of the water'.

Thomas Malthus, an English cleric, was alarmed by such figures. American growth could not, he said, be sustained. It was 'a rapidity of increase probably without parallel in history'. The figures suggested to him that populations grew by doubling – from two to four to eight to more than a thousand in just ten generations – whereas resources increased from two to four to six and so on: 'I think I may fairly make two postulata. First, that food is necessary to the existence of man. Second, that the passion between the sexes is necessary and will remain nearly in its present state . . . I say, that the power of population is infinitely greater than the power in the earth to produce subsistence for man'. His interest was in morals and not in biology. God 'ordained that population should increase much faster than food' and had provided a sense of restraint to prevent it from so doing. Marx and Engels also saw the parallel between the economies of Nature and of man (although neither was fond of Malthus; to Marx he was 'a shameless sycophant of the ruling classes' and his doctrine to Engels a 'vile, infamous theory, a revolting blasphemy against nature and mankind').

Tertullian had, long before, seen what such expansion implied: 'We have grown burdensome to the world . . . nature no longer provides us sustenance. In truth, pestilence and famine and wars and earthquakes must be looked upon as a remedy for nations, a means of pruning the overgrowth of the human race'.

If the burden is reduced by the survival of one set of genes at the expense of others when faced by plague, war or famine, then a nation (or a species) has no choice but to evolve. In 1838, at the age of twenty-nine, Darwin read (for amusement, as he claimed) Malthus on *Population*. Here, he saw, was the basis for a theory of change: that, owing to the struggle for life, any variation, however slight, if it be profitable to any individual in its complex relations to Nature, will tend to its preservation and will be inherited by its offspring. Any species – any country – that fails to adapt will join the pantheon of the extinct.

A struggle for existence inevitably follows from the high rate at which all organic beings tend to increase. We have better evidence on this subject than mere theoretical calculations: namely, the numerous recorded cases of the astonishingly rapid increase of various animals in a state of Nature, when circumstances have been favourable to them. Still more striking is the evidence from our domestic animals of many kinds which have run wild in several parts of the world.

In the sixteenth century, the Spanish colonists of Hispaniola considered quitting their colony because of an outbreak of a fearsome stinging ant. Fortunately, prayers to St Saturnin ensured that the plague subsided. The Englishmen on Barbados a couple of centuries later were visited by the same scourge, and their offer of twenty thousand pounds for a solution worked just as well. On both islands ant numbers decreased, and the animal is now no more than a harmless member of the local fauna. Why it burst into such abundance and then declined, nobody knows; but something in its balance of life had changed.

Cattle were introduced into the Americas in 1493 by Columbus. The overgrowth had at once a chance to hint at what it can do. Two centuries after those first cows, huge herds ran wild in Mexico as a wave of beef moved into ungrazed lands to the north. The Jesuits abandoned a ranch with its five thousand cattle in Argentina in 1638. The animals grew into a swarm. A traveller saw plantations and orchards whose walls were built of skulls stacked 'seven, eight, or nine deep, placed evenly like stones, with the horns projecting'. The same happened with wild horses. At the time of the Gold Rush, thousands were driven to their deaths over the cliffs of Santa Barbara to make room for cattle. Plants can do even better. A pretty aquatic fern, *Salvinia*, has been introduced into much of the tropics from Brazil. Everywhere it has become a pest. It blocks rivers and interferes with water supplies. *Salvinia* reached Australia in 1952. It can double every three days and, by 1977, Lake Moondarra in Northern Queensland was covered by fifty thousand tons of weed.

Any ant, cow, fern or whale will, given time and resources, increase in this way. Most do not. Gilbert White published his *Natural History of Selborne* in 1789 as an account of the animals and plants of his Hampshire village. In it he describes a struggle for existence among its swifts: 'I am now confirmed in the opinion that we have every year the same number of pairs invariably; at least, the result of my enquiry has been exactly the same for a long time past. The number that I constantly find are eight pairs, about half of which nest in the church, and the rest in some of the lowest and meanest thatched cottages. Now, as these eight pairs – allowance being made for accidents – breed yearly eight pairs more, what becomes annually of this increase?'

Today, Selborne has doubled in size. The straw-roofed cottages are gone, or have been reborn as bijou homes whose thatch is held down by mesh that birds cannot penetrate. The church tower was altered in the 1950s and swifts can no longer get in. Even so, their numbers have changed little. In 1983 the village held twelve pairs, which, given its growth, is almost the same density as that recorded by Gilbert White. The birds he saw could, with unrestricted increase, have brought forth ten thousand billion billion billion descendants in the two hundred years since he counted them. Something (not, it seems, a shortage of places to nest) holds them in check.

Eggs or very young animals seem generally to suffer most, but this is not invariably the case. Seedlings, also, are destroyed in vast numbers by various enemies. In a tract of tropical grassland, a hundred and seventy thousand palm seeds were scattered. A third became seedlings, a tenth saplings – but just fifteen reached maturity. Even worse, most young plants and animals must attempt to force themselves into a place already more or less full. Thousands of seeds can germinate on a square yard of bare English ground. A few years later, when a community has become entrenched, but a single one of those thousands will succeed.

The struggle for existence is a term best used in a large and metaphorical sense. It may be of one individual with another of the same species, or with the individuals of distinct species, or with the physical conditions of life. It is the doctrine of Malthus applied with manifold force to the whole animal and vegetable kingdoms. It includes not just survival or speedy growth, but sexual success; and what is fought over can be food or sunshine, a place to live or a mate to fertilize.

The battle is a complex affair, with many players. It may involve distant relatives or close kin. War breaks out between unexpected enemies. Humpback whales carry a thousand pounds of barnacles. The turbulent water around an encrusted whale feeds the passengers but slows down their vehicle. The whale retaliates with skin, grown at a rate three hundred times that of our own as a kind of antifouling to slough off the tiny travellers in an intimate contest between two implausible foes. Often the conflict is at its most bitter when relatives are involved. Britain has two common land snails, much studied by geneticists and hard to distinguish at a glance. Since the 1960s, one has retreated at the expense of the other at a rate great enough to render it extinct within a century.

The struggle may last for years. Even the most peaceful place is full of strife, with any weakness among its inhabitants at once exploited. It can, as in snails, lead to victory or to defeat; or to an uneasy truce. Often, the advantage passes from side to side as a reminder that the war of Nature never stops.

From season to season, the number of snow grouse shot in Canada fluctuates by five times. The rhythm has, say the records of the

Hudson's Bay Company, existed since 1821 (and has in all likelihood gone on for twenty times as long). The cause seems simple. Lynx kill thousands of the birds and as the big cats become common they drive down their prey. The lynx pay the price and many die of starvation. In the Malthusian struggle of cat and bird, it seems that first one prevails and then the other.

The truth is much more complicated. Grouse are not the lynx's favourite food. Instead, the cats prefer snowshoe hares, animals sometimes abundant but in other years almost never seen. The numbers of hares cycle by forty times. The lynx, in return, vary seven times over from peak to trough. What leads to this ten-year rhythm, in perfect time across the whole of Arctic Canada?

Far more is involved than a simple battle for survival between grouse, hare and cat. First, the unfortunate hares pay a psychological price for their predicament. Almost every hare during a population decline is killed by a predator and the few survivors suffer much stress. This reduces their ability to reproduce. The time needed to recover a certain equilibrium after the predators have gone makes for a lag before the next hare boom. Other contestants are also involved. Hares eat the fresh twigs of willow and birch. When these are plentiful, the hares are well fed and have many young, most of whom survive. Soon, their numbers begin to shoot up. Parents, children and grandchildren all begin to gnaw back the twigs. When the food is gone, the population crashes and the hungry remainder are forced into the open, where they become prey for lynx.

The trees can defend themselves. For a couple of years after each plague their new branches are filled with poisons that make them bitter and hard to eat. As a result, hare numbers stay low after a crash. In time, new generations of sweeter twigs appear, the animals return and the cycle re-starts. That of the lynx is a symptom of the hares' own swings in abundance, which turn as much on food as on predators. Grouse are dragged in as the hungry lynx turn to them when their main item of diet disappears. Other reluctant players include squirrels, ptarmigan, coyote, owls, goshawks and ravens, all of whom see-saw in synchrony. Red-backed voles dance to the beat of a different (and as yet unknown) drummer, with cycles of their own. Because the system has not settled into the usual glum stalemate of Nature, the complex and unexpected checks and relations between organic beings which have to struggle together in the same country are revealed in all their intricacy.

Fluctuations of this kind can lead to catastrophe. In 1980, a thousand black long-snouted weevils – a grazer on the *Salvinia* weed in its native Brazil – were released on to the fern-clogged Lake Moondarra. By April 1981 the lake had a hundred million weevils, but in August came disaster. The fern population collapsed from a peak of fifty thousand tons to a remnant of less than a ton. The weevil population crashed in sympathy and, soon, both participants disappeared.

The ancient and intimate dance of plants and animals on the Canadian prairie stops when settlers move in. The confusion of Nature yields to the tedium of the farm and, as fields break up forests, the numbers of hares steady. When a patch of forest has been filled, the excess can move to an empty wood with plenty of young twigs. The circle has been broken by the grand simplification of agriculture and, although the struggle goes on, the evidence is hidden because the numbers of each player do not change.

Such stability is true of most of Nature – but it does not mean peace. Life, like a silent forest, may seem to be in harmony, but often that is no more than chaos postponed. The battle may be long deferred, and it may involve not different kinds – lynx, hare and beech – but the same. Indeed, the struggle almost invariably will be most severe between the individuals of the same species, for they frequent the same districts, require the same food and are exposed to the same dangers.

Chile has huge forests of southern beech, filled with plants that started their lives at the time of Columbus. For mile after mile, trees of the same kind and about the same height cover the land. A beech grove is at first sight as calm a place as any on Earth. Its branches stretch a hundred feet towards the sun and shade out most of the light. Little can survive beneath the canopy – not even beech itself. Most places have no young plants, which is odd in a forest that has lasted for so long. The beeches, in the absence of recruits to replace plants dying of old age, seem doomed. Every tree makes innumerable seeds in its five-hundred-year existence, but for few do any succeed.

Sometimes, though, there is an earthquake, eruption, or storm. Records from the 1700s show that two-thirds of a typical stand is wrecked by gales or fire within three short centuries. Then, Nature erupts and the survivors have a chance. The sun floods in, young plants shoot towards the sun and the struggle starts anew as the winners block light from their inferiors. All the trees in a beech forest are the same age because they all started to grow at the same time. Their conflict may last for just a few months in a hundred years or more but is as vicious as the fight between lynx and hare. What seems a forest primeval is in truth an interval between catastrophes. Battle within battle must ever be recurring with varying success; and yet in the long run the forces are so nicely balanced, that the face of nature remains uniform for long periods of time.

The forest is a cathedral, but it often burns down. Lightning strikes the earth two hundred million times a year and sparks off innumerable fires. The pioneers of North America remarked on the park-like landscape of New England. Its timber was scattered not for aesthetic reasons but because it had been burned by lightning or by Indians. Not before fire control did Longfellow's 'murmuring pines and the hemlocks, bearded with moss, and in garments green' have a chance to grow. Five thousand years ago the jungles of South America were reduced by fire to copses encircled by grassland. Three hundred million years earlier even Ireland was covered with tropical forest. Its rocks are filled with ancient charcoal, the remnant of a forgotten firestorm.

Farmers try to second-guess the struggle for existence when they move plants or animals from their native land to somewhere new. How their subjects survive and what kills them in their novel abode are clues to what they face at home.

Penzance, Nice and Alicante are proud of the palms along their sea-fronts. All are aliens as, apart from a few dwarf specimens in Southern Spain, Europe has no native palm trees. They fail because their shoots are damaged by frost. A single cold night in fifty years – a moment of defeat in Nature's struggle – will stop them.

The denudation of the Cornish Riviera is a trifle compared to the risks faced by coffee-growers. Wild coffee is found nowhere in the world where the mean temperature in any month drops below that of an English spring. A millennium and a half ago, farmers got into the act. In the Ethiopian province of Kaffa, coffee was used to season food. A thousand years later it was roasted, ground and used as a drink by Arabs. Now millions of tons are grown each year. Almost half the crop comes from Brazil and Colombia. For most of the time, conditions are ideal – but, sometimes, it freezes.

The New York Coffee, Sugar, and Cocoa Exchange gambles against the weather, with all the uproar of an 'open outcry' trading floor. The growers and roasters deal in futures, a bet on the price in months or years to come. They hope to reduce their exposure to the risks of cold. Their opponents count on the profit to be made from the changes that follow a sudden frost. Fortunes can be won or lost on a guess about bad weather as the contest between speculators moves back and forth.

The market is evidence of a struggle against the elements. Coffee prices peaked in 1912, 1928, 1953, 1975, 1979, 1981, 1986 and 1995. The cycle is less regular than that of the snowshoe hare because it is driven by chance changes in climate rather than a shifting balance between implacable foes; but its peaks and troughs are just as much a response to stress (in this case mitigated by greed). A frost year would put paid to the crop without human intervention; and most good harvests are followed by a slump brought on by over-optimism and excess production in a buoyant market.

Coffee has survived, but its insecurity in an alien home puts the economy in constant danger. From the 1890s, 'coffee presidents' ruled Brazil. In spite of attempts to control the trade, crop failure, followed by excess output and price collapse, led to unemployment. The revolutions of the 1920s in Brazil and the more recent chaos in Colombia (the latter helped by a war against the coca crop) are a reminder that the natural state of any ecosystem is of convalescence between frightful battles.

For coffee, the weather is a simple enemy – it kills it. For other plants, even when climate, for instance extreme cold, acts directly, it will be the least vigorous, or those which have got least food through the advancing winter, which will suffer most. Many crops struggle more with starvation than with hard weather. Farmers fertilize their fields because more food means better harvests. Often, some of the fertilizer leaks out and its recipients respond. As the River Po spews filth into the Adriatic it releases algae from their struggle. They form mats for a hundred miles down the coast.

Parts of the open ocean are so pure as to support little life. Great quadrants of the Pacific, from the Equator to Antarctica, have almost none of the tiny marine plants – the plankton – that abound elsewhere. The seas have plenty of the nitrogen and phosphorus needed for growth. What the plants search for and fail to find are minute amounts of iron. Without it, the green machinery that soaks up the sun's energy is starved. A thousand pounds of iron salts added to a five-mile square of sea near the Galapagos caused the numbers of plankton to explode. A shortage of a few parts per million of a simple but crucial item limits life over whole oceans. The seas were more productive at the time of the ice ages because of the battle for iron, which, in those cold and dusty times, was blown far out to sea from the parched land.

Much of the struggle for existence – on land as much as in the sea – is for a share of the sun's energy. The sunnier a place, the more kinds of animal and plant it contains; and, within any locality, animals that keep themselves at a constant temperature are less numerous than those that do not. In general, the more energetic a creature, the less common it is – in Britain, for example, land mammals are a hundred times as abundant as birds of the same size, in part because fliers need more energy to stay alive.

The fuel for nature's commerce is not well used. On land, about one part in three hundred is trapped by plants, in the sea even less. The energy flows upwards, but does not go far. Most food chains have four or five steps from bottom – the plankton in the ocean or the grass of an African savannah – to top, be it lion, eagle, or killer whale.

When a deer eats a leaf, it gains less than a tenth of its worth as food. The lion that feeds on the deer does rather better. Even so, more than half the prey's value is wasted. Too many links in the chain leave too little for those at the top. It becomes impossible to add another and to evolve a beast that preys on blue whales, or on lions. Such an animal would have to be fast, fierce, and of necessity rare. Most of the islands of the Outer Hebrides have no eagles as they are not big enough to provide enough food. A bird that ate eagles would need most of Scotland to support itself.

Short though life's chains may be, their links are tightly packed. South Australia was once covered with a spiny scrub called the jarrah. It had to cope with drought and with the giant kangaroos that browsed upon its leaves. A ten-yard square may contain a hundred kinds of plant. Intense competition means that each bush or flower is forced into its own narrow niche and that together, they exploit all that sun, soil and rain can offer. Some are tall, some short; while some roots penetrate deep and others spread out.

Now, all have been replaced by a great field containing a small part of the five hundred trillion wheat plants in the world. Wheat is sheltered from ecological reality. It grows for half the year and wastes most of the sun's energy as it passes through its upright stalks. The fields flood in winter and in summer dry and starve because every plant grows to the same depth. Entire harvests may be destroyed by pests. The jarrah, shaped by generations of struggle, was more resilient. Where fields have been abandoned, the jarrah creeps back. Wheat, in contrast, cannot force its way into the native vegetation without human help.

For jarrah, wheat and giant kangaroos the conflict is obvious. Any plant or animal must avoid its enemies, find a place to live and get enough to eat in order to have a chance to pass on its genes. However, to win those battles is just to set the stage for the greatest contest of all: the battle for sex.

When it comes to that interesting pastime, most members of most species come to grief. When they fail depends on their habits. For elephants, the majority survive, although not all have offspring, but for orchids (which may have a billion seeds) a minute proportion last to have issue of their own. Investment in each stage depends on its chances of success. An acorn is tiny compared to its parent, but a kiwi lays an egg a quarter her own weight.

Malthus himself saw that the struggle for existence involved both survival and sexual success. With his Christian standards, he was a strict examiner. Far better for the excess to fail at once and die young, rather than interfere with God's procreative plan. A boy who 'cannot get subsistence from his parents . . . and if society do not want his labour' should face the facts. 'At Nature's mighty feast there is no vacant cover for him. She tells him to be gone, and will quickly execute her orders'. Any attempt to rig the test by 'unnatural promiscuous concubinage ... violations of the marriage bed, and improper acts to conceal the consequences of irregular connexions' – was a vice. Human nature was, to Malthus, red in tooth and claw, and must be controlled (he favoured 'moral restraint' as an escape, and lived up to it, with a marriage at thirty-eight and just three children). His image of the cruel fate of the surplus population still sums up, in the public mind, the struggle for existence.

Tennyson's sanguinary phrase reflects not a rule of life, but despair at the death of a friend. Nature is often not at all like that. The battle may be for light rather than land, or for traces of metal instead of terrified prey. It may be confined to a few moments in a whole life-time. The war of Nature is not incessant; no fear is felt, death is generally prompt, and the vigorous, the healthy and the happy survive and multiply.

Except, that is, when politics comes in. Then, the struggle is to the end. Mark Twain parodied it: 'I confined an Irish Catholic from Tipperary, and as soon as he seemed tame I added a Scotch Presbyterian from Aberdeen. Next a Turk from Constantinople ... a Buddhist from China; a Brahman from Benares. Finally, a Salvation Army Colonel from Wapping'. A few days later 'there was but a chaos of gory odds and ends of turbans and fezzes and plaids and bones and flesh – not a specimen left alive. These Reasoning Animals had disagreed on a theological detail and carried the matter to a Higher Court'.

Those animals, reason as they might, have often destroyed their resources, and their lives, as the consequences stare them in the face. Fishermen are closer to the wild than are coffee-traders and act – like any predator – only in their own short-term interests. Politicians try to control them, but almost always fail. As in the beech forests of South America, lethal skirmishes are separated by intervals of calm. The annual quota of Alaskan halibut is taken in two days of mad scramble – the Halibut Olympics. For the rest of the year, millions of dollars of capital lie idle and, it seems, at peace. So incensed are its owners that the controls are to be lifted, at whatever cost to the fish.

Such battles go back to the dawn of America. The Pilgrim Fathers obtained a grant of land in North Virginia (now New England) to set up a cod fishery. They had reason to be hopeful. John Cabot, a century and a half earlier, had scooped fish six feet long from the sea in baskets. The Pilgrims were better at dogma than fish and had to pillage the Indians to stay alive. So devoid was the country of their favourite food that they reported with some chagrin that the sole 'dish they could presente their friends with was a lobster'.

In the end, cod triumphed and New England boomed. The court that burned the witches of Salem had a codfish on its seal, the first coins minted in Massachusetts carried its image, and a wooden figure of that noble animal was hung from the roof of the Old State House in Boston. The fishery spread north and soon Cape Cod, the Grand Banks and the Gulf of St Lawrence provided enough fish to feed the United States and much of Europe.

The stock seemed boundless: after all, a female codfish can live for two decades and lay nine million eggs a year. As Alexandre Dumas wrote in his *Grande Dictionnaire de Cuisine*: 'It has been calculated that if no accident prevented the hatching of the eggs and each egg reached maturity, it would take only three years to fill the sea so that you could walk across the Atlantic dryshod on the backs of cod'.

The steam trawler was invented in England in 1881. The breakthrough was followed by Clarence Birdseye's discovery that cod could be frozen, even at sea. Before long, the waters off North America were plundered. The steam-powered harpoon appeared in 1864. The number of whales it killed rose from thirty in that year to sixty-six thousand in 1961.

In 1992, the Canadian government closed the seas off Newfoundland. The stocks had fallen to twenty thousand tons from more than a million. Although the number of fishermen in the world has doubled since the Second World War, most stocks are in decline and a third of the output of the continental shelf already ends up on the table. Whales, too, were hunted to near extinction before the moratorium of 1986. In the 1960s, the Soviet Union built the *Sovetskaya Rossiya*, a whaler the size of an aircraft carrier. They kept the industry alive (and cheated on their quotas) to pay for it. It was pensioned off as a marine slaughterhouse for Australian sheep.

Now, the old Massachusetts whale-port of New Bedford has become, like Ellis Island, a tourist attraction. It celebrates not a triumph over the struggle for existence, but a catastrophe. The cod-fishing villages to the north, in Newfoundland, are an even starker memorial. In spite of a billion dollars in aid, many have disappeared, while others are sunk in poverty. They are a monument to Malthus. He saw that 'The positive checks to population are extremely various ... Under this head, therefore, may be enumerated all unwholesome occupations, severe labour and exposure to the seasons, extreme poverty, bad nursing of children, great towns, excesses of all kinds, the whole train of common diseases and epidemics, wars, plague and famines'. The fishermen of North America faced most of them; and, like their prey, have failed.

CHAPTER IV

NATURAL SELECTION

Natural Selection: the factory for the almost impossible — Its power compared
with man's selection — Illustrations of the action of Natural Selection, from
Manchester to the Caribbean — Speed of action — Its power at all ages and
on both sexes — Evolution's second paper: Sexual Selection — Sex, age and
death — Cuckoldry and the balance of the sexes — On the intercrossing of
individuals — Circumstances favourable to Natural Selection, as illustrated
by agents of disease — Extinction caused by Natural Selection — Divergence
of Character related to the diversity of resources — Action of Natural
Selection on the descendants from a common parent — Superiority of the
work of Nature to that of Man

I ONCE WORKED for a year or so, for what seemed good reasons at the
time, as a fitter's mate in a soap factory on the Wirral Peninsula,
Liverpool's Left Bank. It was a formative episode; and was also, by
chance, my first exposure to the theory of evolution.

To make soap powder, a liquid is blown through a nozzle. As it
streams out, the pressure drops and a cloud of particles forms. These
fall into a tank and after some clandestine coloration and perfumery
are packaged and sold. In my day, thirty years ago, the spray came
through a simple pipe that narrowed from one end to the other. It
did its job quite well, but had problems with changes in the size of
the grains, liquid spilling through or – worst of all – blockages in the
tube.

Those problems have been solved. The success is in the nozzle.
What used to be a simple pipe has become an intricate duct, longer

than before, with many constrictions and chambers. The liquid follows a complex path before it sprays from the hole. Each type of powder has its own nozzle design, which does the job with great efficiency.

What caused such progress? Soap companies hire plenty of scientists, who have long studied what happens when a liquid sprays out to become a powder. The problem is too hard to allow even the finest engineers to do what they enjoy most, to explore the question with mathematics and design the best solution. Because that failed, they tried another approach. It was the key to evolution, design without a designer: the preservation of favourable variations and the rejection of those injurious. It was, in other words, natural selection.

The engineers used the idea that moulds life itself: descent with modification. Take a nozzle that works quite well and make copies, each changed at random. Test them for how well they make powder. Then, impose a struggle for existence by insisting that not all can survive. Many of the altered devices are no better (or worse) than the parental form. They are discarded, but the few able to do a superior job are allowed to reproduce and are copied – but again not perfectly. As generations pass there emerges, as if by magic, a new and efficient pipe of complex and unexpected shape.

Natural selection is a machine that makes almost impossible things. Consider a typical protein such as whale myoglobin. That molecule is but one of seventy thousand or so proteins in the animal's body and contains a hundred and fifty-three units called amino acids. These come in about twenty forms. The number of possible combinations of amino acids in a structure the size of myoglobin is hence twenty raised to the power of a hundred and fifty-three. The figure, ten with about two hundred zeros after it, is beyond imagination and is far more than all the proteins in all the whales, all the animals and all the plants that have ever lived. Such a molecule could never arise by accident. Instead, a rather ordinary device, natural selection, has carved out not just myoglobin but millions of other proteins and the organisms they build.

Selection is simple, efficient and inexorable. George Bernard Shaw did not like it: 'When its whole significance dawns on you, your heart sinks into a heap of sand within you. There is a hideous fatalism about it, a ghastly and damnable reduction of beauty and intelligence, of strength and purpose, of honour and aspiration'. Unfortunately for Shaw, beauty and intelligence have themselves now turned to that fatal device for help. Some artists use a computer to generate altered versions of an original. Then, by aesthetic preference, they select each generation until a work to their taste has emerged. Architects do the same, and machines even write plays (albeit not to Shavian standard).

Like the tapeworm or the tree-kangaroo, the buildings, plots or nozzles made by selection may grate upon the senses of a purist. Even so, they work, and do so without a single blueprint or equation. Instead, they have evolved through descent with modification. Darwin is loose on the shop floor, and industry has become a branch of the biological sciences. Natural selection has returned the compliment.

Illustrations of the action of Natural Selection. In Britain in 1690 three million tons of coal were mined. By 1950 the figure was eighty times higher. Evolution has been busy since the invention of the flying shuttle – the New Machine for Opening and Dressing Wool – two and a half centuries ago. The Industrial Revolution was a great experiment in biology.

In the eighteenth century, Manchester was described by Daniel Defoe as the greatest mere village in England – a market town, but little else. Within a hundred years, its population had multiplied and it had become Cottonopolis. As de Tocqueville put it: 'From this filthy sewer pure gold flows. Here, civilisation works its miracles, and civilized man is turned back into savage'. London, too, in its different way, had grown. The inhabitants of each city and every other had to adapt or die. Many died – but many more evolved to stay alive.

To stand on a clear day by the monument on Highgate Hill, where, five hundred years ago, Dick Whittington was persuaded to turn again, is, sometimes, to take in a splendid view. It extends across London to the Downs, twenty miles to the south. Dick Whittington, in days when the prospect of the city included 'A delightful plain of meadow land, interspersing with flowing streams', saw the same range of low hills in the distance. Nowadays the best to be seen is, too often, the glum towers of the City of London as they peer through a sulphurous haze. Smog is a London word, coined to describe the smoky fog that once killed thousands. Just thirty years ago it let through a third less sun in the East End than in the more affluent west of the city.

1848 was a year of revolutions, brought about by the social changes that followed the spread of manufacturing. In February, in Paris, the populace rose and overthrew the Emperor. A few months later Austria, Germany and Hungary were in tumult. The revolts were a sign of the political storms that were to form the modern world.

In Britain, the main upheaval was among the moths. In that year there was a report from Manchester of a black form in the peppered moth, an animal that had until then been grey. By the time of Darwin's death in 1882, the melanic type was the commonest form in much of industrial Britain. Collectors were quick to notice: 'Near our large towns where there are factories and where vast quantities of soot are day by day poured out from countless chimneys, fouling and polluting the atmosphere with noxious vapours and gasses, this peppered moth has during the last fifty years undergone a remarkable change. The white has entirely disappeared, and the wings have become totally black.' It was the nation's response to the Industrial Revolution.

The force of selection was simple. Dark moths could not be seen by birds against the filthy trees on which they sat and were safer from attack than the original form (which was camouflaged against the lichens of rural woodlands). The dark moths appeared in London by 1897. By then, just one Manchester peppered moth in fifty was of the original type. In rural Britain, blacks were rare or absent (although a few were found in East Anglia, a place much polluted by drifts of smoke from the Midlands). The rest of the world tells the same story – melanics turned up in peppered moths in the industrial parts of Europe, and, from 1906, in North America too. Japanese peppered moths remain white, but they live in the countryside and have not gone into towns. All over the world, natural selection had responded to the new challenge. It caused melanics to spread so fast after 1848 that a simple sum shows them to be half again as able to succeed in cities as was the original form.

With smoke-control, the dark moths began to disappear. As the soapworks were cleaned up, the frequency of blacks in the Wirral fell from 95 per cent in the late 1950s to 5 per cent at the turn of the millennium. Soon, the new form will be gone, and the latest episode in the moth's evolutionary adventures will be over.

London still has black mallards, black pigeons and black squirrels (which became less common in America at the time of its own Industrial Revolution, because dark fur was in fashion). Life in the city affected sex as much as survival. Unlike their grey relatives, black pigeons – the cooing majority in Trafalgar Square – breed throughout the year. Their pigment soaks up more of the sun and persuades the calendar in the brain that winter days are long and that it is time to turn on the testosterone. Cities, unlike their ancestral cliffs, provide enough food to breed in cold weather. As a result, melanics, scarce in the countryside, prevail.

Town-dwellers must deal with a world of change, or die. That is nothing new. The countryside is as merciless as the toughest slum. Every plant or animal descends from ancestors tried by a struggle for existence for three and a half billion years before cities came on the scene. Man selects only for his own good, Nature only for that of the being which she tends (which is why the streets are not full of marauding sheep or tomatoes).

The Industrial Revolution was a test of the theory of evolution. Its results were often unexpected, sometimes unwelcome and always unplanned. So familiar is the testimony of natural selection that we do not recognize the best proof of its power: the fit of life to where it lives. When we see leaf-eating insects green and bark-feeders mottled grey; the alpine ptarmigan white in winter, the red-grouse the colour of heather and the black-grouse that of peaty earth, we must believe that these tints are of service to those birds and insects in preserving them from danger.

Evolution, say its critics, is not a real science, because it cannot predict the future. It has, no doubt, done pretty well at explaining the past. But how can we know what it will do next? Even Thomas Henry Huxley, Darwin's Bulldog, felt that his hero 'does not so much prove that natural selection does occur, as that it must occur, but in fact no other sort of demonstration is attainable'. He was wrong. Now, evolution has joined the scientific mainstream. It designs its own experiments to confirm what mere observation might suggest.

The Caribbean is full of scores of different kinds of lizards called anoles, climbers skilled enough to make their way up a vertical sheet of glass. On each island, the lizards fall into half a dozen physical types, specialized for the forest crown, for trunks or twigs, or for bushes, grass or the ground. Bare rocky places have lizards with shorter legs than those from sites with large trees, because short limbs allow a lizard to be agile at the cost of being slow. A lizard able to sprint along a thick branch will fall off a stick, which is not much help when it tries to catch food or escape enemies.

In 1977 small groups of lizards were moved from their home on a Bahamian island shrouded in thick vegetation to an islet without lizards of its own. There, the cover consisted only of thin twigs. Ten years later, the emigrants had changed. Their limbs were shorter than those of their ancestors and had evolved the pattern of stubby legs on thin branches found throughout the Caribbean. On the new island, natural selection had been at work, daily and hourly scrutinizing every variation, even the slightest, rejecting that which is bad, preserving and adding up all that is good; silently and insensibly working at the improvement of every organic being in relation to its organic and inorganic conditions of life.

Short legs (or black wings) help their owners to stay alive. That is of course important, but selection, in the end, depends only on the relative number of offspring. A long, happy and well-adapted existence, without issue, is of no use at all. What matters is how many genes are passed on. As no animal can do everything, evolution, like economics, is full of compromise.

Any investment – in flesh or in cash – trades off stability against growth. The highest yields are from a risky bet, while security means a lower rate of interest. Because (as Malthusians and financiers each know) compound interest is powerful, any speculation in genetic futures needs to squeeze in as many rounds of interest, with as high a rate of return, as possible. The market, however, calls for a decision between a long life and a fertile one, with success in one vocation paid for by failure at another. Chickens bred for meat lay fewer eggs

and even among the egg-layers those selected to lay young die before
their time so that the number of eggs goes up not at all. A balance of
cost and benefit moulds what any business (evolution included) can
do.

The ecological market determines when plants and animals
should make the biggest investment of all: to reproduce. Variation in
when to breed, in how much risk to take, and in how many young to
produce is potent fuel for evolution. In their happier days, when
some were the size of a man, North Atlantic cod delayed their sex
lives until they were four or five years old. Now, the giants have gone
and the few that escape the net are tiny. With no chance of old age,
evolution favoured those that reproduced as soon as they could.
Those reduced animals have struck a balance between mortality and
sex, and now lay eggs at the age of two.

For cod and all other creatures, sex and death are weighed in the
scales of selection. Trinidadian guppies have different lives in different places. Fish in the lower rivers are much harassed by predators.
As a result, most die young. Because their enemies cannot get above
the waterfalls, guppies in mountain streams are free from attack and
are guaranteed a death deferred. As a result, lower and upper rivers
evolve their own investment strategy. Upstream guppies rejoice in
peace and quiet, while life for their lowland relatives is shorter and
nastier. With so few chances of an erotic encounter before Nemesis
arrives, the downstream fish put all their eggs into one sexual basket.
They mature more rapidly and have more, smaller, young, each with
a lower chance of survival and faster growth, than do fish from
higher up. Fish moved from the lower parts to the headwaters soon
evolve local ways. Within ten years they grow larger, delay sex and
have fewer young. The males adapt to their new and relaxed home
by evolving at ten times the rate of their consorts.

As natural selection adjusts a guppy's existence, it illustrates the
links between sex, age and death. They are universal and unavoidable. The more a male fruit-fly copulates, the sooner it dies, and
the more cones a Douglas fir makes, the slower it can grow. Life
is a gamble, and not just for fish. Any animal must decide whether
to make a large stake in the hope of a pay-off, or to delay a bet in
the hope of better odds. As insurance companies realize, that

logic applies to death as much as to sex. If astrology worked and everyone knew when they would die, nobody could sell any policies. Clients told that not much time is left would buy cover at once, but those promised many happy years would spend the money elsewhere.

Insurance and evolution must each cope with the fact that risk goes up with time. As I write this chapter I have, as a fifty-something male, a chance of around one in a thousand of death before I finish it; and, I have to say, I am not too worried. The chance is hundreds of times bigger than that of a Lottery win, should I be fool enough to buy a ticket. However, the two risks are subtly different. Not every gambler wins the lottery, but all of us (and every fish) will die. A sixteen-year-old has a one in two thousand chance of demise before his next birthday, while for centenarians the figure drops to one in two. Unlike Lottery odds, the yearly chances of death alter with time – for humans, doubling about every eight years.

The worth of a young animal, calculated in the currency of how many young it can have, is, as a result, higher than that of an old one. The force of selection, like the value of a new insurance policy, weakens with age because less time is left for the contract to run (which is why, for downstream guppies, a short life means a heavy investment in premiums). A small increase in the chance of reproduction when young means more than a sexual triumph long delayed. This means that evolution favours youthful vigour at the expense of later decline. Why should it care if the price of sex is to become a burned-out wreck? Any gene able to help its carriers to copy their DNA will spread, however evil its effects, if they are long enough deferred. Selection favours only those who can pass on the genes and cares not at all for those who cannot. Age is a tax on sex, levied by natural selection. It is as much a product of the struggle for existence as are the black wings of the peppered moth.

Sexual Selection. Sex is a marketplace for natural selection. As a merger between genetic enterprises it is, like any other business with two partners, liable to discord. To find a mate, fight off the opposition, suffer sexual congress and pregnancy, and raise a brood, are all expensive and dangerous. Each involves subtle differences in the share-dealing strategy of the parties involved. Sometimes there is sexual war that may lead to traits that seem opposed to the interests of those who bear them. They are a reminder that the biological battle goes much further than the search for food or for light: there is, in addition, an unavoidable struggle to pass on genes.

Evolution is an examination with two papers. To succeed demands a pass in both. The first involves staying alive for long enough to have a chance to breed, while the mark in the second depends on the number of progeny. Malthus assumed that sex was so attractive that, given the chance, all would indulge as much as they could and that birth rates were always near their maximum. Death was what counted. He was wrong. Until the eighteenth century, the English married late – at twenty-five for women and even older for men. The rapid growth in numbers at the beginning of the Industrial Revolution came not from a decline in the power of death, but from an emancipation of the national libido. The age of marriage dropped by several years so that changes in fertility, not in survival, caused the population explosion that gave Darwin his theory.

Inherited differences in sexual ability fuel a subtle and unexpected kind of selection. The drop-outs from the first part of the ecological exam are easy to identify as they are, after all, dead. Variation in sexual success is more difficult to measure, particularly for the male candidates (who tend to disappear before their papers are ready for marking). The battle is of male against male, of female against female, and of either sex against the selfish interests of the other.

The successes and failures of two hundred marked female sparrowhawks in Eskdale in Southern Scotland were followed for a quarter of a century. Like the swifts in Selborne the number of birds in the valley stayed about the same. That stability hides great differences in how well each did in the contest to transmit DNA. Most of the birds are in an evolutionary dead end. Three out of every four females die young, fail to find a mate, or have no progeny; but some triumph in the sexual struggle. A fifth of the young produced nine-tenths of the next generation. A few live for a decade and have two dozen fledglings, while many more die before they have any chicks at all.

For sparrowhawks, life is unfair, with no question of equal shares, be they for food or for sex. Man himself insists on inequality. As a result, wealth – which leads to happy old age and plenty of sex – soon becomes concentrated among a few. Half the private land in Scotland is owned by three hundred and fifty people (in a country where half the population has no landed property at all); and the greatest proprietor of all, at a quarter of a million acres, is the Duke of Buccleuch. The Duke's lesser titles include a couple of Earldoms, a Barony or two – and the Lordship of Eskdale.

How many of the testamentary differences among birds or dukes are due to good genes, how many to good parents and how much to good luck nobody knows. When it comes to the nobility, good (or aristocratic) parents are what count. For the birds of the ducal acres, success may say more about genes. Whatever the reason, in Eskdale or anywhere else, most sparrowhawk – most animal – DNA does not make it to the next generation.

Dukes vary more than duchesses in the ability to make the reproductive grade. The struggle for sexual existence nearly always bears more upon males. They put less effort into each encounter, because sperm are cheaper than eggs and those who make them are not left holding the baby. Males have more chances for sex – but they are limited by the numbers of partners available. The imbalance means that, in general, any female, however unattractive, will find a mate, but that just the best males will win. Males must, as a result, fight for access to females as much as for any other resource in short supply.

Their struggle is manifest in many ways. Often, the winner must hold a territory against his ardent rivals. Shoot a male red-grouse and another moves into his space, proof of how crowded existence must be. If a bird is injected with the male hormone testosterone, the size of his patch increases. Like a body-builder attacked by steroid rage, he displaces his unlucky neighbours from their homes. They may survive, but the battle to transmit their genes is over.

Many of Nature's most attractive features – flowers, birdsong, mandrill bottoms – result from rivalry among male sex cells for access to eggs. Male alligators have been described as fighting, bellowing and whirling round, like Indians in a war-dance, for the possession of the females; male salmons have been seen fighting all day long; male stag-beetles often bear wounds from the huge mandibles of other males. Many are armoured; for the shield may be as important for victory as the sword or the spear. Even the giraffe's neck is involved. If access to high leaves is important, why stretch necks rather than legs – and why do males have longer necks than females? In fact, they fight (and die) over mates by battering each other with their heads, and a long handle makes an effective club. Because their brain is ten feet higher than their heart, giraffes have the highest blood pressure of any mammal. The sexual struggle is to blame.

Some females are firm in monogamy, but many more go in for manifold males. The successful (and anxious) partners follow their mate and beat off the competition and its insistent sperm. Dogs have the unsubtle strategy of copulating for long enough to give their sperm a head start over the next candidate, while mice and many insects insert a plug to keep out a later deposit. Those excluded can fight back, with a hooked penis to unblock the female or a nozzle to siphon out any earlier contribution. The fight goes on after sex, because a sperm has to compete with others donated by the previous swain. A typical mammal has a hundred million sperm in each ejaculate, which leaves plenty of room for an internal struggle. A hidden battle rages inside any female who mates more than once. A quick DNA test shows which partner succeeded as a father. Often, the last male to mate sires all or most of the young. It hence pays a male who mates with a female for the first time to make a lot of sperm to flood out an earlier donation. Ejaculates with a new partner are several times larger than those produced for a familiar mate.

Whoever is on top, the sexual interests of males and females differ. It pays all males to be lazy, selfish and debauched, while any female is better off with an active, helpful and faithful spouse. Such paragons are hard to find and may be less trustworthy than they seem. Certain flies court for days before a male can persuade a female to respond, but once he succeeds, he inserts a cocktail of chemicals. Some force the female to lay eggs sooner than otherwise she would, and some reduce her sexual interest in later males. A female who mates many times dies long before one who mates just once.

Plants do not escape the battle of the sexes. The flower is an organ of allure, evolved to attract the pollinators who move male genes. In flowers that are both male and female, a reduction in size makes little difference to female success, as just a single bee can deliver the crucial genes. It has, in contrast, a great effect on the export of pollen. Plenty of pollinators means lots of male sex cells on the move. Flowers are silent screams of masculine passion.

Often, their cries are unheard. In some orchids, a mere one plant in fifty attracts even a single insect. Pollinators do thirty billion dollars' worth of work for American farmers each year. In orchards, with millions of flowers in bloom at once, sexual competition is so intense that growers hire hives of peripatetic bees to ensure that there are enough to go round. Insecticides have killed half the honeybees in the United States over the past decade or so. Many plants now fail to set seed, and, in the 1970s, the Canadian blueberry crop failed, at enormous cost, because of a lack of pollinators. The sexual struggle in American plants is, as a result, more intense than ever before.

Males are not always the aggressive sex. The female spotted hyena has a structure remarkably like a penis: an enlarged and erectile clitoris, as big as her partner's genital organ. For intercourse, she rolls it back, rather like a man might roll up the sleeve of his shirt. She gives birth through her penis – and it must be a painful business, as around a sixth of all females die when they have their first cubs. When it comes to sex, the female spotted hyena has an unrelaxed existence that calls for some steroid rage – and the appendages that grow from it. Her bizarre organ appears because she makes lots of the male hormone, testosterone. As in the grouse, that useful chemical is involved in a sexual battle. In spotted hyenas, females are in charge. The top bitch passes on her rank to her daughters, who must battle to keep it. Status pays, because females at the top get more meat and have twice as many young as do others. Hyenas have twins, and newborn sisters attack each other within minutes. One is killed, or lives as a cringing inferior who may never have cubs of her own. All this stress leads to the production of male hormones – a great aid to courage – by females. The ladylike penis follows on.

Because of its well-endowed female, Aristotle believed the spotted hyena to be a hermaphrodite. He was wrong, but animals in which boy-girl meets girl-boy are common. They are an experiment in the struggle of the sexes. A long debate determines who will be unlucky and bear the cost of eggs rather than sperm. To mate may take hours, with an elaborate nuptial dance. Careful bargaining ensures a fair division of the bill (although in some slugs the argument goes so far as to involve each partner trying to bite off the other's penis). In sea-slugs the first penis enters, and then the other, while in hermaphrodite fish a series of packets of eggs are traded for an equal number of bundles of sperm.

Sexual conflict – penis-biting included – is as bitter as the struggle for food or for light. It can produce traits that are favoured only because they increase sexual success, even as they reduce the chances of staying alive. Alarmed by the cost of such a structure, Darwin wrote that 'The sight of a feather in a peacock's tail, whenever I gaze at it, makes me sick'. The outlandish organ, and many like it, reduces its bearers' hopes of survival and as a result might seem evidence against evolution. However, the birds' gaudy train does not deny the truth of Darwinism. Instead it is a reminder of the rigours of its second examination paper. Such bizarre decorations are driven by a particular force of selection called sexual selection. That depends, not on a struggle for existence, but on a struggle between the males for possession of the females; the result is not death to the unsuccessful competitor, but few or no offspring.

In sex, who wins is often a matter of status, determined by fights, threats and submissions. The risk is real: among baboons, for example, half of the males who try to take over a group from its owner are seriously injured. There can never be enough status around, because a top dog exists only by virtue of his inferiors. Such battles can be reduced with the use of signals to establish whether a challenge is worthwhile. To sport a badge of rank is cheaper than to face the risk of violent death. The symbols include massive antlers, loud croaks, or characters invisible to the human eye. Male and female bluetits look the same to us, but in the ultraviolet (seen by birds) the males have a conspicuous patch on the head. Male insects make complicated scents, or sound inaudible to ourselves. Many have ornate genitals, studded with knobs and spikes. Such titillators (from the Latin *titillare*, to tickle) were once seen as keys to open a partner's lock and ensure that a female accepted a male of her own kind. They are, however, found only among insects whose females mate many times. Those with faithful females have simple penises. A spike may hence be a statement of excellence, rather than identity, tested by one party before the other is allowed in.

Although sexual selection moulds the public image of biology, natural selection is always on the alert to stop it from going too far. In Uganda in the 1930s, almost every male elephant had tusks, structures evolved (at least in part) as statements of reproductive excellence. Sixty years later, ivory poachers had much reduced the number of animals, with those with the largest tusks at greatest risk. Now, a third of adults are tuskless, because the negative effects of ivory on survival outweighed its role in allure.

However bizarre the behaviour involved, sex is, in the end, a struggle to find someone to fuse with. Sperm and egg do not grow or build a home, but they do unite. All biographies come in two parts. A long and tedious episode of advance from egg to adult is followed by a brief fight to find another cell. Selection is hard at work on both legs of the journey from egg to grave. The opposed priorities of sex and survival explain why the two sexes may look so unalike, not just in trivial characters such as antlers or tails, but in their very essence.

To meet someone else it pays to be common and to move around a lot. As a result, sexual selection favours small, active and abundant sex cells. To get a head start on the way to adulthood needs a large and well-stocked embryo, and natural selection is all in favour of reproductive cells that are big, stuffed with food, and reluctant to waste energy in going places. The balance between the two forces explains the evolution of sperm and egg – and of the sexes themselves.

Males have small sex cells, females large. The difference (the true definition of what males and females actually are) results from an ancient conflict. Long ago (and in some simple organisms today), sex cells were all the same size and fused to make an embryo, well provided with food. Then self-interest made an appearance and one partner cheated by making smaller but more abundant cells. He (for such was, from that moment, his gender) might have hungry young, but there were more of them. His success was limited by the risk of those small cells fusing to make an embryo with too little food to grow at all. Selection favoured union with the large cells of his ally, and so the sexes were born. Males have been frauds since they began.

All males make more sperm than can succeed. This leads to a struggle among their sex cells. Often, a male shoots himself – metaphorically – in the foot as, once an egg has been fertilized, the sperm shut out may well be his own. If he is the only male in the reproductive arena, there is not much point in lots of sperm (or in being big enough to make it). As a result, in most animals (except the small sample – cows, deer or humans – with which we are familiar) males are smaller than females. Sexual selection has insisted on small and active sons, natural selection on large and well-fed daughters. Some males are tiny. The male angler fish is a worm-like appendage attached to his mate. Certain barnacles, indeed, were once thought to be hermaphrodites, but the male is there, snug inside, busy with sperm and not much else. In the vastness of the sea each male has but a small chance of a sexual encounter. The best strategy is to have many small males who stick like glue on the rare occasions when one makes the grade.

Not until the battle escalates, with males rather than sperm at war, do there evolve the virile beasts so beloved of television producers. Fish show how size can matter. Many start as one sex, and change into the other as they grow. Some are male first, some female. Precedence follows a simple rule. In those whose sex cells are ejected into the water (with competition among sperm rather than among males themselves) it pays to be male first, because any male, small as he is, can generate embarrassing amounts of sperm while an older (and larger) female can make more and better-fed eggs. Not unless the battle is among males for possession of the females does it pay to make eggs first and to delay manhood until full size and aggressive potential is reached.

The sexual equation has two terms. Males compete for females, while females choose the males that they find, for some reason, attractive. Male battles are obvious enough, but can the aesthetic tastes of females drive males to such lengths? It may seem childish to attribute any effect to such apparently weak means: but if man can in a short time give elegant carriage and beauty to his bantams, according to his standard of beauty, there is no good reason to doubt that female birds, by selecting, during thousands of generations, the most melodious or beautiful males, according to their standard of beauty, might produce a marked effect.

Females do prefer the frog with the deepest croak, the bower-bird with the most decorated nest, the plant with the most symmetrical flower, and so – endlessly – on. Why? There are many ideas about the evolution of such sexual habits, and perhaps almost as many truths behind them.

The choice may help with a meal, or somewhere to live. Many male insects give their mates a tasty grub before sex. The larger the gift, the more acceptable the partner. Female fruit-flies digest the ejaculate itself to help them make eggs; but other insects ask for more. A certain moth is protected from its enemies by a poison picked up from plants. A male passes on with his sperm a solid dose of the substance, which in turn reaches the eggs. His sex scent (used to attract the females) is based on the poison, and the more scent he can make, the safer his partner and her eggs and the better mate he will be.

Females do not always approve of male ardour, because it wastes time and cuts down their own chance to choose. As well as sex, females have to find food and a place to live and lay eggs. They may be loath to accept a partner. Some female insects evolve sharp spikes to discourage courtship, and squirrels remove and eat the male's copulatory plug as soon as he has left.

Sexual selection leads to a *folie à deux*. If the best choose the brightest, female preference and male trait will evolve together. The system has volatility built in, and once the sexual fuse is lit, an evolutionary explosion is more or less guaranteed. If a sexy male has sexy sons who are chosen by the most fastidious females, the process will run away with itself to give characters as strange as the bluetit's crown. How far it can go is limited by natural selection. Even an ultraviolet coronet is expensive, because it can be seen by sparrowhawks.

Females can, without doubt, choose mates on the basis of what genes a male might carry. Because of the dangers of incest, a brother or a cousin is a less attractive partner than is a non-relative. Whole systems of kin recognition have evolved to avoid such matings. Mice excrete scents in their urine that enable females to avoid their close kin, and plants carry matched cues on pollen and egg to ensure that close relatives never pair. But why should the choice involve a showy crown or a pair of luminous buttocks? What does a female gain from such a well-endowed mate? His characters seem at first sight useless or even perverse. It may be – once more – a matter of marketing. An extravagant signal could be a general statement of a male's ability to cope. A peacock able to manage his accessory must, no doubt, be quite a lusty specimen and could make an excellent father.

As soap-makers know, advertisements are expensive. That is part of their point, for if a company can afford lavish publicity it must be a solid concern. Because cheats devalue the currency, business goes a long way to protect its brands. The Distiller's Company in London – makers of Gordon's Gin and Johnny Walker Whisky – has a 'black museum' of fakes from around the world (the least plausible a shampoo, Johnny Washy's Old Sloshy, which a customer once drank by mistake). The company always sues those who mimic them, and almost always wins. France once had a brutal and much-feared band of bank robbers called the Moustache Gang. Cashiers learned to hand over the money as soon as they appeared. In a few months there appeared squads of counterfeit gangs with the trademark but not the violence. Quite soon, the intimidatory power of upper-lip hair began to wane. In the same way, a feeble male sparrow painted with the black bib of sexual dominance is attacked by other males for his presumption.

If a male can afford an expensive symbol, to pick him as a mate may help, because his young may be as well endowed. In tree-frogs, females prefer males who make a longer call – and they gain nothing but his DNA, because he disappears before the tadpoles hatch. When half the eggs of a female are fertilized by long-call males, and the rest with sperm from those with short calls, the former grow faster. The genes advertised by their father in his lengthy croaks do indeed help his offspring.

The work of natural and of sexual selection stands in great contrast. Natural selection often reaches similar ends with different means. Thus, wasps or snakes that warn of danger each use black and yellow, while both grouse and peppered moths have mottled camouflage. Sexual characters are more capricious. Each haughty

male has a badge unique to his own kind. Some bower-birds are
bright and showy, but others are drab and excite their females with
elaborate shelters instead. There is no general statement of sexual
excellence (apart, perhaps, from size) common to all male birds, or to
all flowers. Perhaps the characters are arbitrary and the ancestral
female's choice was frivolous. The machinery of sexual selection did
the rest.

Males can be beguiled into a signal by the workings of the female
mind. A bird brain is quite naïve. A female oystercatcher given an
egg several times the normal size will abandon her nest to sit on what
seems a manifest fake. The message that says 'egg' overwhelms
what little common sense she has. Female zebra-finches prefer
sex with males with red plastic leg-rings and even have a penchant
for a mate decorated with a white chef's hat. Eccentric as these
quirks seem, they are no more than enhanced versions of what a
male zebra-finch already has: a red bill and a white facial stripe
proclaiming him to belong to the correct stock. Female bias is wired
deep into the brain, poised to send her sexual partners into a dead
end.

For sex, as for business, what matters in the end is not advertising
but sales. Feeble players – the discount stores of the sexual world –
may do as well as their more impressive opponents. The bluegill
sunfish has a society based on lies. Each fish sets off down a path to
one of several sexual styles. At the age of two, a male makes a
momentous decision – whether to become adult or to stay young.
Some mature while small, but others delay adulthood for six years or
more, by which time they are several times larger than their preco-
cious sibs. When maturity comes at last, the slow developers are big
enough to defend a home.

They own grand premises, but they have a problem: their cunning
sibs, who lurk nearby. When a landlord has enticed a female on to his
patch, the rivals dash in and emit semen. It lacks an impressive
wrapper and comes without a guarantee, but is cheap enough to
undercut their rival. When the sexual scoundrels grow too big
to sneak in, they again change their style. They begin to resemble
females, until they can saunter unafraid on to a territory, the sole risk
one of courtship by its besotted holder. When a real female appears a

transvestite's deception pays off. He fertilizes her eggs and makes a hurried exit.

A large male courts and inseminates females far better than does a prowler or cross-dresser. It has, though, taken him years to achieve his sexual peak – and for most of that time his rivals have, in devious ways, been passing on genes. To lurk or lie needs rapid maturity. A cheat's feeble frame and ability to exploit the market is as much a product of sexual selection as the beefy bodies of his rivals.

For a sunfish or a supermarket, the best game to play depends on what everybody else is up to. If most of the competition is territorial, it pays to cheat, because the water is full of potential cuckolds. If impostors become too common, then there will not be enough hosts to go round. It pays to be different and to do what others do not. If a strategy becomes rare it gains an advantage – but that is lost as soon as it becomes common. In time, the system settles down and house-holders and sneaks pass on genes with equal efficiency.

That explains why most animals have equal numbers of each sex. After all, males and females are themselves no more than alternative solutions to the problem of handing on genes. If a single male can fertilize dozens of partners, why the spares? Why not a hundred females and one male? The reason is simple: as in sunfish, the rarer caste is always better off. If so few males exist that each always finds lots of mates, then it pays a parent to have sons. A shortage of females puts an equivalent premium on daughters. Soon, things come to a balance, with equal investment into the transfer of genes through each gender.

Sexual selection helps explain the evolution of characters that at first sight make no sense. Unfortunately, it can also be a general excuse. It is unwise to attribute all sexual differences to this agency: for we see peculiarities arising and becoming attached to the male sex in our domestic animals (the wattle in male carrier-pigeons, horn-like protruberances in the cocks of certain fowls, and so on) which we cannot believe to be either useful to the males in battle or attractive to the females.

We see analogous cases under Nature. The penis raises man above the primates. Our organ is, in comparative terms, huge. The gorilla has a guilty secret: its one-and-a-quarter-inch member. The chimpanzee, a copulator of gigantic appetite (a male manages hundreds of sexual encounters with dozens of females each year), does little better.

Why do men have such large genitals? Some call on sexual selection as the cause of penis expansion. Like the peacock's tail, it simply ran away with itself. A penis may not cost as much in metabolic terms as does a showy tail, but think how useful it would be if all that too solid flesh were to be recycled into brain or muscle. Our unique organ might all be a matter of female choice. A natural experiment seems to test the idea. In New Guinea, men of the Ketengban tribe enhance what Nature has provided with a pointed gourd – a phallocarp. Without this additional foot or so of re-assurance, they say, they feel naked. Perhaps this is what the penis would evolve into were it not held back by some contrary force. It is, above all, a signal of virility.

In most places, of course, the penis is not a signal of social or genetic excellence at all, because so few get to see it. The organ is not flaunted in the sexual marketplace, but hidden away until the last possible moment. Even Don Giovanni did not wear a phallocarp. The Duke of Edinburgh, on a visit to New Guinea long ago, was given one, but has always refused to reveal whether he put it on. Most Western males follow in the ducal path. They are proud of their appendage, but in private. As far as is known, societies that expose or conceal that useful organ have no difference in penis size, but as both exist it is fatally easy to appeal to the option that fits the theory. Of course, things might have been different in the days before trousers, and *Homo erectus* may have lived up to his name in more ways than one. To test the idea that today's male accessory results from sexual selection long ago is even harder than to establish whether it is at work now.

The dangerous flexibility of the penis argument warns of the risks of searching too hard for the hand of selection. It has long been fashionable among evolutionists to mock the claims of the Rev. William Paley, whose *Natural Theology* (and its watch found upon a heath as proof of a divine watchmaker), published in 1802, multi-plied examples of the perfection of the body as proof of the existence of a Creator. Too often, enthusiasts for evolution do the same as they hail every quirk among plants or animals as evidence for its power. Sometimes they are right; but to assume that everything must be adapted simply because it evolved is to practise theology rather than science. The feats of selection are such that blind faith in its abilities is not needed.

On the Intercrossing of Individuals. Sexual selection – with all its expense – depends, of course, on the presence of two sexes in the first place. Most animals fulfil their destiny with that curious strategy. The division of reproductive labour has so many drawbacks that it is hard to see why it has not been quashed. Pious Jewish men in the second century chanted three times a day: 'Praise be to God that He has not created me a woman!' In fact the real question is why they (or anybody else) were created as men. Why should any female bother with such creatures? They seem almost useless. To give up males could at once double a female's output of daughters, who would copy her DNA; and each child would contain just her own genes, undiluted by those of a stranger.

Even so, when it comes to chastity (or at least to asexual reproduction), most plants and animals refuse to go the whole hog. Many try to escape sex, but they usually fail. Some creatures indulge but once a year, with long periods of abstinence as they make copies of themselves. Others are hermaphrodites, while some persist for countless generations without males at all. For most creatures, however, it seems that a cross with another individual is occasionally – perhaps at very long intervals – indispensable. The persistence of sex is one of the puzzles of biology. As is the case for the tail of the peacock, the theory of sex is well ahead of the facts needed to support it.

Many hermaphrodites – slugs, for instance – fall into what Woody Allen called 'sex with someone you really love': they mate with themselves. As a result, their progeny are liable to inherit two copies of the same gene and are more alike than average. If the process goes on for long enough, all the members of a line become, in effect, identical twins. On the way, genes are exposed that are best kept hidden. Although the survivors are purged of inherited weakness they have lost their variety.

For British slugs, sex stops at Preston. North of there, the familiar large slug of gardens and wild places retires from the sexual arena and takes incest to its logical conclusion. The slug of Scotland and Scandinavia is in effect a single strain of billions of identical animals. In southern parts, the slugs of Welsh mountain-tops or of Dartmoor (and, oddly enough, of parts of Cambridge) take the same reproductive route. The pattern of less sex in cold places is found in many creatures. It hints at one reason why sex might maintain itself: faced with the predictable enemies of frost and starvation it is better to evolve a single set of hardy genes that are never broken up by admixture with others. In the steamier parts of the world – Cheshire, or the Amazon Basin – the adversaries come from biology rather than physics; they are other animals (parasites included). These can themselves evolve, and a constant production of new genetic combinations through sex is essential if their victims are to have a chance.

Sex is not just an escape from the painful slowness of evolution, but an ingenious way to make scapegoats who, by their own sacrifice, save others from having to atone for their genetical sins. Several bad genes may come together in one shuffle of the reproductive cards. All are disposed of at the cost of a single bearer's death. In Levitical style, he carries the faults of many with him as he goes. In genetics, as in life, sex and guilt are close companions; and each is just as far from a convincing explanation.

Circumstances favourable to Natural Selection. What helps natural selection? A large amount of inheritable and diversified variability is favourable. A large number of individuals, by giving a better chance for the appearance within any given period of profitable variations, will compensate for a lesser amount of variability in each individual and is an extremely important element of success; and a large continental area, which will exist for long periods in a broken condition, will be the most favourable for the production of many new forms of life.

In other words, bacteria are bound to win their war against medicine. Nowhere else does the evolutionary battle take place in an arena where, in effect, one player holds all the cards. Bacteria show what natural selection needs and what it can do when it gets the chance.

The Murray Collection is a series of reference strains of harmful bacteria gathered between 1914 and 1950 and kept in suspended animation ever since. Every strain is, when reanimated with warmth and food, susceptible to every one of the dozens of antibiotics used today. They are a reminder of what a revolution those drugs made. Before the Second World War, wards were filled with patients close to a horrible death from infections of the blood. After penicillin, they could be cured with a few injections. Those glorious days will soon be over, because of evolution.

Antibiotics have become a human right. Penicillin was first used in the 1940s. Its history is an object-lesson in natural selection. The first resistant strain was found within a year of its use and soon spread. Twenty years ago, the drug could kill the bacterium that causes meningitis. In many places – the United States and France included – three-quarters can now defy it. The more the drugs are used, the more resistance spreads. In Norway, where antibiotics are controlled, one strain in five hundred of the septicaemia bug is resistant to more than one drug, while in Greece, where such remedies are available over the counter, half the strains are.

Natural selection, with the help of stupidity, has triumphed over medicine. Twelve million doses of antibiotic are given each year in the United States to fight colds or sore throats, against which they do not work. Plenty of European countries use more than a ton a day, most of which is wasted. In the Third World, things are even worse, and in Kenya, powerful drugs like tetracycline and ampicillin are sold on the street. Farmers who add the chemicals to animal feed pour yet more into the environment. Their 'growth promoters' include drugs that might be needed in medicine. To put them into food could almost have been designed to speed evolution. Again, Africa leads the way. It is easier to add a powder than to clean up a farm, and Kenyan chicken guts are filled with bacteria resistant to tetracycline. Now, fruit trees are sprayed to kill their bugs, and salmon farmers use drugs by the sack. The latest twist is to sell plastic food bags treated with anti-bacterial substances as defence against a non-existent threat to health.

Why have the bugs done so well? First, there are a lot of them about. Just a tenth of the cells of our bodies are human. Most of the rest belong to bacteria (although a few fungi, mites and worms leaven the mix). When things are good, the inhabitants of British guts double in number every twenty minutes or so, compared to the sixty years that it takes the population of these islands, even in expansive times, to do the same. With a world population of around ten with thirty zeros after it, bacteria are so common that the most improbable events are, in effect, bound to happen. Mutation is almost guaranteed and one individual among billions is certain to draw the successful ticket in the genetic lottery. That explains why many of them rarely indulge in sex (although some – the agent of gonorrhoea included – are as sexual as anything else). For most bacteria there is no need to exchange genes with another when, quite soon, the same one will turn up in your own family.

When necessary, they cheat to ensure a supply of new mutations. DNA is supported by a mechanism that cuts down the number of mistakes each time it is copied. As soon as things get tough our enemies have a clever stratagem. They circumvent their own repair machinery and, as a result, increase the mutation rate in a last-ditch attempt to generate variants that may save them. Some drugs spark off this emergency response and help their targets to defend themselves.

Even so, the rate of change per gene is small, at about one in ten million. All new variants are at first rare and most disappear, purely by accident, before selection notices them. There were, no doubt, hundreds of moths with brand new genes for black wings in nineteenth-century Manchester. The majority failed, for reasons unrelated to pollution – they starved, or drowned in a Mancunian downpour. The chances of success for any new gene are smaller than they seem. The peppered moth was common enough to cope with the accidental death of most of its new melanics (although it took twenty years for the gene to succeed). For the billions of bacteria in a single decent-sized pustule, the chance loss of a few mutations is less of a worry.

Geography, too, plays a part in the evolutionary equation. If everybody lives in the same place and faces the same challenges, then selection is absolute: a new mutation succeeds, or it fails and the whole population dies. In a divided group, a few individuals may find a safe refuge in which to await a favourable change.

New York in the 1980s suffered an outbreak of tuberculosis, concentrated among the poor. Black men had a rate of infection fifty times the national average. A lengthy course of the right medicine can cure the disease. In New York, though, resistance became impossible to contain. Most patients failed to complete their treatment: a single dose made them feel better, there were side-effects, and the lives of many were in such chaos that they could not manage a course of therapy. In Harlem, only a few patients took more than a few pills before giving up.

As a result, drug resistance flourished in the poorer boroughs. It soon spread to the affluent parts of the city, in which the disease had seemed to be defeated. A reservoir of resistance, maintained by low doses of drugs, was enough to overcome all efforts to get rid of the disease. Many cities are great continents of people, but those of the United States are divided by social barriers, with success on one of their many islands of humanity nullified by failure in another. In places in which the population is treated as a whole, tuberculosis has been conquered.

The spread of bacterial genes is helped by sex as much as by politics. Although some rarely indulge, others enjoy it on demand, and in many ingenious ways. For them, venereal disease evolved early on. Infectious third parties called plasmids, sections of mobile DNA inserted into the genetic material, are (rather like AIDS viruses in human cells) multiplied each time their hosts divide. Some can hop from host to host, carrying resistance genes as they go. In time, a single plasmid may accumulate many such genes and become invaluable to its carriers.

The Murray Collection had plenty of plasmids, but no resistance genes at all. Their descendants are full of genes that enable them to cope with several antibiotics at once. Many of the multiple-resistance plasmids were first seen in hospitals, but are now everywhere. In Madagascar a single strain of plague bacillus can resist ampicillin, chloramphenicol, streptomycin, spectinomycin, kanamycin, tetracycline and sulphonamides. All seven resistances are carried on a single short length of mobile DNA. With sixty million air travellers a year, such elements can move at some speed. A resistant agent of pneumonia, first seen in Spain, was found within five years in the United States, Korea and South Africa. Even worse, plasmids can hop between species. Some found in harmless denizens of guts have entered pathogens (such as the agent of gonorrhoea) and, at a stroke, render them safe from several drugs.

Bacteria show how natural selection builds its defences. Evolution is happy to pick up and use whatever is at hand. It presses new mutations into service as they arise, and is just as ready to make do with what is already around. Resistant bacteria may break down an antibiotic, block its entry, pump it out, store it where it will do no harm, or change the shape of its target molecule. Sometimes, complete new pieces of biochemical equipment evolve, but more often workaday genes are pressed into service. Tetracycline can be coped with in twenty different ways and penicillin in almost as many. The most effective destroyers of drugs are ordinary enzymes made in huge amounts by resistant strains. The bugs pay a price, as their economy is so crippled by the need to fight off the enemy that they become bacterial drug addicts, unable to survive without the poison given to exterminate them.

The last new class of such drugs was discovered twenty years ago and no more are on the horizon. Medicine's finest days may soon be over, but antibiotics, in their brief flowering, have revealed as can nothing else what evolution needs to do its finest work.

Extinction. The creative force of evolution has a dark side, for life today was earned at the cost of the death of almost all that went before. The idea of a past now gone alarms fundamentalists because it casts doubt on the perfection of God's plan. Thomas Jefferson was so concerned that he told the explorers Lewis and Clark to keep an eye open for mastodons as they travelled through their new continent. Charles Lyell, the geologist whose work formed Darwin's views, also denied the idea of loss. Instead, he envisaged a time when 'the pterodactyl might flit again through umbrageous groves of tree-ferns'. Extinction, however, is a crucial part of the evolutionary machine and is as inevitable as is the origin of species.

Because of the conservation movement, such catastrophes have become impossible to ignore. Public concern about the fate of the planet suffers from overkill. Twenty years of lament about the imminent disaster has led to a general view that things cannot be as bad as is painted. Many who once cared about the environment now share a Voltairean sentiment that the easiest way out of the crisis might be to strangle the last panda with the guts of the last blue whale. However, today's cataclysm is no different from many others (and is far smaller than some of the accidents that befell ancient life). Few plants and animals last for long. The descendants of a very few, transformed by natural selection, make up the world today. Some survivors manage to remain unchanged for tens of millions of years, but for most, death soon follows birth.

Why do plants or animals meet their end? If they arise by the slow process of natural selection, why should they not disappear for the same reason? If each area is already fully stocked with inhabitants, it follows that as each selected and favoured form increases in number, so will the less favoured forms decrease and become rare. As better adapted successors emerge, others are forced into a corner. They are driven back in the great battle for life until, at last, they are gone. Like the British Motor Corporation, they have been driven to ruin by their better-adapted successors.

The evidence is in the fossil record. Although there have been some pulses in the past, with more new forms arising at certain times, the big picture is consistent. Over the past five hundred million years, through all its ecological alarms and excursions, new kinds appeared at an almost constant rate. A survey of tens of thousands of marine animals over that time gives a rate of four hundred and fifty new species a year. The world is more or less full. Any new species must push out a predecessor to have a chance and will, sooner or later, be squeezed out in its turn.

Sometimes, a few survivors of an earlier age remain as relics of an ancient race. Lungfish first appeared about four hundred million years ago. They were an active and diverse group that adapted themselves to whatever they were faced with. A cadet branch of the family played an important part in the transition of vertebrates from water to land. Then, all of a sudden, lungfish evolution slowed down, while around them that of other vertebrates exploded. Now a mere half-dozen kinds are left. They live a glum existence in the lakes and rivers of Africa, South America and Australia, shrouded in mud for much of the year. They are living fossils, reminders of a universe now lost.

Any form represented by few individuals will, during fluctuations in the seasons or in the number of its enemies, run a good chance of utter extinction. As new species in the course of time are formed through natural selection, others will become rarer. Then, bad luck begins to play a part. Although time and the gradual appearance of new and better forms kills most of them, chance is also important. Bighorn sheep in the Rockies have been studied for almost a century: and in that time all groups of fifty or fewer animals became extinct, while nearly all those with more than a hundred survived. When gambling with nature, it pays to have a strong hand.

It is hard to be sure about extinction. Nobody, after all, writes to the newspapers about the last cuckoo of spring. The youngest fossil of the coelacanth, a fish at one time thought to be important in the origin of land animals, is eighty million years old, and it was once presumed to be gone for ever. An example of that remarkable beast was caught off the coast of South Africa in 1938. For a time coelacanths seemed to have a population of a few hundred and to teeter on the edge of demise, but, sixty years on, a specimen on sale in an Indonesian market led to the discovery of many more, thousands of miles away off the coast of Sulawesi. Coelacanths may be common all around the Indian Ocean.

The same uncertainty applies in other places. About half a million kinds of beetle are known but, because most were found by gassing tropical trees, many are recorded from just a single location. An absence on a second visit may not mean that they have gone, but that they are on another tree. Species thought extinct quite often re-appear. In California, where there has been much concern about the loss of Mediterranean plants, more kinds supposed to have been driven out have been rediscovered than have in fact disappeared. Even so, when it comes to the destruction of what evolution has made, we live in interesting times. They prove how fast selection can carry out its baleful work as soon as it gets the chance. All over the world, life has been swept away, as if by some murderous pestilence. That pestilence is man and his hangers-on.

At any moment, a hundred thousand people are suspended over the Atlantic. Some smuggle alien plants and animals, but many more have seeds, insects and more in their turn-ups or their baggage. Other creatures travel in soil or packing-crates, or are introduced for food or ornament. Many of the travellers make a home in their new world and drive out the natives. The worst culprits are those – like men or rats or hedgehogs – able to eat many things. If they choose to graze on the eggs of some rare bird, they can drive it to extinction while they sustain themselves on other food. The cats on Marion Island in the sub-Antarctic killed four hundred thousand seabirds a year before they themselves were wiped out by a virus. On Stephens Island in the Cook Strait, in 1894, the lighthouse keeper's cat brought in the first known specimen of the Stephens Island wren. Its descendants ate the last one.

The Galapagos were uninhabited when first seen by Europeans in 1535. Half a million fossil bones of reptiles and birds have been found in the lava tubes scattered across the landscape. In the eight thousand years before the Spaniards arrived, each island lost, at most, three of its native vertebrates. In the four centuries since the onslaught began, the rate of loss has gone up by a hundred times. Cats, dogs, goats, pigs and ants have done huge damage. Wasps were introduced in the 1990s and have spread. By 1998, five hundred different foreign plants were on the islands. The guava is a pest, as is a vine called the Curse of India. Blackberries and passionflowers are on the march. As they grow they shade out botanical equivalents of Darwin's finches. The cacti are unique to the archipelago, but many have been strangled by vines. The giant tortoises and land iguanas that feed on them are damaged in their turn. The Floreana flax, evolved on a single island, now consists of a mere eight plants in the wild and three more in cultivation. With a growth rate in the human population of 10 per cent a year the future is bleak.

The emergency goes further than the Galapagos. Half of all kinds of bird in the world may disappear within the next three hundred years, a rate of loss thousands of times that in most of the fossil record. About one in twenty of all the people who have ever existed are alive today, compared to a mere one in a thousand of the different kinds of animal and plant. The fate of the defunct is as much testimony to the force of natural selection as is the triumph of the survivors. Biologists often bemoan the ecological crisis – the loss of the cod, the whale or the dodo – as something outside the normal world of evolution. It is not: extinction has happened millions of times before and is an exciting opportunity for science. The years since Darwin have relived an evolutionary experiment that shows how the more adapted will always drive out those less able to cope.

Divergence of Character. Because all animals must compete with their relatives, evolution favours things that differ from each other. The more it can do so, the less each of its products is forced to depend on an asset in short supply. The more diversified the descendants from any one species become in structure, constitution and habits, by so much will they be better enabled to seize on many and widely diversified places in the polity of nature.

Canadian lakes are full of sticklebacks. They must be recent arrivals, for the whole country was covered by ice until a few thousand years ago. Most sticklebacks live in the sea or in estuaries, and the fish have invaded fresh water many times. They come in two forms, sometimes found in the same lake. The first has a stocky body and lives in the shallows, where it eats grubs, while the other grows longer and has a smaller mouth suited to open water and a diet of surface swimmers. The two live in different patches of an ecological quilt. Each is better off in its own habitat, and each originated not long ago, each within its own lake system. Fish of different shape are already reluctant to interbreed. Each is, if not yet a separate species, well on the way to an identity of its own.

Small lakes contain but one kind of fish, of intermediate form. When put into competition with the specialists, they fail. As soon as the chance arises the sticklebacks split into two types with their own peculiar habits and structure and exploit more of what the lake can offer. Canadian salmon, too, come in distinct versions, one – the sockeye – migrating to the sea before spawning, the other – the smaller kokanee – staying in the lake for its whole life. Sockeyes moved to empty lakes quickly evolve a new and reduced version that stays at home. This, too, is evidence of the pressure to divide.

Natural selection has, built in, what may be called the principle of divergence, causing differences, at first barely appreciable, steadily to increase, and breeds to diverge in character both from each other and from their common parent. The process is captured as a snapshot when, as sometimes happens in snails, water-fleas and plants, variation that circulates through a population is partitioned into a series of asexual lines. What was once a variable group is split into a series of distinct clones, each of which contains part of the diversity of their sexual parent.

A certain New Zealand freshwater snail exists in both sexual and asexual forms. Within a lake, the sexuals are accompanied by dozens of daughter clones. The sexual form, in all its diversity, is found in most places. Each clone, in contrast, lives at a certain depth and on certain plants. It is frozen into a separate place in the economy, unable to invade the equally cramped niche next door. Its narrow life is testament to the fact that competition will generally be most severe between those forms which are most nearly related to each other. The snails show how finely divided the evolutionary cake may be.

The snail was introduced into Europe in the 1880s, on a load of ship's ballast dumped into the Thames, close to where the dismal suburb of Thamesmead now stands. It spread to fill most of the lakes and rivers of Europe. It lacks any form of sex and has just the three clones first introduced, rather than the hundreds found in its native land. They divide up their waters quite amicably, and each occupies a wide range of habitats, from ponds to estuaries. Without an infinity of competitors they can afford to spread out and are not crowded into a smaller and smaller space by their kin. Like any business, life must diversify its manufactures, or fail. As a result, evolution – like capitalism – has to run to stay in the same place. If the young overtake their parents, the parents have no choice but to find another trade, or die.

That brutal fact launched the Industrial Revolution and drives the economies of today. Commerce depends, like life itself, on a constant input of energy. In modern London or Manchester, it flows not from the coal-fired stations that poisoned the Victorians and their moths, but from boilers fueled by uranium, gas or oil. Evolution is not mocked. Pylons carry the electricity across the land. Each is protected by a layer of zinc, which drips on to the ground when it rains. Zinc, like copper or lead, is poisonous to plants, but some have genes able to deal with it. Natural selection has come up with the same response again and again: under most pylons is a patch of grass with genes for zinc tolerance. Some populations lack the right genes and cannot grow, but in most places they cope with ease. In the archipelago of steel that fills all cities an experiment has been repeated thousands of times.

Zinc faces evolution with nothing new. Man has long spread his poisons. Four thousand years ago the production of lead reached a peak not matched until the Industrial Revolution. The records of ice and peat show that in 1979, the height of its use in petrol, the air had fifteen hundred times the background level. As a result, every road is lined by a swathe of lead-tolerant vegetation. Many soils (such as those of the Lizard Peninsula) are in any case awash with metal. Dozens of plants have evolved to cope, some able to deal with amounts sixty times more than those lethal for others. For selection, a pylon is a minor provocation.

Cities, like pylons, do not last. The first was founded but ten thousand years ago, and many have come and gone since then. Sometimes, the sole evidence of their passing lies in evolution. Fourteenth-century Africa had a culture based on copper, mined from the deposits – still the largest in the world – of what is now Zaire and Zambia. Around today's mines and smelters the soil is so full of metal that only plants with genes for tolerance can survive. One, the copper flower, grows in dense violet clumps on the most polluted soils of all.

Patches of that plant are found far from any habitation. They are the tombstones of lost villages, the remnants of a forgotten Industrial Revolution. Hundreds of copper crosses, used as money by the miners, are buried beneath the violet blooms. The genes of the copper flower are monuments to those who made the coins: all else has disappeared. They are a reminder of how fleeting are the wishes and efforts of man, how short his time, and in consequence how poor his products, compared with those accumulated by nature during whole geological periods.

Natural selection is no more than a machine. What it makes depends on what it has to work with and where it started. Evolution does its job as well as it needs to, and no more. Sometimes, as in the balance of the numbers of each sex, it does it well; but often it is satisfied with what seems slapdash. Most of its products do not last. Who could ever have designed a tree-kangaroo? Clumsy as the animal may seem, it is infinitely better adapted to the most complex conditions of life, and plainly bears the stamp of far higher workmanship, than anything achieved by man.

Industry has begun to notice the superiority of Nature. Nowadays, a billion dollars' worth of copper a year is extracted not with furnaces, but with a bacterium able to break down ore and to release the metal. The bug obtains its energy by chemical means and may drive smelters to extinction. Tolerant plants can accumulate a hundredth of their own weight in zinc or nickel; and, in Africa, if given enough fertilizer, can be a crop twice as valuable as wheat. It is even possible to harvest plants from goldmines to reap their treasure.

Man, in his factories for copper or soap, has begun to use the instrument that shapes biology, but has rarely matched the work of Nature. In spite of Otto Lilienthal's great work *Bird Flight as the Basis for Aviation* and the long – and lethal – series of birdlike gliders and flappers that followed, today's aeroplanes do not have feathers. A few useful ideas have been lifted: the inventor of the tunnelling shield based it on the shipworm that chews wood and passes the waste through its body, while the 1874 patent for barbed wire stated that the invention was designed to look like a thorn hedge. Like thorns themselves, the new product diversified into (among many others) Griswold's Savage, Blake's Body Grip and Brink's Stinger. The patentee almost lost his millions because rivals claimed that he had invented nothing, but merely copied the living world.

Nature, though, starts in a different place and uses materials quite unlike those available to man. It cannot smelt copper or make crosses, but, with what it has, it works miracles. Man, his machines and Darwin's idea may – given a few million years – do almost as well.

Summary of Chapter. If during the long course of ages and under varying
   conditions of life, organic beings vary at all in the several parts of their

organisation, and I think this cannot be disputed; if there be, owing to the high geometrical powers of increase of each species, at some age, season, or year, a severe struggle for life, and this certainly cannot be disputed; then, considering the infinite complexity of the relations of all organic beings to each other and to their conditions of existence, causing an infinite diversity in structure, constitution, and habits, to be advantageous to them, I think it would be a most extraordinary fact if no variation ever had occurred useful to each being's own welfare, in the same way as so many variations have occurred useful to man. But if variations useful to any organic being do occur, assuredly individuals thus characterised will have the best chance of being preserved in the struggle for life; and from the strong principle of inheritance they will tend to produce offspring similarly characterised. This principle of preservation, I have called, for the sake of brevity, Natural Selection. Natural selection, on the principle of qualities being inherited at corresponding ages, can modify the egg, seed, or young, as easily as the adult. Amongst many animals, sexual selection will give its aid to ordinary selection, by assuring to the most vigorous and best adapted males the greatest number of offspring. Sexual selection will also give characters useful to the males alone, in their struggles with other males.

Whether natural selection has really thus acted in nature, in modifying and adapting the various forms of life to their several conditions and stations, must be judged of by the general tenour and balance of evidence given in the following chapters. But we already see how it entails extinction; and how largely extinction has acted in the world's history, geology plainly declares. Natural selection, also, leads to divergence of character; for more living beings can be supported on the same area the more they diverge in structure, habits, and constitution, of which we see proof by looking at the inhabitants of any small spot or at naturalised productions. Therefore during the modification of the descendants of any one species, and during the incessant struggle of all species to increase in numbers, the more diversified these descendants become, the better will be their chance of succeeding in the battle of life. Thus the small differences distinguishing varieties of the same species, will steadily tend to increase till they come to equal the greater differences between species of the same genus, or even of distinct genera.

We have seen that it is the common, the widely-diffused, and widely-ranging species, belonging to the larger genera, which vary most; and these will tend to transmit to their modified offspring that superiority which now makes them dominant in their own countries. Natural selection, as has just been remarked, leads to divergence of character and to much extinction of the less improved and inter-mediate forms of life. On these principles, I believe, the nature of the affinities of all organic beings may be explained. It is a truly wonder-ful fact the wonder of which we are apt to overlook from familiarity that all animals and all plants throughout all time and space should be related to each other in group subordinate to group, in the manner which we everywhere behold; namely, varieties of the same species most closely related together, species of the same genus less closely and unequally related together, forming sections and sub-genera, species of distinct genera much less closely related, and genera related in different degrees, forming sub-families, families, orders, sub-classes, and classes. The several subordinate groups in any class cannot be ranked in a single file, but seem rather to be clustered round points, and these round other points, and so on in almost end-less cycles. On the view that each species has been independently created, I can see no explanation of this great fact in the classification of all organic beings; but, to the best of my judgment, it is explained through inheritance and the complex action of natural selection, entailing extinction and divergence of character.

The affinities of all the beings of the same class have sometimes been represented by a great tree. I believe this simile largely speaks the truth. The green and budding twigs may represent existing species; and those produced during each former year may represent the long succession of extinct species. At each period of growth all the growing twigs have tried to branch out on all sides, and to overtop and kill the surrounding twigs and branches, in the same manner as species and groups of species have tried to overmaster other species in the great battle for life. The limbs divided into great branches, and these into lesser and lesser branches, were themselves once, when the tree was small, budding twigs; and this connexion of the former and present buds by ramifying branches may well represent the classification of all extinct and living species in groups subordinate to groups. Of the many twigs which flourished when the tree was a mere bush, only two or three, now grown into great branches, yet survive and bear all the other branches; so with the species which lived during long-past geological periods, very few now have living and modified descendants. From the first growth of the tree, many a limb and branch has decayed and dropped off; and these lost branches of various sizes may represent those whole orders, families, and genera which have now no living representatives, and which are known to us only from having been found in a fossil state. As we here and there see a thin straggling branch springing from a fork low down in a tree, and which by some chance has been favoured and is still alive on its summit, so we occasionally see an animal like the Ornithorhynchus or Lepidosiren, which in some small degree connects by its affinities two large branches of life, and which has apparently been saved from fatal competition by having inhabited a protected station. As buds give rise by growth to fresh buds, and these, if vigorous, branch out and overtop on all sides many a feebler branch, so by generation I believe it has been with the great Tree of Life, which fills with its dead and broken branches the crust of the earth, and covers the surface with its ever branching and beautiful ramifications.

CHAPTER V

LAWS OF VARIATION

Heredity, its myths and errors — Effects of use and disuse; heredity as memory — The inheritance of privilege — Acclimatization, to heat and to poisons — Atavism and the failure of the average — Genes as particles rather than fluids — Mendel and the physical basis of inheritance — Mutation and the rate of evolution — Correlation of growth; genes and development — The simplicity of Mendel's law contrasted with the complexity of the real world — Genetics the foundation of the theory of evolution — Summary

GEORGE SPENCER HAD the misfortune to live in Connecticut at the height of the Puritan obsession with sex. In a childhood accident he had lost an eye. When a one-eyed pig was born in the town, the culprit seemed obvious and Spencer was accused of bestiality. Terrified (for this was a capital offence) he first admitted the crime, but then withdrew. Two witnesses were, the law said, needed. So anxious were the magistrates to hang him that his confession was accepted as the first and the pig as the second.

Sex and confusion have long been bedfellows. For ten thousand years of success for plant and animal breeders, its practice was quite divorced from its theory. The laws of inheritance were quite unknown, and notions now seen as absurd – the transfer of eyes from pig to man included – were believed by everyone. Many of those ideas were misguided, but some contain enough truth to explain (and perhaps to excuse) the morass in which biology wallowed for so long.

It is easy to multiply (and to mock) curious beliefs about inheritance. His ignorance of the subject worried Darwin and led him, in his later years, to complicate and confuse his ideas on evolution. Now genetics has become the science that catches the collective imagination as does no other. To the public it seems beautifully simple, but it is not.

The laws of genetics are as elegant as is the idea of natural selection. Mendel started a science that still rests upon his simple rules. Like evolution itself, it has become more complicated since it began, and Darwin's perplexity makes more sense today than it did fifty years ago. Many of his difficulties now look like an honest attempt to find simple patterns in complicated situations. This chapter is less faithful to its original than are others, if only because Darwin got it so wrong. Even so, with hindsight, *The Origin* points at problems about heredity that are still scarcely understood.

Effects of Use and Disuse. The notion of the inheritance of acquired characters does away with any need for a theory of evolution. In melanic moths the camouflaged young were once said to result from a 'powerful impression on females during the all important period of life, viz., that of propagation, coupled with an instinctive provision for the protection of its future progeny'. The Italian botanist Odoardo Beccari went further. He speculated about a 'plasmative epoch', a time when Nature was malleable: 'Had man been associated with the dog during the plasmative epoch, I believe that to the expression of our face and to the sound of our voice there would have been aroused in the dog, owing to the attention with which he listens to us and observes us, analogous movements in its vocal organs, which, instead of expressing themselves by inarticulate sounds, would have enabled it to talk and to learn a language'. Our own dreams of flight are a relic of days when all animals fantasized about it, and some took off to become the birds. Humans followed the same rules: and the noble Madeleine d'Auvermont assured her son's succession by her claim that she had become pregnant when her husband was away, just by thinking about him.

Before Mendel, all heredity was (as the French aristocracy noticed) memory. The idea seemed to make perfect sense. The ostrich is exposed to danger from which it cannot escape by flight, but by kicking it can defend itself from enemies as well as any of the smaller quadrupeds. We may imagine that the early progenitor of the ostrich had habits like those of a bustard, and that as natural selection increased in successive generations the size and weight of its body, its legs were used more, and its wings less, until they became incapable of flight. Such a notion is easy to contemplate: but it is wrong.

Nature has plenty of instances of use, disuse and inheritance. Blacksmiths have thicker arms than bank clerks, but migratory birds put both of them into the shade. Some birds double in size before their journeys (and, unlike any human, increase the volume of their testicles by a hundred times in spring). Such characters are not themselves passed to the next generation. The young are heirs to an ability to grow large organs, rather than to the structures themselves. Fat parents have fat children, in the main, not because stoutness is in the genes, but because they feed their offspring with a diet like their own. Fat people have fat cats, too, but nobody blames that on DNA.

A moment's thought shows that the inheritance of acquired characters must be common. Parents and their young nearly always share environments as much as they do genes. If identical plants are grown in pots of soil from a few inches apart in a forest, the differences in habitat cause large changes in size and shape. As seeds fall close to their parent, the fate of a young plant depends both on its heritage and on where it grew. Although such effects last only a few generations, biologists are nowadays wary about giving exclusive authority to genes. The conflict between nurture and nature has lost much of its meaning. The attributes of most interest to evolution – size, shape, or behaviour – are influenced by both. Those who first asked big questions about how species originate failed, in part, because much of the answer lay in a smaller question about inheritance within each one.

The idea that a character acquired in an animal's lifetime can be handed on was once anathema, dismissed with a story about Jews and foreskins in the first lecture of every genetics course. It is now commonplace; but the notion is a detail on the edifice of genetics and not as its foundation.

All children get more than genes from their parents. The songs of birds and the teamwork of lions are each passed down the generations by education. Macaque society is based on rank. Every female monkey knows her place (although, now and again, she makes a furtive challenge in the hope of promotion). Her status comes from her mother, who helps her daughters to lord it over those lower in the pecking order. Any female from a noble line, feeble though she is, ranks above anyone from a lower stratum of society. Rank does not travel in the genes, but in the mind. The hierarchy set by custom can last for years – and has a physical as well as a mental effect, as low-grade animals have more heart disease than those higher in the social scale. All this might seem a slight exception to the great Mendelian truth, but is a small part of the great range of characters whose inheritance is in some way acquired.

Nutrients are passed to the next generation in egg or seed, diseases strike before birth, and a mother's diet affects her young. Chemicals, as well as culture, are passed between generations (as doctors who deal with babies born addicted to heroin know). To give a newborn female mouse a hefty dose of thyroid hormone depresses her ability to make the right amount of that crucial substance – and the effect is transmitted to her offspring, who grow within her damaged body.

To breed from large or small mice can make one line twice the size of another through artificial selection of the most conventional kind. However, a mouse's size also depends on its food. An animal with a good diet will be larger than its twin on starvation rations. As a result, for young animals, what matters most is how well fed their mother might be. Her environment influences their fate. As might be expected, in inbred mice, large and well-fed mothers have more young than their identical but small and hungry sisters. However, the young of large mothers tend to be small, because they grow up in a crowded womb and must struggle for what milk is available. Because they are so tiny their own few offspring grow up uncrowded, well fed – and large. However good a mouse's genes, the environment is enough to defeat them. The antagonism between nature and nurture controls their fate. Such complexity baffled those who tried to work out laws of inheritance from the experience of animal breeders.

Plants are much the same. When flax plants are given fertilizer they grow faster and have more branches with larger leaves. Simple enough; but when the offspring of such plants are grown alongside others whose parents were less fortunate, they too are more branched and leafy. The effect persists for several generations. The reason is straightforward. When a plant is well fed, certain genes multiply and help it deal with the extra food. Some of the copies are passed into the egg and reach the next generation, who benefit from their parents' happy lives.

Genes, too, have a memory of who transmitted them. It can make a difference whether they are passed on by a male or a female. Any father is anxious to persuade his partner to invest as much as possible in his children, while her own priority is to minimize the amount taken in the hope of more children later, perhaps by a different mate. A certain mouse gene, when it goes wrong, interferes with a mother's ability to succour her young. Its harmful effects are greater if it passes through the father. Perhaps he marks the gene to ensure the maximum of care. Each female carrier removes the imprint and sheds the burden of excess responsibility for her offspring. All this and more does away with the idea that the genes are sacred, safe from the insidious effects of the outside world.

Acclimatization. Heat up a fruit-fly, and it may die; but if allowed to recover and heated again it will cope better with the second shock. Plants, too, can resist heat or cold if they are warmed or cooled before the main challenge. They adapt to their new conditions with special proteins that fight stress. These are switched on when danger threatens, to be ready when it appears in its full ferocity.

Not all defences need be kept at full alert at every moment. Instead, natural selection keeps much of its armour in reserve: with its troops stood down until needed. A tan is no use at Christmas, but it does not take much sunshine when spring comes to prepare the body for the next hot day. In the same way, a sunbather who downs a stiff drink after a long sober winter switches on a set of enzymes able to deal with it; which means that, in time, more and more alcohol is needed to put him into the right mood. The first drag at a cigarette does the same – and the tobacco hornworm, one of the few insects able to eat the plant, recoils in horror at its first juvenile taste of the bitter leaves. Not until its anti-nicotine enzymes have been activated can it settle down to its natural diet. Tobacco itself, in retaliation, turns on a whole set of poisons after a leaf has been damaged, to warn off later browsers.

None of this is much use when the stress is not there; and the ability to respond when called for is as honed by natural selection as is the response itself. Anyone interested in, say, the inheritance of dark skin, and unaware of the role of sunlight, would find it hard to sort out why some people are brown and some not. Acclimatization even has an effect on body structure. Flies given a sudden burst of high temperature as pupae have many deformities as adults. The errors are a side-effect. The proteins that rush to the aid of the heat-stressed cell and ready it for another bout have a second job; to insulate the body against the effects of genetic damage. They act as a scaffold during development and contain minor flaws to ensure that a perfect fly emerges. At times of danger, the need to acclimatize takes their support away and the body reveals its inborn weaknesses.

Distinct species present analogous variations; and a variety of one species often assumes some of the characters of an allied species, or reverts to some of the characters of an early progenitor. Most children are the average of their parents, and a fat pig mated with a thin one tends to have young of intermediate size. There are exceptions to this otherwise persuasive observation. It was a very surprising fact that characters should reappear after having been lost for many, perhaps for hundreds of generations. But when a breed has been crossed only once by some other one, the offspring occasionally show a tendency to revert in character to the foreign breed. That single ugly fact, the reappearance of a character in a lineage after it had seemed lost, was the foundation of the scientific study of inheritance.

Genetics is to biology what atomic theory is to physics. Its principle is clear: that inheritance is based on particles and not on fluids. Instead of the essence of each parent mixing, with each child the blend of those who made him, information is passed on as a series of units. The bodies of successive generations transport them through time, so that a long-lost character may emerge in a distant descendant. The genes themselves may be older than the species that bear them.

Mendel was both lucky and a genius. His luck was favoured by a prepared mind: by his insight that to understand inheritance it was necessary to study simple characters in a simple organism. His genius was to separate the products of inheritance from the mechanism of heredity. He chose to work on peas, which have an odd but useful sexual system. Like many garden plants, they exist as a series of inbred families (or 'pure lines'). These have been kept separate for so long that, within a line, every plant is identical, but among lines the plants are distinct. The lines diverge in many traits – flower colour, plant height, pea shape and the colour of the pea included. What is more, peas are at once male and female, and (unlike many of that ilk) can fertilize themselves.

His experiments now seem simple, but nobody, in ten thousand years of agriculture, had tried them. Mendel took male sex cells – pollen – from one pure line and used them to fertilize the eggs of another. He looked at the various pairs of traits used to distinguish each inbred family. For example, pea colour in different lines was either yellow or green.

A plant from a yellow pure line crossed with another in which all peas were green gave only offspring with yellow peas. That was itself remarkable. It at once disproved the notion of blending inheritance, because all the seeds looked like those of one parent and were not the average of the two. To cross those progeny among themselves gave another useful result. Green and yellow peas each appeared in the next generation; and whatever had made the plants green was restored after it had lain hidden in a plant whose own seeds were yellow. The agent of inheritance – the gene – had, it seemed, an existence separate from that of its vehicle, the plant.

Mendel's ratios were always (given the accidents of sampling) the same. In this second generation, there were three yellow peas to one green. From this, Mendel deduced that the units of inheritance came in two forms, or 'alleles'. Body cells contain a pair of alleles for each character, while each pollen or egg cell receives just a single one. They combine in different ways: two yellows, a yellow and a green, or two greens. One, the dominant allele, can conceal the presence of its recessive partner. A yellow pea can be made with two yellow alleles, or a yellow allele and a recessive green. A recessive allele must be present in double copy to show itself, and all green peas have two green alleles.

The yellow peas in Mendel's first generation descended from parents of different colour. Each has a single copy of the yellow allele, matched with a single copy of the green. When those hybrid plants were intercrossed, simple arithmetic gives a ratio of a quarter with two yellow alleles, a quarter with two greens, and a half with one of each allele; to give a proportion of three yellow to one green plant. Inheritance, he thought, was explained.

Mendel's logic applied to every character, from flower colour to plant height. Even better, the pattern for each was independent of those of the others. It made no difference to the three to one rule for yellow and green peas if the two stocks also differed in flower colour. Genetics seemed simple. It was based on independent particles, each coding for a single attribute and each following its own path down the generations. The idea explained how a character long lost could reappear: it had been hidden by a matched copy of the dominant allele. A rare recessive allele almost always suffers such a fate, as only in those unusual cases in which two parents each carry a hidden copy is there a chance of its appearing in their progeny.

Mendel saw the value of his work to the theory of evolution (as is manifest in the marginal notes to his own copy of *The Origin*). Although his research was known to many biologists of his day it was, alas, ignored as – at best – a discovery of interest only to those concerned with peas. Later, Mendel went on to study inheritance in hawkweeds, but was baffled by the plant's (then unknown) ability to reproduce without sex. He could get no results as elegant as those with peas. Discouraged, he gave up and became an administrator.

Once, Darwin almost got it right. He noticed that the young from a cross between two different stocks of pigeons were uniform, but that when these mongrels were crossed for several generations then hardly two were alike. Mendelism is, we now know, at work, as the stocks differ in several genes that later come together in many ways. For once, Darwin's insight failed him.

The importance of Mendel's work was at last realized in 1900. Crosses were soon made on a variety of plants and animals. In most cases they followed his laws. Inheritance was, it appeared, quite simple. Of course, it was not, and (as often happens in biology) the more that is known the more there seems to be left to find out.

Genetics still has its basis in Mendelism, but on that firm foundation has grown a complex and often enigmatic structure.

Sometimes, his famous ratios can shift. A cross between two Manx cats gives not a three to one proportion in the next generation, but two animals without tails to each one tailed. The ratio emerges because a double dose of the Manx allele (unlike a double dose of the allele for yellow pea colour) kills the one in four kittens unlucky enough to inherit it. Other odd patterns appear as different genes band together – as they must – to build a living creature. Some abolish the effects of a whole string of others. A white cat, like an albino human or white whale, has a biochemical quirk. It masks the presence of genes that would, given the chance, make a patterned coat or a dark eye. In other cases, genes co-operate rather than compete. Persian cats have their elegant coats because of an inherited variant that reduces the intensity of the coat pigment laid down by quite a different gene. When it comes to attributes such as size, shape or behaviour, many genes of large or small effect are involved.

Quite soon after the rediscovery of Mendel's work, there emerged an important exception to his laws. Some characters are, it appeared, not independent. Instead, certain combinations tend to pass down the generations together. The link between the fellow-travellers is sometimes broken, but there remains an incomplete association between them. All genes fall into one of several groups that are passed on in consort. Only members of different groups follow Mendel's rule.

The discovery was the key to the physical apparatus of in-heritance. In fruit-flies four such 'linkage groups' emerged from crosses. They match the four pairs of chromosomes, dense bodies in the cell nucleus. These were the tangible signs of Mendel's magical particles. If two genes were on the same chromosome they could be passed on together. If they were not, they were inherited as autonomous units. The association was not absolute because the chromosomes themselves were broken up and rejoined in new combinations each time sperm or egg was made. It did not take long to realize that the more a pair of loci deviated from independence, the closer together they must be. The first genetic map was made by comparing the tendency of such pairs to stick together as the generations succeeded.

Chromosomes are complicated things, made up of hundreds of molecules. One element, DNA, seemed an unlikely candidate as the vehicle of inheritance. It had a mere four different units (or 'bases') and had been dismissed as 'the stupid molecule' as a result. A bold experiment in which DNA was transferred between bacteria with colonies of different shape showed it to be the crucial agent. The story of the famous double helix and of the code for the structure of proteins is part of the cultural inheritance of the twentieth century.

Now, all kinds of marvellous technology are used to read the message of the DNA. The structure of proteins can be deduced from its sequence, and this in turn gives a hint as to what each one does. The order of the bases has now been established for many creatures. Already, many viruses and bacteria, together with yeast, have had their genetic message laid bare. A small worm – similar in its fundamentals to a man or an oak – has had all ninety-seven million of its units read off. The human map itself, three billion bases long, will be complete soon after the millennium (and less than fifty years after the discovery of the double helix).

For most things more complex than bacteria, maps of the DNA do not much resemble a chart based on crosses. Instead, the genes are full of waste and redundancy. Some are interrupted by strings of material that appear to code for nothing. All this is, perhaps, less remarkable to those who do not come (as did many of the pioneers) from physics or chemistry, but from biology. Anyone used to the muddle and waste of evolution, its products cobbled together over long ages of expedience, is not surprised to see in the genes themselves the same history of uneasy compromise. The genome is as complicated, makeshift and imperfect as the creatures it builds.

Genetics is the science of difference. Variety is the raw material of evolution, used up as natural selection takes its course. Once it has been consumed, the Darwinian machine comes to a stop. Diversity is renewed by chemical errors – mutations – made as DNA is copied. Geneticists were once so impressed by mutation as to suggest that new forms arise not through the accumulation of small changes but in great leaps. Evolution was due to the instability of genes and genetics had, perhaps, destroyed Darwin's idea.

It had not: mutation is the fuel rather than the engine of biological advance. The process involves mechanisms undreamed of in the science's first days.

Some mutations are simple. They are no more than a change in the genetic alphabet that alters the properties of a gene, or stops it altogether. A single shift can persuade a growing protein to stop dead and, by so doing, kill or much modify those who inherit it. Other mutations arise from a sudden duplication or deletion of genetic material or of whole chromosomes.

Mutation could once be studied only in bacteria or in fruit-flies. Now, those in humans are as accessible as those in any other species. The bones of the Russian Royal Family, killed in 1917, were identified through the fit of their mitochondrial genes to those of today's aristocrats (the Duke of Edinburgh included). Embarrassingly enough, the Tsar's own DNA did not quite match that of his presumed relatives, because he carried two distinct types. One must have arisen in his own lifetime. The United States Army records the genes of its soldiers, and sometimes has cause to compare them with those of their relatives. Again change is rapid, with a mitochondrial mutation in every forty parent–child comparisons.

Much of the genetic damage lies at the feet of age and sex. Men are defined by their ownership of a single gene. It has the modest task of persuading the early embryo to divide faster than it might. The early stimulus sets it off down the path of masculinity. The gene sits on a certain chromosome, the Y, which has few other jobs to do. It is so small that it does not mask genes carried on its partner, the X. Females have two X chromosomes, which behave much like all others. In males, though, every gene on the single X – whether dominant or recessive in females – shows its effects (which is why there are more colour-blind men than women: the gene, a recessive, is on the crucial chromosome).

As a result, any new mutations in a man are likely to bear more heavily not upon his sons (who receive only his Y) but upon his daughters, who get one of their two X chromosomes from him. There lay the first clue about sex and age. Among European royalty, the daughters of old fathers die earlier than do those with younger sires. The difference is as much as two years for a fifty-year-old compared to a thirty-year-old father. For sons, parental age made no difference.

Males, it seems, are not just conduits for genes between females. Instead, they are responsible for most mutations. Males make sperm all the time, while females make their eggs early on, releasing them when needed. As a result, more cell divisions take place between the sperm in one generation and that in the next than between egg and egg. In humans, a mother uses an egg separated by a couple of dozen cell generations from the egg that made her. A father, in contrast, makes sperm separated by hundreds of sperm generations from the cell responsible for his own existence. The chance of error increases each time a cell divides.

Older fathers (with many divisions behind each elderly sperm) have a rate of error twenty times greater than do females. Males are the source of much of the raw material of evolution. To compare the males and females of related species of mammal reveals that the Y chromosome – which spends its time in males alone – changes much faster than does the X. In birds, in contrast (in which females rather than males have an equivalent of the Y), the Y chromosome evolves at normal speed. Female birds show that the accumulation of genetic change arises not from how sex is determined, but from masculinity itself.

Mutations are not just simple faults in a rigid set of commands but a flexible and inconstant system that works to its own rules. Most creatures have a complex system of enzymes that repair DNA, which is such a lengthy and unstable chemical that it would decay without constant help. They evolved, in effect, to reduce mutation rate: indeed, if it could, natural selection might act to eliminate it, halting evolution altogether.

In spite of the apparent chaos in the DNA, the rate of change for individual genes is quite small, at about one in a million per generation. The figure seems tiny, but in total is quite large. London has about two million cats and each cat perhaps seventy thousand genes. There may hence be a hundred thousand genetic changes each year in that city alone. Worldwide, any mutation is almost a certainty. If it is useful it will at once be picked up by natural selection and will spread.

Does genetic accident limit the rate of evolution? In some senses, it must. Pigs, after all, have not mutated to make wings. To increase its rate can sometimes improve the ability to respond to a challenge. Plant-breeders know this, and irradiate their seeds in the hope of turning up new and useful forms. In the same way, fruit-flies respond better to selection when genes that push up the mutation rate are introduced. However, evolution sometimes has to wait before natural selection can get to work. If a mutation does not happen, it becomes impotent. Tsetse-flies are susceptible to most insecticides because they have not come up with the genes to deal with them. Some agents of disease, too, have failed to evolve their way around medical advance, although some day they may. Syphilis was for many years easy to treat with penicillin, but a resistant strain has now been found in Africa. No doubt it will spread.

Resistance to insecticides, now ubiquitous, did not begin for a couple of years after the first use of DDT. The brief respite before the pests could fight back was due to the time it took to come up with the mutations needed for selection to work its magic. As soon as one appeared it spread, showing that evolution was indeed limited by its absence. One fly evolved a sudden resistance to a certain chemical. Each of its defiant billions, from Pakistan to California, carries the same genetic change, with the same length of DNA around it. Each copy must descend from the same error within a single animal, and evolution seized the mutation as soon as it arrived.

Mutation alone is not enough. On a farm sprayed with a dozen pesticides, to have new genes able to protect against one or two – or eleven – is not much help. Twelve changes in a row in the same family line are too much to ask even of insects, but the situation is saved by the most fundamental of all laws of variation: sex. Most of genetics is no more than the scientific study of that eccentric pastime. Sex makes offspring unlike either parent because they contain new combinations of genes. It allows a favourable alteration in one family to get together with another in a separate line. Without it, there would be a long wait in each lineage for the second one to turn up.

In an all-female mosquito, the only safe individuals would be the direct descendants of the first to strike lucky, rather than the multitude of otherwise unrelated animals into which it can spread through sex. Free exchange between families has enabled some insects to resist twenty poisons at once. Each new mutation appeared in a different line, and in distant parts of the world, but soon got together. They outwit the best the chemical industry can do.

Correlation of Growth. Inheritance is full of links among disconnected characters. Darwin noted them, but was baffled by what seemed a series of incoherent facts. What can be more singular than the relation between blue eyes and deafness in cats, and the tortoiseshell colour with the female sex; or, again, the relation between the hair and teeth in the naked Turkish dog? It can hardly be accidental that if we pick out the two orders of mammalia which are most abnormal in their dermal coverings – whales, and armadillos, scaly anteaters and so on – that these are likewise the most abnormal in their teeth.

As *The Origin* notes, and as Goethe expressed it, 'in order to spend on one side, nature is forced to economise on the other side'. Nobody can do everything. As a result, to breed from animals or crops desirable for one reason often leads to failure in another part. An attempt to breed mice as large as rats began in the 1930s and lasted for thirty years. It was abandoned with an animal less than twice the size of a typical mouse, because the much-selected beast was sterile. Much later, genes for growth hormone were engineered into mice, and the rat-sized mouse became a reality. The failure of the mouse-breeders involved an unexpected correlation of growth among the genes responsible for size and for sex and is a reminder of how little such traits are understood.

These patterns show, in their several ways, how one change can alter many characters at once. In the 1940s, a goat without forelegs was born in Russia. It learned to stand upright, and developed large hind legs, a curved spine, an oval chest cavity and a thick neck. A single accident during development had led to all these changes. Whether genes were involved in the fate of the unfortunate goat nobody knows, but plenty of small changes in DNA can have equally large and unexpected effects.

Often, the error takes place early on. Blue-eyed cats are deaf because those at first sight unrelated characters share an embryonic pathway. The gene involved makes the dark pigment – melanin – that is responsible for skin colour. Melanin is found in many other parts of the body, brain, eyes and ears included. Black cats are full of it, while pale animals have less. Those with no pigment at all are white, with blue eyes. Melanin plays an unexpected part in the brain, for it guides the cells responsible for certain nerve pathways to their correct places. As a result, a shortage of melanin gives a cat a white coat and blue eyes – and makes it deaf.

The bald and edentulous Turkish dogs tell another part of the genetic story. A joint loss of hair and teeth is found in many animals (such as the Chinese crested dog once popular, for symbolic reasons, with striptease artists). Darwin himself, fifteen years after *The Origin*, wrote of 'a Hindoo family in Scinde, in which ten men, in the course of four generations, were furnished, in both jaws taken together, with only four small and weak incisor teeth and with eight posterior molars. The men thus affected have very little hair on the body, and become bald early in life. They also suffer much during hot weather from excessive dryness of the skin ... Though the daughters in the above family are never affected, they transmit the tendency to their sons: and no case has occurred of a son transmitting it to his sons.'

The problem comes because both teeth and hair are derived from the same tissue: damage it, and each one suffers. Sweat glands are involved, too, which is why those with the condition are uncomfortable in hot weather. The gene makes a protein that exchanges signals between skin cells as they develop. Tabby mice, as it happens, have the same error, and it has been detected in a hairless dog. Darwin's insight was to note the pattern of inheritance of this strange mixture of attributes. They are passed through females but seen most often in males. The pattern arises, we now know, because the gene is carried on the X chromosome. The hairless sons of Scinde are another hint about how close he came to the mechanism of heredity.

His tie between tortoiseshell cats and the female sex is also explained. The gene affects hair colour, and codes either for orange or for black. It is carried on the X chromosome. A male, with his single X, has either the orange or the black allele and is an un-remarkable animal. Many females carry both the black and the orange version of the gene. They are not pure orange or pure black, as might be expected. Instead, in some of their skin cells the black allele is switched on, and in others its alternative. This gives the tortoiseshell its pattern of patches of different coloured hairs. The reason lies, again, in development, as female cells use only one of their two X chromosomes in any tissue, switched on at random.

In this age of faith in the power of genes, the public is as gullible about the wonders of DNA as it once was towards the one-eyed pigs of Connecticut. Most ancient ideas are simply wrong. Nonetheless, many apparently eccentric notions – use and disuse, acclimatiz-ation, the correlation of different parts of the same animal, the reappearance of characters long lost – turn out, in the light of modern biology, to have a basis in fact. Filled with complexities and exceptions as it is, genetics remains as the rock upon which the whole edifice of evolution rests.

Summary. Our ignorance of the laws of variation is profound. Not in one case out of a hundred can we pretend to assign any reason why this or that part differs, more or less, from the same part in the parents. But whenever we have the means of instituting a comparison, the same laws appear to have acted in producing the lesser differences between varieties of the same species, and the greater differences between species of the same genus. The external conditions of life, as climate and food, &c., seem to have induced some slight modifi-cations. Habit in producing constitutional differences, and use in strengthening, and disuse in weakening and diminishing organs, seem to have been more potent in their effects. Homologous parts tend to vary in the same way, and homologous parts tend to cohere. Modifications in hard parts and in external parts sometimes affect softer and internal parts. When one part is largely developed, perhaps it tends to draw nourishment from the adjoining parts; and every part of the structure which can be saved without detriment to the indiv-idual, will be saved. Changes of structure at an early age will generally affect parts subsequently developed; and there are very many other correlations of growth, the nature of which we are utterly unable to understand. Multiple parts are variable in number and in structure, perhaps arising from such parts not having been closely specialised to any particular function, so that their modifications have not been closely checked by natural selection. It is probably from this same cause that organic beings low in the scale of nature are more variable than those which have their whole organisation more specialized, and are higher in the scale. Rudimentary organs, from being useless, will be disregarded by natural selection, and hence probably are variable. Specific characters – that is, the characters which have come to differ since the several species of the same genus branched off from a common parent – are more variable than generic characters, or those which have long been inherited, and have not differed within this same period. In these remarks we have referred to special parts or organs being still variable, because they have recently varied and thus come to differ; but we have also seen in the second Chapter that the same principle applies to the whole in-dividual; for in a district where many species of any genus are found – that is, where there has been much former variation and differenti-ation, or where the manufactory of new specific forms has been actively at work – there, on an average, we now find most varieties or incipient species. Secondary sexual characters are highly variable, and such characters differ much in the species of the same group. Variability in the same parts of the organisation has generally been taken advantage of in giving secondary sexual differences to the sexes of the same species, and specific differences to the several species of the same genus. Any part or organ developed to an extraordinary size or in an extraordinary manner, in comparison with the same part or organ in the allied species, must have gone through an extraordinary amount of modification since the genus arose; and thus we can under-stand why it should often still be variable in a much higher degree than other parts; for variation is a long-continued and slow process, and natural selection will in such cases not as yet have had time to overcome the tendency to further variability and to reversion to a less modified state. But when a species with any extraordinarily-developed organ has become the parent of many modified descendants – which on my view must be a very slow process, requiring a long lapse of time – in this case, natural selection may readily have succeeded in giving a fixed character to the organ, in however extraordinary a manner it may be developed. Species inheriting nearly the same constitution from a common parent and exposed to similar influences will naturally tend to present analogous variations, and these same species may occasionally revert to some of the characters of their ancient progenitors. Although new and important modifications may not arise from reversion and analogous variation, such modifications will add to the beautiful and harmonious diversity of nature.

Whatever the cause may be of each slight difference in the offspring from their parents – and a cause for each must exist – it is the steady accumulation, through natural selection, of such differences, when beneficial to the individual, that gives rise to all the more important modifications of structure, by which the innumerable beings on the face of this earth are enabled to struggle with each other, and the best adapted to survive.

CHAPTER VI

DIFFICULTIES ON THEORY

Difficulties on the theory of descent with modification — Absence or rarity of transitional varieties — The fate of hybrids — Transitions in habits of life and the origin of flight — A change of diet and a new existence — Organs of extreme perfection — Organs of little apparent importance, from caves to Everest — The neutral theory of molecular evolution — The startling structure of the genome — Partial, profligate and promiscuous DNA — The confederacy of life

LONG BEFORE HAVING arrived at this part of my work, a crowd of difficulties will have occurred to the reader. Most have been noticed (and used) by anti-evolutionists since the subject began. The lack of intermediates between species, groups with forms distinct from their relatives, animals and plants of strange and unique habits; all seem hard to explain on the theory of slow and gradual change. Even worse, there have evolved some structures that appear to play no part in the body's economy and – another problem – complex organs that seem flawless. Can instincts be acquired and modified through natural selection? What shall we say to so marvellous an instinct as that which leads the bee to make cells, which has practically antici-pated the discoveries of profound mathematicians? And how can we account for species, when crossed, being sterile and producing sterile offspring, whereas, when varieties are crossed, their fertility is un-impaired? Behaviour and sterility deserve chapters of their own: but all these apparent difficulties, much used as fuel for the creationist diatribe, are, in truth, each evidence for evolution.

On the absence or rarity of transitional varieties. Existence is divided into its many kinds, and seems to have been since it began. Why is Nature not all in confusion instead of the species being, as we see them, well defined? The primeval soup may have been a simple liquid but has today become closer to minestrone. Why is life so lumpy?

Nobody, say the anti-evolutionists, has ever seen a species arise. That, as it happens, is not true, but it is hard to deny that few creatures appear to be in transition between one form and the next. Why should this be? It has to do with natural selection and with the replacement of the old by the new.

Evolution is, for most of the time, a race to stay in the same place. The worst enemies of any animal are among its relatives and descendants, who need the same things and may have evolved better ways to get them. Unless a parent can keep up with its children, its fate is sealed. Most cannot, and disappear. As a result, at any time, just the tips of the twigs of any evolutionary tree are on view. The branches have been replaced by something better. Because the past has been wiped out by the present, the ancients usually leave no clue to what their fate might have been. Where are the animals halfway between whales and hippopotami, or house-flies and fruit-flies? As innumerable transitional forms must have existed, why do we not find them embedded in countless numbers in the crust of the earth?

The problem of the missing links arises in part because it is hard to see who they were. The dinosaur whose descendants gave rise to birds had arms a mere fraction longer than its doomed sibs and the same may be true of some organ of some bird today that might be a step towards a new and dramatic form of future life. Even so, neither the fossil record nor the modern world is full of creatures caught in the act of a change in lifestyle. Why should this be, as so many must have made a shift from one existence to another?

Biologists are only too used to the criticism that part of an eye or an ear is of no use and that evolution is as a result disproved. The argument denies utility to a plate camera, a carbon microphone, or the Maniac computer of the 1950s that used a room full of valves to provide little more power than a digital watch. That today's versions are better than what went before says little about what they were worth in the days of Daguerre, Edison, or Alan Turing. For life, the problem of the intermediates is more subtle: the eye, the voice or the brain were not designed from scratch, but had to get from where they were to where they are, step by step, while still doing their original job.

It may be hard to get from one form to another without a middleman who is worse at the old task even though his descendants are better at a new one. Evolution often faces the mountaineer's dilemma. Few peaks are a straight slog upwards to the summit. Instead a climber has to lose his hard-won gains by crossing a valley before he can reach the next high point. Plenty of tasty butterflies gain protection from predators because they mimic, almost exactly, the bright warning patterns of unrelated species filled with poison. By so doing they flourish, but how could their camouflaged ancestor have taken the first step towards his false advertisement? Any gene that made a savoury insect easier to see must, it seems, be disadvantageous; even if, in time, it leads to a new form of protection. Quite how the insects traversed the valley of death – in a sudden leap, with a single gene pushing them most of the way, or by small changes getting together almost by accident – is not clear. What is certain is that the intermediates were worse off than either their camouflaged ancestors or their dishonest descendants and must certainly have disappeared.

From time to time, a natural experiment hints at how such forms meet their demise. When species that diverged not long ago form hybrids, animals appear that are halfway between two well-adapted forms. They hint at the dangers of obsolescence. Such creatures are doomed to death or a narrow existence because all else is denied to them. Suppose that three varieties of sheep are kept: one adapted to an extensive mountainous region; a second to a narrow, hilly tract; and a third to wide plains at the base. The great holders on the mountains or on the plains will improve their breeds more quickly; and consequently the improved mountain or plain breed will soon take the place of the less improved hill breed. The intermediates, of sheep or anything else, are squeezed out.

Take the crow, denizen of mountain and plain (and devourer of lambs' eyes). Most crows are black; they are the carrion crow – 'that loathsome beast, which cries against the rain' – of England, Wales and much of Europe. The crows of Scotland, Scandinavia and Eastern Europe (hooded crows, as they are called) differ from their loathsome cousins as they have pale grey sides. Apart from its flanks, the hooded looks almost the same as its sombre kin (although it prefers to live in rather different places).

Where the two birds meet, they hybridize, to give a zone full of birds halfway between the carrion and the hooded kind. The hybrids are restricted to a narrow band between their improved descendants. In Scotland, the hoodeds take the high ground, the carrions the low, while the intermediates are confined, like the sheep of the foothills, to a narrow strip in between. They have no chance of supplanting either common form.

Victorian Cambridgeshire was full of hooded crows (as noted by Charles Kingsley, the first person to turn Darwinism to theological ends). In *The Water Babies*, the crows kill one of their number because she will not steal eggs: 'They are true republicans, these hoodies ... so that for any freedom of speech, thought, or action, which is allowed among them, they might as well be American citizens of the new school'. By the 1950s, those radical birds had retreated to the north of a line from Glasgow to Aberdeen. Forty years on, the hooded has been pushed further back, to Inverness. As the carrion crow expands its range, the hooded retreats, because the two kinds are so similar that they cannot live together. The hybrid zone, too, is on the move, but it stays as narrow as it ever was and is, as before, confined to the foothills. The intermediates relive the fate of an ancient and supplanted bird. They have no future because carrion and hooded crows are already so close in their needs that no other animal can squeeze in.

What about the greater steps on the evolutionary road? Why are there so few forms in transition not just between low and high ground but from land to air? Again, selection has done its inexorable work in getting rid of the intermediates. When we see any structure highly perfected for any particular habit, as the wings of a bird for flight, we should bear in mind that animals displaying early trans-itional grades of the structure will seldom continue to exist to the present day, for they will have been supplanted by the very process of natural selection. Each new form will tend to take the place of, and finally to exterminate, its own less improved parent. If the improvement is great it will not take long to complete the move from old to new. The chances of survival for any ancestor are small indeed.

Most animals able to fly do the job quite well. Those who hesitated on the boundary of the new medium have gone, but it is possible to guess at what they were like. For bats or birds, the task is simple. Their wings are modified arms. Plenty of animals behave as the fore-fathers of the eagle or the vampire bat must have done. They glide from high to low, to save energy and to avoid the dangers of the ground. Lizards, frogs, rodents and even a remarkable flying lemur (at first misclassified as a bat rather than a primate) all go in for it. A flap of skin between fore- and hind-legs, enlarged feet or a flattened body all help. Part of an aerofoil, even in the form of enormous toes, is a great deal of use. To change a forelimb into an organ of flight is no great task. Each bone in a bird's or a bat's wing has its match in the rabbit's foreleg or the whale's flipper.

However, nothing today looks like a creature halfway to a bird or bat. Bats are, the molecules show, related to rabbits, an eminently ter-restrial group, but there are no living hints at what the ancient rabbit-bat might have looked like. In the family of squirrels, though, we have the finest gradation from animals with their tails only slightly flattened, and from others with the skin on their flanks rather full, to the so-called flying squirrels that have their limbs and even the base of the tail united by a broad expanse of skin, which serves as a parachute and allows them to glide through the air to an astonishing distance from tree to tree. The bat wing-membrane still reveals traces of an apparatus originally constructed for gliding. If such a sequence from land-bound life to expert flight can be found among different animals today, why should there not have been such a gradual change among the ancestors of today's bats or birds? Any improvement in the ability to glide or flap would soon put paid to a less effective foregoer.

Some vertebrates tried other routes into the air. In 1910 a German miner found a beautiful fossil of a chicken-sized aerial reptile. One expert thought this improbable and removed the wings as mere remnants of a dead fish. This was indeed the earliest vertebrate to fly. It used a flawed but ingenious stratagem, twenty million years before the next one left the ground. Bony rods stiffened a pair of curved wings made from skin flaps. These opened like a Japanese fan and allowed the animal to glide a hundred times its own length, with an approach quite unlike that of bats or birds. The first modified fore-limbs, attached by joints and muscles to the skeleton, had the potential to evolve into elegant pinions and, on the way, to replace all earlier forms. The Japanese fan went nowhere – because, unlike a modified arm, there was nowhere much to go.

Most flying animals are not birds, but insects. Like angels, they grow wings without losing their arms. The structure of an insect wing gives no clue about what its ancestor may have been, and insects have no equivalent of the squirrels to suggest what their predecessors lived through.

The first airborne insects to be preserved as fossils, more than three hundred million years ago, were already blessed with magnificent wings that carry most of the struts and aerofoils used by modern dragonflies to fly with such skill. The distant and unknown parents of those aeronauts may have had simple skin folds that helped their bearers to glide, or clumps of hair that kept them afloat. Perhaps, instead, the first step was via a central-heating radiator. To aim a plate at the sun soaks up energy, to turn away loses it. In time the solar panel grew to become useful in another sphere. The earliest wings could even have emerged from flaps used by aquatic insects to absorb oxygen and hijacked by evolution as aids to flight; but it is hard to see how an animal that lives in water took to the air.

All those routes to a new world have analogues – the curse of evolution – today. Crickets leap upwards and glide to ground. For tiny insects, the air is viscous enough to allow them to drift through the summer sky supported by tufts of bristles. Butterflies, in contrast, open their wings to the sun to warm up and take off. Any organ able to do such things may have allowed a feeble flight. Which one actually did the job once rested on speculation.

DNA proves what the first fliers were. Shrimps are distant relatives of insects. A search through their genes reveals a set almost identical to those that initiate the series of steps that makes the wing of a fruit-fly. The shrimp versions are active not in the body wall (as might be true if the glider, balloonist or central-heating theories were right), but in a set of specialized limbs used as gills. The wing, they show, must have evolved from a jointed leg, first used not for waving, but to stop its owner from drowning. A few animals still follow the ancient ways. Those primitive beasts, the stoneflies, develop in streams and emerge as adults to the surface. They use raised appendages based on gills to sail or to skate until they reach land. That, no doubt, is what the first wings looked like, and there, among the stoneflies, they remain, as implausible steps on the road to flight.

On the origin and transitions of organic beings with peculiar habits and structure. Natural selection can make new organs and new forms to replace what went before. The theory of evolution has to do much more than that. It must explain how some animals adopt habits quite different from those of their relatives. Such dramatic moves could not, it seems, be achieved by gradual change, as there must be a shift in the many characters that separate the old version from the new.

We sometimes see individuals of a species following habits widely different from those of others and might expect that such individuals would occasionally have given rise to new species, having anomalous habits, and with their structure either slightly or considerably modified from that of their proper type. Plenty of animals take up unlikely opportunities as they arise. To imagine from the behaviour of an eccentric bear that a whole race could, in time, be rendered more and more aquatic, till a beast was produced as monstrous as a whale is a little much to ask even of natural selection. Even so, a small shift in behaviour can have large effects on the evolution of those who make it. Of cases of changed habitats it will suffice merely to allude to the many British insects which now feed on exotic plants. Each has had to adapt to new food and to a new place to mate and to spend its days.

The Europeans who settled the United States saw the chances on offer. Apples were planted wherever they would grow and soon developed into huge orchards. In the 1860s, in the valley of the

Hudson, the crop was attacked by a new pest, the apple maggot fly. It was at first assumed to be a European immigrant but was, in fact, a native that had changed its habits. A local insect was able to evolve, within a century, a new calling. It now does millions of dollars of damage each year.

The apple pest is an altered form of the hawthorn fly. The first of those animals to visit the new host found a mountain of food. A simple shift led to a long chain of consequences and to a step towards a new species with an identity of its own. It was the latest of many in its family. One fly made the crucial move; and billions of its descendants profited.

The flies mate where they feed, which itself cuts off the apple visitors from their ancestors. They were faced with new challenges to which they had to adapt. Apples differ in many ways from hawthorn. They appear earlier in the year, the fruit takes a whole hot summer to ripen and has its own defensive chemicals. All this faces an insect used to other food with new difficulties. To make up for that, a maggot deep inside an apple is safe from the parasitic wasps that attack those on the small fruits of the original host and, as an added bonus, it does not need to compete with the many local insects who enjoy a meal of hawthorn. As a result, there has been much evolution among the descendants of the first hawthorn fly to take to a new diet, and the apple pest has now altered so much that it almost never meets its forefathers. Given the choice, it will fly to an apple rather than a hawthorn and – although it has no great genetic differences from its ancestor – the apple flies emerge two weeks before their parental form. This reduces the chances of sex (although, in the lab, the two pests still fall on each other with enthusiasm).

Other, more distant, relatives hint at a similar history. One, the dogwood fly, mates with the apple maggot in the laboratory, but not in nature. The blueberry maggot fly never breeds with any other and artificial hybrids are unfit. Yet another lives on wild laurels, is unable to cross with any of its kin, and contains genes found in none of them. For each member of the group, a small change of habit was the first step to an identity of its own. Now, the hawthorn fly is testing its ability to live on cultivated cherries. If it succeeds (and the United States Department of Agriculture is doing its best to stop it), that native American will assume yet another personality.

Many insects have evolved alongside the plants upon which they feed. Indeed, a plant diet drove them to diversify. As many as three hundred and thirty thousand kinds of beetles are known, far more than any other group. They feed, in the main, on flowering plants, themselves evolved from ancestors without flowers. A pedigree based on genes shows that the beetles followed their food. Those near its root still eat conifers or cycads, ancient and conservative plants, much older than anything with flowers. For the first hundred and fifty million years of beetle history, not much happened. Not until the plants themselves blossomed could the beetles diversify their habits and set off, with the flowers, down a tangle of evolutionary roads.

Sometimes, the footprint of the crucial individual with a new lifestyle remains in its descendants. Like the wheel or the thermionic valve, it shows how a single idea can lead to a whole range of new models.

The world is full of poisons. Some, like deadly nightshade, may kill us, but others (such as the nicotine in tobacco) we have learned to love. The body, too, generates wastes that must be made safe. A specialized group of genes does the job. They began with the first poison of all. Oxygen appeared in the atmosphere about two billion years ago. It was lethal to the life of those days (as it still is to the bacteria that have not evolved to deal with it). Anything able to remove the deadly gas was picked up by natural selection. There soon evolved a protein that attached the oxygen molecule to other chemicals and reduced its malign effects.

Then, long ago, there was a neat inversion of evolutionary logic. Oxygen became friend rather than foe and most organisms began to use it to fuel their lives. The defensive protein was utilized not to make oxygen safe, but to add it to noxious substances and to render them harmless. One change of habit generated a family of genes that branched into thousands of different forms. Every mammal has at least two hundred versions of the protective molecule, insects many more. Without their defences they would at once fall victim to the chemical world in which they live.

Some of the new proteins result from an arms race between animals and plants. Every new plant poison was met by an animal molecule adapted to deal with it and, as one party changed its habits, its opponent followed. Some plants are ahead in the race (which is why we do not eat rhubarb leaves), but some animals can attack a plant denied to others because they alone can handle what it makes. The tobacco budworm can break down nicotine, and flourishes on tobacco plants, a single leaf of which will kill a man.

Swallowtail butterflies are spectacular, their bright colours improving the forest's gloom and the glare of the savannah. Each of the two hundred different kinds has a host-plant of its own, kept more or less to itself because other insects avoid them. All their hosts make a toxin called coumarin. This keeps most grazers off, but is defeated by the evolutionary shift that came to the swallowtails. The evidence of an ancient transition is in the DNA. Although parts of the gene's structure have changed, the section able to attack coumarin has not altered over millions of years. A simple change in an ancient insect opened the gates of an evolutionary citadel. It allowed hundreds of new kinds of butterfly to evolve and to exploit a new and diverse set of circumstances.

Whole groups of animals may undergo unexpected shifts in routine and evolve in directions quite different from their ancestors. Can a more striking instance of adaptation be given than that of a woodpecker for climbing trees and for seizing insects in the chinks of the bark? Yet in North America there are woodpeckers which feed largely on fruit, and others with elongated wings which chase insects on the wing; and on the plains of La Plata, where not a tree grows, there is a woodpecker, which in every essential part of its organization, even in its colouring, in the harsh tone of its voice, and undulatory flight, tells its close blood-relationship to our common species; yet it is a woodpecker which never climbs a tree.

Swallowtails, woodpeckers and many other creatures all show how nature can change in an arbitrary way. For each, the world now has a set of products that it did not know it needed. Evolution, given the chance of a better hole, always goes to it. If nobody else can get in, so much the better. A peculiar habit, or a novel structure, opens a world of opportunity that can be exploited in a myriad ways.

Organs of extreme perfection and complication. The perfection of life has often been used to prove the existence of a Creator. Griffith

Hughes, Rector of St Lucy's, Barbados, put the case in his *Natural History of Barbados* of 1750. In the island's animals he 'traced the Workmanship of a Divine Architecture ... without Defect, without Superfluity, exactly fitted and enabled to answer the various Purposes of their Creator, to minister to the Delight and Service of Man, and to contribute to the Beauty and Harmony of the universal System ... For instance, the Potato-Louse, which is so small that it is scarce discernible! ... Yet every Part that is necessary to animal Life is as truly found in one of them as in Behometh or Leviathan ... What less than infinite Wisdom and Power, could dispose a little Portion of Matter, almost too small to be viewed by the naked Eye, into that infinite Variety?'

The notion that existence is so flawless as to need a designer was much taken up by the Victorians and is still trotted out today. The logic is empty. Perfection is relative and, for potato-lice or anything else, depends on subjective judgement. The eye, the ear – even the toenails – all rebut the 'argument from design', as the claim of Griffith Hughes and his many successors is known.

Although the belief that an organ so perfect as the eye could have been formed by natural selection is more than enough to stagger anyone; yet in the case of any organ, if we know of a long series of gradations in complexity, each good for its possessor, then, under changing conditions of life, there is no logical impossibility in the acquirement of any conceivable degree of perfection through natural selection.

The American politician William Jennings Bryan – the victor in the trial of John Scopes, convicted in 1925 for teaching evolution – liked to mock the subject: 'A piece of pigment or, some say, a freckle appeared upon the skin of an animal that had no eyes. That piece of pigment or freckle converged the rays of the sun upon that spot and when the little animal felt the heat on that spot it turned the spot to the sun to get more heat. The increased heat irritated the skin – so the evolutionists guess, and a nerve appeared there and out of the nerve came the eye ... Can you beat it? And it happened not once, but twice!'

In fact, the eye happened not twice but fifty times, and the problem of how to extract information from light has been solved in a dozen ways. The eye is as intricate as it needs to be, and no more. Its apparent perfection does not destroy but upholds the theory of evolution. There are many sequences of eyes in different creatures. Each hints at the stages that even an organ as complex as our own must have passed through before it gained the moderate abilities it can claim today.

Many animals have a single lens used to focus light on to a plate able to convert it into nerve impulses. Humans, worms, jellyfish, snails and spiders all do the job in much the same way. As paparazzi know, the bigger the lens the better it sees; and mice have eyes larger in relation to their body size than are our own. Even a tiny device does quite well. The simple eye of a spider, like the complex organ of the mouse, tells enemy from friend at thirty times its bearer's body length.

All eyes reflect what history has demanded and are restricted by what it provides. The human eye is complex enough; with a hundred million rods, used in dim light, and three million cones, responsible for colour vision. Each rod contains thousands of proteins that transform light into signals via a molecule that crosses the membrane in a sevenfold zigzag. Three colour-sensitive pigments pick up the red, green and blue elements of a scene. Our eye is imperfect; but, fortunately, we are used to what it cannot do. The world has plenty of white flowers, but only to us. Bees can see in the ultraviolet, and to them the plants are full of detail. All cameras correct for the coloured fringes that surround an image passed through a lens; our own sneaks around the problem with a shortage of blue-light receptors in the centre of its field (a fact noted by the Impressionists, who blurred their blue flowers).

In spite of such tinkering, natural selection will not produce absolute perfection, nor do we always meet, as far as we can judge, with this high standard under Nature. Everything has been modified, but not perfected, for its present purpose. Any structure that evolves has to cope with its past. Because the chief part of the organization of every being is simply due to inheritance it is, for most of the time, impossible to get to one place from another. Every animal is limited in what it can do by what it starts with.

The eye is a servant of that inflexible rule. That of mammals has a weakness that has dogged it since its earliest days. It began as a patch of cells and was later formed into a cup or a camera. As a result, the light must pass through the wires that take information to the brain before it reaches the retina. No camera that put the sensitive part of the film on the wrong side and then had to compromise to cope would sell. The feeblest designer could improve it (which is why we have spectacles and microscopes).

Any insect would be astonished by our ability to see. Their eyes are built with not one but hundreds of lenses, each of which concentrates light upon a sensor. That set of tiny and cheap cameras is a forceful statement of what evolution cannot do. Because of where they began, insect eyes are limited in what they are. They specialize in the big picture and are no good at details. Insects are Nature's victims. As any movement could mean death, they have a bird's-eye view of the world, every object in their sights, any activity at once detected. However, the most suspicious insect is ten times less able to identify the fine points of an adversary than is a spider with a crude but more effective organ of vision. Their world-view was described by the first scientist to take photographs through insect eyes as 'a picture about as good as if executed in rather coarse wool-work and viewed at the distance of a foot'.

Whatever the limitations of its raw material, natural selection has improved their eyesight as far as it can. Nocturnal insects have large lenses that increase sensitivity by a hundred times, while dragonflies have more cameras, with a patch of tight-packed small units able to pick up prey against the sky. Bees go on long journeys, and have an upright strip of sensors adapted to the vertical world of trees and branches. The world of water-skaters, by contrast, is flat and their eyes have a horizontal band, suited for the watery plain upon which they swim. Sex comes in, too. Male flies have acute vision to stalk a potential mate to whom her swain is a distant blur. Even the molecules change to fit. Light causes ions to rush across a membrane and sparks off a train of impulses passed to the brain. Fast fliers, faced with a stream of new information, have an ion channel able to respond at once, while that of their relatives with a more tranquil existence is slower.

As insects battle to improve a feeble design, evolution does its best; but its best is not very impressive. The eye of the dragonfly or the water-skater has triumphed, but only because all its competitors are worse. For sight, excellence is in the eye of the beholder.

In the context of evolution, perfection is not necessary. If the eye were only a hymn to the supreme powers of a deity called natural selection it would be no more persuasive as evidence than was William Paley's celebrated watch as proof of the existence of God. His book multiplied examples of flawless design, and, with no other idea of whence it came, turned to a Great Designer. Unfortunately for him, the song of the eye has many discordant notes. They show it to be not the work of some great composer but of an insensible drudge: an instrument, like all others, built by a tinkerer rather than by a trained engineer.

Organs of little apparent importance. Natural selection can give rise to organs of extreme complexity (eyes included) and has often done so. Any increase in the ability to see is useful and is at once seized upon. But what of organs of little apparent use? How can evolution explain the origin of simple parts, of which the importance does not seem sufficient to cause the preservation of successively varying individuals?

Part of the answer is: important to whom? We might not be able to see the point of a structure, but it may be crucial to its owner. The giraffe has a tiny tail, which looks like an artificial fly-flapper. It seems at first incredible that this could have been adapted for its present purpose by successive slight modifications, each better and better, for so trifling an object as driving away flies.

Flies suck blood and carry parasites. The tsetse-fly makes it impossible to raise cattle in much of southern and central Africa. Its attacks on livestock are so fierce that it once forced humans, all over Africa, to carry their own loads as no pack animals could survive. The flies still cause a billion dollars' worth of damage a year. Whether a cow is black, cream-coloured or patchy might be thought a most trivial character – but some breeds are more liable to attack because tsetse are attracted to large blocks of continuous colour and avoid animals with patterns. Even the limited protection given by a flapper is of some help. The tail, the coat, or even the eyelashes of the cow or the giraffe are, trivial as they appear to us, crucial to its defences.

Scientists often dismiss organs as unimportant because they have not bothered to find out what they do. Once, many of the endocrine glands of the body – the pineal, or the thymus – were shrugged off as mere useless relics, rather than as the masters of its internal economy. Even the eye has unexpected tasks. It is used to see with, but it does much more. Many cave animals are blind. Few things seem less useful than an organ of sight in a place with no light. In the dark, the eye at once loses its importance – or so it might seem. In fact, darkness reveals uses that are otherwise invisible. The blind mole-rat of Israel has the smallest eyes of any mammal, sealed beneath the skin. Even the parts of its brain most involved in vision are much reduced. However, its organ, diminished as it is, is still an eye. It has kept the remnants of sight in a place where it can see nothing.

Some of the eye's nerves go, not to the visual centres of the brain, but to the hypothalamus, a place much involved in the control of temperature, of feeding and of sex. A single three-second burst of illumination will set the brain clock, and in mole-rats enough light crosses the eyelids in the brief moments when they kick earth out of their burrows to tell them how long the day is and at what time of year to breed. Even the sightless need the remnants of what once allowed them to see. To destroy a blind eye disrupts the rhythm of existence. The organ has, it seems, powers revealed only after generations of darkness.

Even so, plenty of cave animals have the remnants of eyes that seem to be of no use at all. In many fish, even the nerve connection with the brain has gone. In some of the crabs the foot-stalk for the eye remains, though the eye is gone; the stand for the telescope is there, though the telescope with its glasses has been lost. These organs may be real relics, of no importance; but how is it possible to prove a lack of use?

Nowhere is it easier to dismiss the value of any structure than in DNA itself. Great tracts seem to have no function and the molecule has millions of sites that differ at random from one individual to the next – or so it appears. It is easy (and may often be fair) to see most of its changes as beside the evolutionary point. Can all the millions of differences between two mice or two fruit-flies – or the many more that separate insects from mammals – have evolved through the struggle for existence? Perhaps, some say, most of the DNA is an organ of small importance, whose presence is not noticed by Darwin's machine. If so, much of life is neutral ground upon which natural selection enacts its few rare struggles.

That view is supported by a surprising fact: that the parts of the body that vary most are those that appear to be least important. Blood clots are made when small proteins link together in response to damage. For much of the time each unit floats in the plasma, its ability to bind checked by a short piece that blocks the crucial site. After a cut, the plug is snipped out, the molecules link up and the clot forms. Most of the protein does not vary at all, but the stopper, with its simple job, is filled with diversity. Natural selection surely cannot act to retain differences in the part of the molecule with the least exacting task. Most of the changes in the stopper probably have no effect on how it works and merely accumulate with time. In the same way, in DNA as a whole, the parts that make no protein vary more than those that do, and the more embedded in the machinery of the cell any protein may be, the less variable its gene. Diversity, it seems, builds up where it does no harm, but is excluded from places where it might cause trouble.

The champions of Darwinism find it painful to admit that most variation under Nature is a spectator at the evolutionary play. Even so, the idea is now accepted, almost by default. The molecular clock and the trees used to link distant beings assume that change in DNA measures no more than the passage of time. As most clocks and trees make at least some sense, perhaps the belief is fair.

However, the clock itself hints that to dismiss most diversity as random noise may be a mistake. Most mammals are blessed with a personal parasite. The United States has many species of pocket-gophers, each with its own unique louse. Host and dependant evolved together, but the amount of DNA change is ten times greater from louse to louse than from gopher to gopher. Animals and plants measure out their lives not in years, but in generations; and the faster clock in the louse is in part because it squeezes in more sex in a shorter time than does its host. That explains some, but not all, of the difference. In sharks, a sluggish group, the clock ticks slowly; in bony fish with the same length of life at greater speed; and in the primates (ourselves included) for some reason it slows down. Something has changed the clocks; but what it is, nobody knows.

All selection – on behaviour, on colour or on the structure of proteins – is, in the end, on DNA. Almost never do we know where in that giant molecule it is at work. Most of the genetic material does a job that is quite obscure and in our ignorance it becomes as easy to dismiss the value of any change as it was to belittle the tail of the giraffe. Sometimes, though, a molecule's structure and function come together, to show how dangerous it can be to deny use to any character until it has been studied in detail.

The red pigment of the blood, haemoglobin, has many tasks. Its most important job is to pick up oxygen in the lungs and move it around the body. The molecule consists of two pairs of protein chains, folded around each other. The oxygen fits into a cleft between them and escapes as the chains shift. The bar-headed goose migrates across the summit of Everest at more than five miles above sea-level, a place where humans die within a few hours. Andean birds spend much of the year in conditions almost as tough. Each bird's haemoglobin stays in its oxygen-binding shape long after that of others has flipped. The bar-headed goose alters a single DNA letter, which removes a crucial contact between the chains and makes the protein more eager to bind oxygen. The Andean goose alters the same junction to the same effect, but with a change in the opposite chain. Andean vultures become avid for the life-giving molecule with quite a different set of mutations. The three subtle pieces of genetic engineering are each as direct a response to natural selection as are the wings of the melanic moth. Without an intimate knowledge of haemoglobin it would be easy to dismiss them as random noise.

Even so, one bar-headed goose differs from the next, or from its Andean cousins, in millions of sites throughout the DNA. Can selection have crafted them all? It seems improbable. For most of the time natural selection must act as a policeman rather than as an architect. It does not adapt every molecule to each shift in the environment, but spends much of its efforts on a purge of mutations that interfere with the smooth operation of the body. After all, a random change rarely improves a device that works well. To hit a heart–lung machine with a hammer does not often increase the oxygen supply (although, sometimes, it might). In evolution, most changes are for the worse, and most selection acts to prevent modification and not to promote it.

To keep a police force on the alert – and to punish those it does not approve of – is expensive. If every one of the millions of inherited differences influenced their carriers' ability to stay alive or to reproduce, almost everyone would be unlucky in what they drew; one of their genes would fail the test and the rest (advantageous as they might be) would pay the price. Such a stringent application of the Darwinian rules might make it impossible for a population to sustain itself. Many biologists hence assume that variations neither useful nor injurious would not be affected by natural selection, and would be left a fluctuating element. Most diversity, on this neutral view of the world, is mere froth on the surface of the great Darwinian sea: random noise in a system whose important parts are fine-tuned by selection.

But, others counter, if selection works on what seem at first sight trivial characters like the blind eye of a mole-rat, it could influence all diversity, slight as its importance might appear. So far (apart from the odd exception, such as the blood proteins of mountain birds) there is little evidence one way or another. Once, the action of selection was denied in things that seemed as trifling as the colour of moth wings or the structure of haemoglobin. Now, those are seen as the great proofs of its power. Perhaps the same will happen to all changes at the DNA level – but perhaps not.

Haemoglobin has many messages for evolution. It bears on all the supposed difficulties of descent with modification and shows most of them to be false. The molecule and its fellows also raise problems that are hard to accommodate within a Darwinian framework. Perhaps the greater number are only apparent, and those that are real are not fatal: but the natural history of DNA needs explanations beyond those of *The Origin of Species*. In spite of the fuss about its supposed defects, that great work has done well at explaining the diversity of plants and animals. It does rather worse when it comes to the structure of genes.

A whale has three thousand million DNA bases, but a certain salamander less than an inch long has twenty times as many. Other animals, at first sight little different from those stuffed with nucleic acid, have shed such genetic paraphernalia altogether. The fugu, or puffer fish, has all its genes packed into a length of DNA an eighth of our own. Other sea creatures show how far a cell can be stripped down. Small green algae – whole nucleated cells, with all their machinery – contain a mere three hundred and eighty thousand DNA bases. Their genes are jammed together and even overlap, with the end of one marking the start of the next. That, it seems, is the minimum needed to make a cell. Why does a mammal need ten thousand times as much and the salamander twenty times more? Salamanders are more complicated than seaweeds, but surely not that much.

There seems to be a pressure for change within the genes themselves. Evolution once seemed to inhabit the Malthusian world of 'severe labour . . . excesses of all kinds . . . wars, plague and famines'. The language might be modified to fit plants and animals rather than the undeserving poor, but natural selection still involved a struggle between individuals for existence or for sex. Now, biology's attention is being drawn more and more to another conflict: to the war beneath the skin, to the struggles among genes themselves. DNA has, it appears, its own agenda that may conflict with the interests of its carriers. Some of the molecular battles can be explained in familiar terms, but some follow rules that seem at first sight quite alien to Darwinism.

In the 1960s, geneticists noticed an odd result in their crosses. When fruit-flies from the United States were mated with others from Europe, their daughters were sterile, their sons reshuffled their genes in frenzied fashion, and many new mutations appeared. The same happened when American stocks collected thirty years earlier and kept in the lab since then were crossed with modern flies from the same place. Odder still, there was sexual inequality, as crosses in which American (or older) females were mated with European (or newer) males showed the effect, but nothing happened the other way round. All this was, it transpired, the result of an upheaval deep within the genes.

An extra piece of DNA – a moveable element, three thousand bases long – had invaded the Americas. It hopped around and sometimes damaged other genes. Flies long exposed to the intruder were better able to cope than those attacked but recently. The harm they cause stops the elements from increasing above fifty or so copies, most of which stay where they are because they have lost parts of their structure.

Some of the many invaders have a life of their own, while others sit in sullen silence as a reminder that although DNA is a medium it need not be a message. Some are coded by RNA, others by DNA. Some can copy themselves and make a protein, but others are so corrupted that they can no longer do anything. Most of the genetic tumult does no good to its bearers. One mouse mutation in ten is due to mobile DNA, and a certain mouse gene exists as a single functional copy — and two hundred useless versions damaged and multiplied by interfering outsiders. A fifth of the DNA of most mammals is made up of wrecks of genes or of those that ruined them. That must, at least, be a nuisance to copy.

Just a thirtieth of our own DNA makes a protein. Much of the rest consists of thousands – or millions – of repeats of the same sequences, degraded by the accidents of time. Many human mutations (like those of fruit-flies) result from the unwelcome insertion of a mobile element into a piece of working DNA. Genes are, it seems, mere oases of sense in a desert of nonsense.

The genes themselves are marked by the fingerprints of ancient visitors. For many, the code for the protein chain is interrupted by lengths of DNA that make no sense. That meaningless message is read off, only to be edited out so that the genetic gibberish has no effect on the protein itself. It is a fossil of a biological gatecrasher whose ruins remain as flaws in the crafted products of natural selection. Nineteen out of every twenty mammalian genes contain such interruptions, and some have far more useless material than they do coding sections. Other creatures avoid such junk with no apparent difficulty. The fugu analogue to the gene for Huntington's disease (an illness of nerve damage and premature death) is far smaller than its human equivalent, which is full of redundant DNA. Flies have fewer trespassers in the genes, and bacteria none at all.

All this can be reconciled with a modified version of orthodox theory. After all, an organism whose retina is inside-out has no right to complain about the oddities of its genes. Eyes and genes each evolve through a series of compromises. Tapeworms hijack the machinery of the gut, which has to cope with its unwelcome visitor – why should not transposable elements do the same with the cell?

Darwinism is not mocked by movable DNA, but it still has a lot of explaining to do. Physics, after relativity, could no longer account for the behaviour of atom, apple and universe in the rational terms of Newton. Within genetics, too, is hidden a world of turmoil that may need new kinds of answers. Evolution lacks its Einstein – and, some say, does not need one – but its theories, successful as they have been, are sorely tested by the natural history of the genes.

Some segments of DNA have an inbuilt tendency to alter in number. That process can make Darwinian sense. Aphids resistant to pesticides do the job with a vast increase in the dose of the genes used to break the chemicals down, as a result of natural selection of the most ordinary kind. Duplication may also be favoured because one copy can concentrate on part of the task in hand while another sets off on its own. The haemoglobins live as families, changed versions of an original, each now a specialist at a different job. Long ago, there was but a single transport molecule. It resembled the myoglobin that stores oxygen in muscle. The primeval globin doubled up several times to give two great groups of genes. One member of each family makes the haemoglobin found in adults while others produce versions useful in the foetus, which must cajole oxygen from its mother's blood. Each, distinct though its task now is, retains the stamp of its shared ancestry.

Sometimes, however, multiplication seems to do nothing useful, but to emerge only from the instability of the DNA. Certain sets of its letters increase and decrease in number with no apparent benefit to their owners. As they do, they generate the unique 'DNA finger-prints' used to identify criminals. Their message means nothing, and such repeats form great tracts of featureless waste that separate the working genes.

Descent with modification predicts that two groups apart for many years will, on the average, differ more than two that split not long ago. For bones, or behaviour, or blood proteins, the rule works well.

In some repeated parts of the DNA it fails so badly as to give pause to Darwin's most dedicated supporters.

Consider the case of the midwife toad, notorious for its failure to disprove Mendelism when the inheritance of an acquired character (a blackened foot) was found to turn on the furtive injection of ink by a corrupt biologist. Now the toad has become an icon for those who hope to transcend Darwin.

The toad has a gene family that makes parts of the machinery used to translate the DNA message. Each of the hundreds of copies carried by every animal has exactly the same sequence (which is odd, given the tendency of any multiplied gene to diverge from its fellows). The midwife toad has a blood relation, the northern midwife, much alike in appearance. That too has a gene family with an identical job (as do most frogs and toads). Each is derived, no doubt, from some ancient shared ancestor.

The pattern of the DNA is quite unexpected. Every duplicate of the much-repeated gene in the midwife toad has precisely the same structure as every other; while the northern midwife has just as many copies, all identical – but all different from those of its relative. How can this pattern of absolute identity of repeats within a species, but fixed differences between them, arise from simple change over time? Why has there been no divergence among copies within the midwife or its cousin, given that they have altered so much since they split? All this is not a mere aberration among toads. The two common European mice each contain twenty thousand copies of a particular section of DNA. It is the same in all its details within each one, but different between them, although they severed relations a mere few thousand years ago. There has, it seems, been a sort of ethnic cleansing of the genes. Just one version is allowed to remain within a species, whatever happens when different species are compared.

The dissidents can be purged in many ways, forcible conversion included. DNA is surrounded by a priesthood of enzymes anxious to correct its smallest errors. Without them, it would soon fail. If the repair enzymes are biased in their belief about what the correct message should be, then that version of the genetic creed is bound to take over. Other methods of genetic purification can homogenize a DNA sequence. In places with many copies of a particular string of letters, the segments tend to mispair, rather like the teeth of a zip-fastener done up too quickly. One version may have a built-in tendency to increase at the expense of the other and to drive it out.

Some of this behaviour can be persuaded into a Darwinian strait-jacket, but some is harder to subdue. It hints that genes have an evolutionary agenda of their own. Perhaps, to certain parts of the DNA, species are no more than a place to live, great continents of animals linked by sex. Different species (such as southern and northern midwife toads) may look much the same to the outsider, but from the molecule's point of view each is an island isolated from its neighbours by a sexual barrier. As a result, each evolves to its own internal rules. Perhaps this is why the amount of DNA varies so much between relatives – by as much as three times when it comes to pairs of salamanders otherwise hard to tell apart.

Startling as this might be for traditional Darwinians, other patterns are even more unexpected. Certain genes make great leaps across the living world. Haemoglobin itself crops up in unexpected places. A few insects (such as the midge larvae found in stagnant water) have a version of the molecule and are able, as a result, to take up habits widely different from their allies. How did a protein from mammals get into an insect? The family tree of fruit-flies and their jumping genes prove that transposable elements entered the laboratory fly from a quite unrelated tropical fly. As the two did not meet until Europeans (and their flies) filled the New World, the invasion was recent. Perhaps, long ago, a similar accident took haemoglobin as a passenger on a DNA element passed from mammal to midge.

Vertebrate DNA also has a curious distribution. A certain piece is found in sheep, goats and cows – and in several snakes and a couple of lizards. No other animal has anything like it. An ancestor of the boa-constrictor, the viper and the rattlesnake was the source from which this nomadic gene got into the progenitor of today's farm animals forty million years ago. Within both snakes and mammals, changes in its structure fit the standard patterns of relatedness, so that the transfer must have happened just once. Rattlesnakes and sheep pale when compared to plants. A piece of mobile DNA hidden within a gene for part of the metabolic machinery has mounted an assault on a whole range of vegetation. It started off in a fungus, but has invaded a thousand or more hosts, from coffee to foxgloves to bananas, picked off at random from the three hundred thousand kinds of plant with flowers.

Although the idea that genes can move between such different places is unexpected enough, such long-distance commerce is everywhere. The influx of drug-resistance genes into harmful bacteria from harmless forms within the gut is the latest of many exchanges among scattered parts of the universe of life.

Every plant or animal cell has many mitochondria, the generators of power. Each has its own genes, distinct from those in the cell nucleus. Mitochondrial DNA in animals is not a string of millions of bases but a small closed circle. When put on to the grand evolutionary tree, it groups not with animals, but with bacteria (whose own genes are arranged in a loop). An elephant's mitochondria are, as a result, more similar to the inhabitants of its guts than they are to most of its own DNA.

Elephant cells, like those of all other advanced creatures, descend, it seems, from an ancient society with two members. The genetic structure of the mitochondrion is close to that of the agent of typhus, the disease that killed twenty million people at the time of the First World War. Typhus is transmitted by lice and can live only inside cells. As is true for mitochondria, its agent is unable to make its own amino acids, but feeds off its host. Long ago, it seems, the two simple organisms patched up their differences and lived in harmony. In time, the treaty ripened into an indissoluble bond, and each party found it impossible to live without the other.

A simple alliance has now widened into a federation. Different parts of the cells of trees or tigers trace their ancestry to a whole range of ancient beings. The whips and lashes with which cells row themselves about came from another bacterial group, the ancestors of the bacterium that causes syphilis. In the same way, the green elements of plant cells descend from bacteria able to use energy from sunlight. Other parts of the cellular machinery, internal pipework included, were also once independent, although their own genes are now lost to the nucleus. Perhaps the nucleus itself was once a separate organism, engulfed by another. After all, it contains both DNA and

RNA, the latter used as the genetic material by many creatures today.

The first hint of how wide the confederacy of life might spread came from bacteria themselves. One kind common in our own guts has had its entire complement of DNA bases laid out. Great sections of its genes speak a language different from the rest, with a mix of DNA bases that reveals a separate ancestry. A fifth of that creature's genes come from elsewhere among the bugs.

The emerging insight into the molecular anatomy of life shows that, three hundred million years ago, gene exchange was universal. All genomes of all higher creatures are a patchwork of parts that started in different places and retain traces of a bastard ancestry from the earliest times. The structure of seven hundred genes is known from a wide enough range of beings, from bacteria to yeast and worms, to trace the distant past. They group not by those who bear them, but by what they do. One set – whether it finds itself in plants, animals or bacteria – organizes, operates and edits the information kept in DNA. The other does household tasks such as repair, food preparation, waste disposal and moving around. The information branch resembles the genes of simple bugs that pump out methane, while the rest of the genetic material has been assembled from that group and from other sources within the bacteria and their relatives. The housekeeping genes have been hopping about almost since life began, while the data processors are less mobile (perhaps because they have to communicate with others). Even so, life is much more fluid than it once seemed. Trees, by most definitions, have a trunk; but there are plenty of exceptions. Banyans, for instance, have stems that run into each other to make a tangled mass. Trees of genes look much the same. They show that not only is the cell a coalition, but the genes themselves descend from separate founders and have shuffled around in a way unimagined before the advent of molecular biology.

The structure of DNA raises problems so grave for the theory of evolution that it is hard to reflect on them without being staggered. Genetics shows how Darwinism can explain what seems at first inexplicable. It is also a useful reminder that a science without difficulties is not a science at all.

Summary of Chapter. We have in this chapter discussed some of the difficulties and objections which may be urged against my theory. Many of them are very grave; but I think that in the discussion light has been thrown on several facts, which on the theory of independent acts of creation are utterly obscure. We have seen that species at any one period are not indefinitely variable, and are not linked together by a multitude of intermediate gradations, partly because the process of natural selection will always be very slow, and will act, at any one time, only on a very few forms; and partly because the very process of natural selection almost implies the continual supplanting and extinction of preceding and intermediate gradations. Closely allied species, now living on a continuous area, must often have been formed when the area was not continuous, and when the conditions of life did not insensibly graduate away from one part to another. When two varieties are formed in two districts of a continuous area, an intermediate variety will often be formed, fitted for an intermediate zone; but from reasons assigned, the intermediate variety will usually exist in lesser numbers than the two forms which it connects; consequently the two latter, during the course of further modification, from existing in greater numbers, will have a great advantage over the less numerous intermediate variety, and will thus generally succeed in supplanting and exterminating it.

We have seen in this chapter how cautious we should be in concluding that the most different habits of life could not graduate into each other; that a bat, for instance, could not have been formed by natural selection from an animal which at first could only glide through the air.

We have seen that a species may under new conditions of life change its habits, or have diversified habits, with some habits very unlike those of its nearest congeners. Hence we can understand, bearing in mind that each organic being is trying to live wherever it can live, how it has arisen that there are upland geese with webbed feet, ground woodpeckers, diving thrushes, and petrels with the habits of auks.

Although the belief that an organ so perfect as the eye could have been formed by natural selection, is more than enough to stagger any one; yet in the case of any organ, if we know of a long series of gradations in complexity, each good for its possessor, then, under changing conditions of life, there is no logical impossibility in the acquirement of any conceivable degree of perfection through natural selection. In the cases in which we know of no intermediate or transitional states, we should be very cautious in concluding that none could have existed, for the homologies of many organs and their intermediate states show that wonderful metamorphoses in function are at least possible. For instance, a swim-bladder has apparently been converted into an air-breathing lung. The same organ having per-formed simultaneously very different functions, and then having been specialised for one function; and two very distinct organs having performed at the same time the same function, the one having been perfected whilst aided by the other, must often have largely facilitated transitions.

We are far too ignorant, in almost every case, to be enabled to assert that any part or organ is so unimportant for the welfare of a species, that modifications in its structure could not have been slowly accumulated by means of natural selection. But we may confidently believe that many modifications, wholly due to the laws of growth, and at first in no way advantageous to a species, have been sub-sequently taken advantage of by the still further modified descendants of this species. We may, also, believe that a part formerly of high importance has often been retained (as the tail of an aquatic animal by its terrestrial descendants), though it has become of such small importance that it could not, in its present state, have been acquired by natural selection – a power which acts solely by the preservation of profitable variations in the struggle for life.

Natural selection will produce nothing in one species for the ex-clusive good or injury of another; though it may well produce parts, organs, and excretions highly useful or even indispensable, or highly injurious to another species, but in all cases at the same time useful to the owner. Natural selection in each well-stocked country, must act chiefly through the competition of the inhabitants one with another, and consequently will produce perfection, or strength in the battle for life, only according to the standard of that country. Hence the inhabitants of one country, generally the smaller one, will often yield, as we see they do yield, to the inhabitants of another and generally larger country. For in the larger country there will have existed more individuals, and more diversified forms, and the competition will have been severer, and thus the standard of perfection will have been rendered higher. Natural selection will not necessarily produce absolute perfection; nor, as far as we can judge by our limited faculties, can absolute perfection be everywhere found.

On the theory of natural selection we can clearly understand the full meaning of that old canon in natural history, 'Natura non facit saltum.' This canon, if we look only to the present inhabitants of the world, is not strictly correct, but if we include all those of past times, it must by my theory be strictly true.

It is generally acknowledged that all organic beings have been formed on two great laws – Unity of Type, and the Conditions of Existence. By unity of type is meant that fundamental agreement in structure, which we see in organic beings of the same class, and which is quite independent of their habits of life. On my theory, unity of type is explained by unity of descent. The expression of conditions of existence, so often insisted on by the illustrious Cuvier, is fully embraced by the principle of natural selection. For natural selection acts by either now adapting the varying parts of each being to its organic and inorganic conditions of life; or by having adapted them during long-past periods of time: the adaptations being aided in some cases by use and disuse, being slightly affected by the direct action of the external conditions of life, and being in all cases subjected to the several laws of growth. Hence, in fact, the law of the Conditions of Existence is the higher law; as it includes, through the inheritance of former adaptations, that of Unity of Type.

CHAPTER VII

INSTINCT

Instincts and habits, inborn and learned — Conflict, co-operation and compromise — Slave-making instincts — Natural instincts of the cuckoo and ostrich — Simple rules make complex habits: cell-making and social habits of the hive-bee — Gradual shifts to new societies — Cannibalism — Neuter or sterile birds, rats and insects — The evolution of behaviour through natural selection and kinship — Summary

THE AMERICAN KENNEL Club standard for the Chinese Shar-Pei has as its ideal an animal 'regal, alert, intelligent, dignified, lordly, scowling, sober and snobbish'. The Rottweiler, in contrast, should be 'calm, confident and courageous . . . with a self-assured aloofness that does not lend itself to immediate and indiscriminate friendships'.

Breeds differ in how highly strung they are, how much they snap at children and in their fondness for barking. When it comes to how easy each type is to housebreak or how much they enjoy simple destruction, all are about the same. Everyone knows that terriers are excitable, that pitbulls bite and that every pup urinates on the carpet. Temperament, like size or shape, is in the genes. It is known that a cross with a bulldog has affected for many generations the courage and obstinacy of greyhounds; and a cross with a greyhound has given to a whole family of shepherd dogs a tendency to hunt hares.

An action, which we ourselves should require experience to enable us to perform, when performed by an animal – more especially by a very young one, without any experience – and when performed by many individuals in the same way, without their knowing for what purpose it is performed, is usually said to be instinctive.

Instincts are of the highest importance to each animal. They show remarkable adaptations – the migration of birds, the attraction of bees to certain flowers, and the slave-making castes of ants. If behaviour varies and is inherited, it has no choice but to evolve. Apparent miracles of complexity in how animals act need no explanation beyond those that apply to what they are. Behaviour often reveals the force of natural selection in a way that mere appearance does not.

Can habits or instincts, intricate and flexible as they may be, follow the same evolutionary rules as colour, shape or size? That might seem unlikely, for behaviour is often learned or comes from a simple reflex. In the end, however, every action of every animal is a product of genes, however much the environment determines what those genes might do. There is no mystery about such things. Behaviour comes from brains; and brains, like hearts or kidneys, are made by genes. Thomas Henry Huxley, in a stark appraisal of the human condition, claimed that the brain secretes thought, as the liver bile. Our own brain is made of millions of cells, each obedient to the message of its DNA. Within it, half the genes of the body are at work. They are as open to change as are others with less noble tasks. Genes make brains; and brains make behaviour. Thus is instinct transmitted across the generations.

Many animals show inherited differences in how they comport themselves. Even bacteria 'behave', in some general sense, as they interact with each other, with their food or with their host. One soil microbe congregates into swarms and, when things get tough, makes a fruiting body, parts of which disperse to form a new colony. Like a pack of lions or a flock of starlings, the bacteria show social behaviour, with co-operation among different individuals, each with the potential to pass on its own genes. Within a species, behaviour may be as variable as is shape or size. Some wild mice have a gene for 'neophobia' – fear of the new, be it food or nest-hole – that is not found in their fellows. Snails of different colour spend more or less time in daylight because darker individuals heat up more in the sun, and melanic moths prefer to sit in shadier and more camouflaged places than do their light-coloured sibs. Such traits can be embraced by natural selection as much as can an increase in a molecule's capacity to bind oxygen.

Not all behaviour is in the genes, because what they do is modified by habit. Any dog can learn to stand on its hind legs. It is not done well only because bipedalism is laid over an ancestral wish to stay on all fours. Birds can do much more: a certain American nutcracker stores as many as thirty thousand seeds a year, each one in a hole in a tree. Every bird learns the map of its homeland and can return to a seed six months after it stowed it away. In the same way, parrots can be trained to talk and monkeys to manipulate signs in a strained and distant imitation of human grammar. A little dose of judgement or reason comes into play even in animals very low in the scale of Nature.

Take, as an instance, the sea-anemone, an animal that seems safe from most forms of intellect because it lacks a brain. Anemones are among the simplest of all animals, mere sacks of flesh with a fringe of tentacles. They release larvae into the water. These settle, divide and grow into colonies of identical partners. One Pacific kind lives in clonal groups separated from each other by strips of bare rock. Within a colony all is peaceful, but among them rages constant war.

All anemones have sting cells (some of which secrete a poison that can kill a man). The wars of the clones are unending. Each has its own personality. Some are aggressive, while others are calmer but respond at once to attack. Some clones do not fight back, but instead throw more soldiers into the front line as their members are killed. Colonies are able to settle only next to those against whom they have some chance in a fight. In time, a resentful truce emerges and battle starts only when a newcomer arrives.

Anemone society – brainless as it is – shows how all animals, even the lowliest, have an existence beyond that written into DNA. For some creatures, habit rules most of what they do. Each of the thirteen species of Galapagos finch is born with a fixed taste for its own food, be it fruits, or seeds, or insects, picked from the ground or branches, or gouged out with a cactus spine. One is a vampire, drawing blood as it pecks at the backs of seabirds. All members of each species go for more or less the same behaviour, whatever it might be. The Cocos Island finch, in its home three hundred miles off South America, has a different attitude to life. Individual birds vary in diet almost as much as do separate species of their famous relatives. Some eat insects, others nectar, or fruit or seeds, or even snails and small lizards. Each sticks to what it likes.

Cocos Island, unlike the Galapagos, contains but a single kind of finch; but it is an animal with a talent for discovery. Each bird picks up its tastes in its first days. It copies its elders and, by holding to its own task, reduces the amount it has to remember. It inherits not a fixed set of actions, but the ability to adapt to many. With a change of habit, a single species fills as many gaps in the economy of nature as does a whole group on the Galapagos themselves.

Learning, knowledge, habit, tradition – or whatever else it might be called – can itself be inherited. For much of the year, humpback whales live as two separate populations around the Poles, but all move to the Equator to breed. The northern and southern groups mate at random as they frolic in the tropical waters, but the calves follow their mothers to one Pole or the other and learn to feed where she takes them. A history held in the mind rather than in the genes splits them to the opposite ends of the earth.

Sometimes, a new habit can be seen as it begins. Somewhere, a bluetit was the first to open the top of a milk bottle, and on a Japanese beach forty years ago a macaque washed his sandy food before he ate it. Both ideas spread, as have many others. They allowed those who used them to move into a new part of the ecological economy. Cultural damage can do as much harm as does gene mutation. The golden-lion tamarins reintroduced into Brazilian forests assumed that all branches would hold their weight (as had their zoo's familiar ropes) and, even worse, the tiny monkeys had no fear of snakes in their new home.

The culture of the bluetit or the tamarin comes from its history and its ability to learn. How they behave is built upon what evolution has provided (which, in the brain of the monkey, is quite a lot). In humans, biology grants even more, but the same laws apply. If Mozart, instead of playing the pianoforte at three years old with wonderfully little practice, had played a tune with no practice at all, he might truly be said to have done so instinctively. Of course, he did not. His ability to learn the piano came from his ancestors. He played as he did because he was Mozart. Genes set the limits even to genius.

Although that is not a term to be applied to animals, what they do is also constrained by what they inherit. Even a brain tiny in comparison to that of Mozart can achieve remarkable things.

Slave-making instinct. One of the most powerful testaments to natural selection is to observe that the instinct of each species is good for itself, but has never, as far as we can judge, been produced for the exclusive good of others. Some animals may seem to be Good Samaritans; but in truth they never are. Even the many cases of apparent co-operation – plants and their pollinators, the green algae that live inside corals, the mitochondria within all plant and animal cells – evolved through a mutual manipulation that can soon go wrong. The balance of cost and benefit is always calculated, and although many creatures suffer at the hands (or teeth) of their enemies, none selflessly help their friends.

Some parasites admit their lowly status and creep with regret into the body of their host. Others persuade onlookers that theirs is the noble role; that they are not exploiters but heroes, who rule (rather than cheat) their victims. Yet others set up societies in which two or more creatures seem to live in harmony. Different though each appears, all are somewhere on the path that links conflict to co-operation. Where they end up depends on a struggle among instincts as vicious as that against the elements.

Malthusian theory, inexorable as it was in condemning the poor to starve, saw the common good as superior to self-interest. Darwin's view was simpler and more ruthless. To him, evolution had no commonwealth; self-interest is what matters. He was right. There is no charity in nature.

The ants run the gamut from conflict to conquest to apparent co-operation. A few creep into the colonies of others to cajole their hosts into a free meal. With no work to do, the hangers-on become feeble, with thin skins and mouths reduced to tubes. Some even evolve to be, like fleas, tiny travellers on the skin of a larger ant. Others are more confident. Certain ants raid the nests of others and seize them as slaves. The ant is absolutely dependent on its slaves; without their aid, the species would certainly become extinct in a single year. The males and fertile females do no work. The workers or sterile females, though most energetic and courageous in capturing slaves, do no other work. They are incapable of making their own nests, or of feeding their larvae. When the old nest is found inconvenient, and they have to migrate, it is the slaves which determine the migration, and actually carry their masters in their jaws.

Energetic and courageous though the slave-makers might appear, they are parasites – for they could not live without their lackeys. Slave-makers, with their fearsome jaws, look more like victors than beggars, but are in fact as much dependent on their victims as are their feeble relatives. Some ants enslave their own kind. Thus, honeypot ants often attack neighbours and carry off their workers. More often, they steal from other species; the amazon ants, for instance (widespread in Europe) thieving from the red ants around them. The colony is attacked, its guards killed and the cocoons taken back home. There, when they hatch, they labour for their masters, unaware that they are helping to pass on foreign genes. Some slave-makers are more ingenious. Rather than risk an attack, they spray the slave colony with an alarm signal and, as its guards flee in panic, steal the cocoons or pick up the scent of their vassals and persuade them to think of themselves as members of their owner's family. One American driver of slaves takes advantage of two of its neighbours: a small kind to act as a worker and a larger insect to serve as its soldiers. That happens in Europe, too, but there those enslaved fight among themselves, the stronger species ejecting the weak.

Slaves and their owners have obvious conflicts of interest, as one gains at the expense of the other. Other relationships seem at first sight to be benign. Ants and aphids have evolved to depend on each other; the ants gain from the sweet secretions of the aphids (and even 'milk' them to obtain it) and the aphids of those species, lacking soldiers of their own, trust in the jaws of the ants to keep their enemies at bay. In fact, neither party is a philanthropist. The aphids provide not just sugars, but vital nutrients that the ants would other-wise lack. Their excreta – the 'manna' of the Bible are nine-tenths sugar (and, as bees take advantage of what is left by ants, much of our own honey is made from their droppings). So close is the relationship between ant and aphid that some queen ants carry aphids with them when they found a new colony, to be sure of their source of food. The aphids, in turn, have bacteria in their guts that help them digest food. Kill them off with antibiotics, and the aphids – and their ants – will die.

The ants have other allies. Some plants help them with modified thorns in which they nest (and, by stinging, keep out other insects anxious to eat the leaves). A network of apparent kindness is in fact a complicated web of mutual exploitation.

Ants have progressed from slavery into domestication. The New World has hundreds of species that depend on fungus gardens, carefully tended by special castes. The first farmers appeared fifty million years ago. They fertilize the small mushrooms in their nests and pick out weeds or spray them with herbicides. When a queen sets off to found a new colony she carries a small packet of fungi to set up a garden of her own. The farmers depend on their harvest to such a degree that they have lost their own stomach enzymes and depend on the fungi to digest food. Genes show that every ant colony has its own clone of fungi, distinct from that on the neighbouring farm, and that some clones are used by different ant species. Ancient though it is, the relationship seems to be a free and open one, with both parties benefiting from it. Some of the farmed mushrooms are identical to others that live free and have been picked up only recently.

Friendship is, of its nature, unobtrusive. When two parties live in harmony, neither makes much fuss and a relationship may pass unobserved. Lichens, consortia of fungi and algae, were once thought to be single creatures. Only when the relationship ends and each struggles to gain the upper hand is attention drawn to the conflict. When the cost of co-operation outweighs its benefits, society soon breaks down. The soil bacteria that behave so harmoniously in their native home can grow in liquid culture in the laboratory. Within a thousand generations, the animals no longer meet to feed or share a reproductive structure. Instead they dwell as sturdy loners, each happy to live and breed in independence. The genes for co-operation have been replaced by others that do the job more cheaply.

Why should a slave submit to its master, or a fungus allow itself to become a commodity? The balance is finely adjusted. Clones of identical fungi are passed down from parent colony to offspring, reducing the chance of conflict in a well-adjusted system. Both parties have reduced their options through their mutual dependence, which helps it to persist. Some fungi, however, creep in as weeds and are suppressed with a special chemical that kills them alone. In other creatures the relationship is even closer and seems permanent – the wasps that pollinate figs, unlike ants and their fungi, have a family tree almost identical to that of their host, with hundreds of species of fig and of wasps evolving in parallel. Quite why the wasps, which breed far faster than their figs, agree to the relationship is not clear, but it has lasted for forty million years.

Slavery goes well beyond the insects. Plenty of animals persuade others to bring up their young. About one kind of bird in a hundred lays its eggs in the nest of another (and some fish even dump their eggs in the mouths of others who tend them). A battle – a microcosm of many other struggles – rages between exploiter and exploited, with sometimes one party ahead and sometimes the other. Because it is un-resolved it hints at the tactics involved before a compromise is reached. In birds, the habit has evolved at least half a dozen times (cuckoo-ducks, for instance, take advantage of seagulls). It can cost the host a lot – a whole brood pushed out or stabbed to death by a sharp beak. Some parasites, such as the cowbirds of Africa and North America, dump their eggs on any of dozens of kinds of host and almost every nest may be filled with unwelcome visitors. Others, the cuckoos included, are more refined birds and concentrate on just a few victims.

A tiny reed warbler, faced with a cuckoo, feeds a chick much larger than herself. She defers to this foreign adolescent although it does not resemble her own young. Her true offspring, once satiated, stop begging; but the cuckoo is more persistent and forces her to keep up the food supply. The alien, when it demands food, makes a noise not at all like that of a single warbler chick but one that is close to the cacophony of a whole nest of young warblers.

As in the ants and their fungi, the evolutionary arms-races of the avian world go on at full speed. North America has just one common parasitic bird, Africa thirty or more. Almost all the perching birds of Africa can identify (and eject) foreign eggs, while almost none of their American equivalents are able to do so. The process can work in reverse. The village weaver-bird in its native Africa rejects most cuckoo eggs, but since it was introduced into the Caribbean (which has no cuckoos) in the eighteenth century it has lost its ability to judge. Education also plays a part, as first-time parents are happier to accept a cuckoo's egg than are their more cynical elders who have brought up a brood without an intruder.

The parasites can fight back, with a range of eggs that mimic those of their chosen host. The huge range of the common cuckoo, from Africa to Asia, contains dozens of races, each specialized to lay a particular egg. Some hosts are able to identify the foreigner, and eject it or abandon the nest, but others (such as the dunnock, which finds it difficult to tell even black and white eggs apart) are less fastidious. In most places, the cuckoo mimics its hosts: blue eggs for the redstart, say, or spotted grey, like those of the reed warbler.

Sometimes, the parasites turn their attention to the most obvious victim. The occasional habit of birds laying their eggs in the nests of other birds of the same species is not very uncommon, and this perhaps explains the origin of a singular instinct in the ostriches. For several hen ostriches unite and lay a few eggs first in one nest and then in another. This instinct, however, of the American ostrich – the rhea – has not as yet been perfected, for a surprising number of eggs lie strewn over the plains. Half a dozen females share a nest, a simple hollow within the square mile or so of a male's territory. One female is the boss. She mates with that male and lays and guards her own eggs. Then, the extra females – some of them mated by the fortunate male – wander in and add to the pile. They act like huge and flightless cuckoos, as they stroll off and take no part in parental care. So well do the parasites perform that a nest may hold forty eggs – far too many for a lone bird to mind.

The nest-holder is no fool. She can tell her own eggs from those of the spongers, and keeps them in the middle of the group, with the aliens pushed to the edge. The foreigners pay the price: just half of the parasites' eggs hatch, while almost all those of the owner do. At three pounds, an ostrich egg is a good meal, and plenty are eaten; but those near the edge go first because not even a hungry jackal can manage more than one. The rootless females dump their eggs because they have failed to find a territory, or have lost their own nest. As a result, it pays them to lay even if most of the eggs are eaten. The holder suffers their attentions not because they share her genes, but because of the help it gives to her own eggs. No benevolence is required.

Cell-making instinct of the Hive Bee. The beehive is a metaphor for human society. Shakespeare was impressed: bees are 'Creatures that by a rule in nature teach/ the act of order to a peopled kingdom./ They have a king, and officers of sorts:/ Where some, like magistrates, correct at home;/ Others, like merchants, venture trade abroad;/ Others, like soldiers, armed in their stings,/ Make boot upon the summer's velvet buds;/ Which pillage they with merry march bring home/ To the tent-royal of their emperor:/ Who, busied in his majesty, surveys/ The singing masons building roofs of gold;/ The civil citizens kneading up the honey;/ The poor mechanic porters crowding in/ Their heavy burdens at his narrow gate;/ The sad-eyed justice, with his surly hum,/ Delivering o'er to executors pale/ The lazy yawning drone . . .'

It does not take much to cause bees or any other animals to act in a complicated, or even an elegant, way. Three simple rules are called for: variation, inheritance and natural selection. The variation may be in the genes or in past experience and the vehicle of inheritance can be DNA or memory, but whatever the machinery, a complex pattern of behaviour soon evolves. Bee society also shows how the principles of economics can be modified by kinship and lead to actions that at first sight seems to violate the very laws of evolution.

Shakespeare tells better than most the story of the beehive. He is mistaken only in the sex of the ruler. The monarch of the hive is not a male but a single female, a fact not realized until just before Shakespeare's death: 'We must not call the Queen "Rex", the Bee-state is an Amazonian or feminine Kingdom' (a notion rejected by the bee-master to King Charles II on the grounds that it denied the divine right of kings)

The queen, who lays nearly all the eggs, is helped by several thousand female workers, accompanied at times by a lesser number of males called drones. She depends on the workers for food and can lay two thousand eggs a day. The workers construct the nest and feed the young, either with the secretions of a special gland, or with honey or pollen. Some are builders, others undertakers who throw out the dead. When the queen dies, another is elected by feeding her with a royal jelly that transforms a lucky larva's status. Once a year, the males have a chance for sex. When their job is done they are judged useless, and killed or thrown out.

For bees and other insects, social life pays. Just one insect in fifty lives in such a way, but those that do flourish. The Brazilian rainforest contains as much biomass in the form of ants (a highly social group) as in birds, mammals, snakes and lizards combined. The habit is fairly new. The insects appeared about half a billion years ago, but stayed solitary until the termites turned up three hundred million years later. Ants evolved a hundred million years ago. Ancient though it is, the balance of each society is still ready to tip whenever circumstances allow.

As in Adam Smith's hypothetical factory for the manufacture of pins (in which each workman took up his own task and much improved its productivity), an ant's nest, a termite mound or hive has division of labour. It is hard to examine the exquisite structure of a comb, so beautifully adapted to its end, without enthusiastic admiration. How could such strange behaviour evolve? Elephants never forget – but bees? Their brains are tiny: why are they so smart?

Karl Marx got it (as usual) more or less right: 'what distinguishes the worst architect from the best of bees is this, that the architect raises his structure in imagination before he erects it in reality'. His statement brings out the pragmatic nature of evolution, in a hive or anywhere else. Bee society has no plans. Instead, its complexity comes from some elementary rules. All that is needed is for each bee to know what goes on in its neighbourhood, to pass its knowledge to others, and to have a threshold at which it changes from one behaviour to another. The hive is an information society. Any action generates feedback and there soon emerges a system that far transcends the actions of those within it.

Such laws determine the colony's architecture. It has been remarked that a skilful workman, with fitting tools and measures, would find it very difficult to make cells of wax of the true form, though this is perfectly effected by a crowd of bees working in a dark hive. Grant whatever instincts you please, and it seems at first quite inconceivable how they can make all the necessary angles and planes, or even perceive when they are correctly made. But the difficulty is not nearly so great as it at first appears: all this beautiful work can be shown to follow from a few very simple instincts.

Each hive – with its hundred thousand cells – has three concentric regions. A central brood area full of larvae is surrounded by a rim of pollen cells, and then by a wall of cells filled with honey. The design might, perhaps, result from some hidden blueprint of what a hive should look like, deep within each bee's brain. The truth asks less of insect intellect.

When an empty comb is put into a hive, the cells are at first filled at random. Then, in time, order begins to emerge. The queen wanders about as she lays her thousands of eggs. She follows a simple rule – lay an egg close to a cell that is already full – and does it with such consistency that, within a few hours, 70 per cent of her products are placed next to an occupied cell. This soon leads to a mass of eggs near the centre, where she will more often re-cross her path.

The workers have laws of their own. They store pollen and honey at random, but remove it more quickly from cells near a larva. Pollen is used at ten times the rate of honey and in places with more brood cells it runs out faster. This leads to more turnover in a pollen cell. Cells with honey, the long-term reserve, are soon blocked. The single place left for a pollen delivery is near the brood, where demand soon empties each cavity for re-use. As a result, to make the three-level pattern of the comb all that is needed is for each bee to test the contents of the cell next door. Soon, there appears a structure that seems – but is not – well designed. Its simple rules mean that small shifts in behaviour cause large changes in the system as a whole.

The hunt for food shows how information and stupidity work together to make what seems like intelligence. A hive has thousands of foragers who comb the landscape for miles around. They come in two forms. Scouts buzz around alone, while the more abundant recruits, guided by the scouts' dance, sally out when needed. Once a rich source is found, scores of bees appear, as if by magic, while a smaller patch is visited by fewer diners. The regulations are simple. If a scout finds a small meal as it flies about, it eats it and goes on its way. If the item is large enough, it goes back home and, with its famous waggle dance, tells its fellows where the food is. They respond and, quite automatically, more join in as each traveller returns with the good news. When the food has gone, its appeal diminishes, the dance stops, the crowd disperses and a new hunt begins.

The better the source, the longer the waggling goes on. Each bee reacts at once to a dance. As a result, long bouts attract many foragers while short ones persuade only a few bees to leave home. When hungry times set in, the scouts become less fastidious and give long dances even for poor food. All this makes the colony at once efficient and flexible.

The cybernetic hive can get quite complicated. The workers divide their labour by age and by experience. The youngest bees clean out the cells and nurse the brood. Some then become undertakers and eject the dead, while others venture out as hunters. Among the foragers some concentrate on nectar, others on pollen. Every bee has a set task that changes with age. As the foragers grow older they move from a juvenile taste for sweet nectar to a more refined preference for pollen. Modest changes in the conduct of single bees have major effects on society. Undertakers take up the pastime in middle age, but their juniors will join in to help clean up after a disaster. A shortage of pollen means that the young turn to it and more is gathered. If not enough nectar can be found, then older bees stick to their juvenile habits and continue to search for the sweet substance.

Genes also play a part. In some colonies, queens mate with several males. Quite why is not clear (although it might help protect against diseases that would kill all members of a group of identical animals), but their behaviour means that several lines of descent are present. DNA tests show that, within a hive, scouts and recruits are dissimilar, so that instinct as well as habit rules their behaviour. In the same way, some bees are born with a taste for pollen, others for nectar.

What seems so complex – bee society – obeys simple rules. They allow each colony to be efficient but adaptable. A shift in preference adjusts the economy to cope with whatever hits it. Genetic diversity means that evolution can seize an opportunity as soon as it arises. The most wonderful of all known instincts, that of the hive bee, can thus be explained by natural selection having taken advantage of numerous, successive, slight modifications of simpler instincts.

For bees, co-operation certainly pays in economic terms – but is it enough to explain the most remarkable property of all, the surrender of sex to a single female? To do so is the mark of the highest form of society. Queen bees and ants gain a great deal from their elevated role. They are the longest-lived and most fecund of insects, with some ant queens lasting thirty years. Those of African driver ants may each give birth to three hundred million young, almost all of them sterile workers who pillage the forest and bring back food for the family. Any queen also produces some males, and a few virgin queens of her own who fly forth in search of a possible new home – although almost none succeed.

All this is fine for the ruling female, but what about the workers? How can natural selection favour sterility – apparently the deadest of evolutionary ends? That is a special difficulty, which might appear insuperable and actually fatal to the whole theory. By good fortune, nature has preserved some hints as to how barrenness can be preferred. The answer lies in economics, modified by family interest.

A series of stages from selfishness to co-operation remains as a reminder of the steps taken on the road from solitary insect to the Byzantine world of the hive. Ninety-five per cent of the twenty thousand different kinds of bee dwell not in colonies but in proud independence, in burrows, empty snail shells, or plant stems. Each carpenter bee, for instance, makes a small nest inside a twig or shoot, lays an egg therein, and provides it with a ball of food. A bigger ball means a better chance for her hungry larva.

Force two females to share a nest, and things become more complicated. Independence at once gives way to collaboration. The household chores are portioned out as one bee stays close to the nest and guards it, while the other – with no task left to do at home – spends her time in the search for rations. Are bees quite so quick-witted? Indeed they are not, but a division of labour does not need much logic. All that is required is that once a task is over another should loom larger in a bee's mind. A head start for one of the partners in a certain department – as a nest-guard, say – means that the other sets to at what is left. The mental rule of doing what is left undone leads the bees, with no conscious effort, to work together for the good of the larva. Who hunts and who guards may turn on tiny differences in size or habit, but once the decision is made, the rest must follow. With a series of such choices a society soon emerges.

In a carpenter bee from Israel, most females set up home in pairs. The partners are almost always mother and daughter. The mother is the reproductive boss. She eats the eggs laid by her daughter, whose sole job is to raise her own young sisters. The system turns on property rights and social convention, because the bee who gets there first (always, needless to say, the mother) lords it over the next to arrive. But how can her daughter's actions be sustained? A small genetic change might let her have grubs of her own. Indeed, in some years daughters set up in independence. Why, most of the time, are they so generous?

It all has to do with kinship, and a shortage of places to live. Except in the best years, good nest sites are hard to find, and any shelter is at risk of a takeover by a rival. A home with a guard is safer and has, on average, more young. The daughter gains because, although by guarding she loses any hope of progeny of her own, her help gives her mother a chance to lay another egg. As a result, copies of her own genes, contained as they are within her young sister, profit.

Evolution favours teamwork, not through goodwill but because of increased efficiency in multiplying DNA. Selection always acts through genes. As mothers and daughters have such a high proportion in common, the balance of profit, loss and risk means that, for most of the time, it pays the daughter to help her parent rather than to strike out on her own. When the economic equilibrium changes (as it does in a particularly favourable year) she at once abandons her mother and makes a home of her own.

Hive-bee society, with its single dominant female, began in much the same way, with added property rights as represented by the hive. Once capital is inherited, a complex society makes even more sense. The first females to take possession can control their sibs and force them to support their young.

Bees are a reminder that individuals are just part of the evolutionary equation. As all in the end die, the fate of the genes is detached from those who transmit them. Most members of most species passed away long ago. Their job as conduits through which genes travel through time is over. Natural selection acts through the medium of DNA, rather than only on the flesh of those who bear it. In other words, it is interested in kinship.

Animal breeders know that well. They prefer to breed from the best – the fattest pig, perhaps. However, a pig can be fat for two reasons: good food, or good genes. To breed from an animal obese just because it ate well is a waste of time. It is, experience shows, better to pick out not the grossest of all pigs but those of more modest form from fatter families. Their household may not contain the most corpulent beasts of all but, as a group, they are better endowed in their genes than are most. The brothers and sisters of moderately chubby pigs may not themselves be unduly large, but to breed from them is a better way to make progress than just to select the few prize specimens out of many. Selection on the genes, and not the individuals, does the job: a thin pig of good family may well have fat genes struggling to get out.

The key to success – in bees as much as in pigs – lies in the blood. The closer the affinity, the better the value of one individual predicts that of another. The second cousin of a fat pig is a less credible candidate for genetic greatness than is his brother. Kin selection, as it is called, is as important in the real world as on the farm. It explains behaviour as strange – and more – as that of the beehive.

Kinship can pay an individual to reduce its own chances if it improves the prospects of other members of its family to a sufficient extent. That can lead to the evolution of animals that cannot reproduce; and to others that kill their own kind. Such actions may, to the casual or the creationist glance, appear to fly in the face of reason; to be evidence against evolution itself. Strange though instincts may appear, not one has been produced for the sole benefit of other animals. If it had, it could not be explained by Darwinism. Generosity or selfishness emerge from natural selection. Once attention is directed to the genes as well as to their bearers, such eccentric behaviour makes perfect sense.

There is more to life than kinship. Some animals care little for the fate of those who share their heritage. The nine-banded armadillo always has identical quadruplets. They behave to each other as they do to a foreign armadillo, in spite of their genetic similarity. Lizards that gave up sex long ago are as aggressive to their clone-mates as are sexual lizards who are not much related; and some bees treat their kin in the same way as anyone else. Perhaps these animals live in places where no help is needed; but perhaps the opportunity for sharing has never turned up. Benevolence needs a gene as much as black wings do and, if it does not arise, then such behaviour will not evolve.

The laws of the animal world are ruthless. Plenty of parents kill their children, and plenty of children murder their sibs. For them, the economic part of the argument looms large and selfishness pays: if only one can survive, shared genes do not much matter. Sometimes, though, three or four of their offspring can pull through; and then the erstwhile murderers can behave as gently as anyone else – at least, to those who share their blood.

Many creatures grow up with their sibs: in a nest, in the corpse of the mouse in which a burying-beetle lays her eggs, attached to their mother's many teats, or as seeds in a common fruit. Quite often, a mother has more offspring than have a realistic chance of clinging to life. Perhaps, in an exceptional year, several might; or a few may replace brothers or sisters who die by accident. Most of the time, though, the fate of the extras is as a meal for their luckier kin, their nursery more an ambush than a sanctuary. The surplus young are a biological insurance policy, and when times look good their parents begrudge the cost of cover. Often, the firstborn are favoured from the moment of birth – or before. In some mammals, different parts of the uterus have more blood vessels, allowing the lucky embryos to develop more quickly. Birds pass testosterone to their eggs and those with more of the hormone grow more quickly as chicks. Piglets are born with sharp teeth. Nipples nearer the mother's head give more milk, and the larger young seize these and fight off their sibs. Again, the relentless rules of cost and benefit determine who lives and who dies. Both are important, but inheritance is often the key. Among bears, lions and howler monkeys, many young are cared for and survive – unless another male takes over the group, when most are killed by the alien intruder.

Animals are forced to live with their relatives for many reasons. In most cases, ecology is to blame. With nowhere to go, the only choice is to stay at home. For a termite, or a bark-beetle, home is a rotten log. It may be crowded, but to set off into the unknown in the hope of setting up another branch of the family business is risky. Other animals live in burrows that take years to build, or in a colony filled with thousands of workmen, so that a pioneer has almost no chance of a residence of his own.

Once confined with one's relatives, kinship and wealth loom large. Like families gathered at Christmas, social animals are poised between co-operation and conflict. The more distant the tie, and the cheaper the gift, the more the chance of a quarrel. A distant aunt is welcomed if she is rich enough, but a cousin will be spurned if he drinks more than he brings.

Shared blood can be a great help in passing on genes – and in persuading some of their carriers to accept the supreme sacrifice. Take, for example, the aphid. Most aphids, most of the time, are celibate. Sex happens just once a year, on a tree. In some, the fertilized female lays her eggs in a gall, a growth in the wood used as a retreat in which her asexual progeny – identical copies of herself – develop. As the weather warms, these multiply, without benefit of males, into millions.

Although aphids seem innocuous enough, some have soldiers to defend their homes. These animals have jaws, or horns, or claws, or sharp mouth parts. They may be small, but are fierce enough to drive off mice and even to bite through human skin. All are sterile. They give up their own reproductive future to protect that of their gall-mates.

The explanation lies in genetics. All descendants of a female are identical twins. As a soldier shares all its genes with those it guards, it makes perfect sense for it to be sacrificed if, by so doing, it increases the chances of its sibs. Soldiers, like the teeth of tigers, are a defence organ, a specialized part of a mass of genetically identical cells. In a few aphids, a single gall contains the products of several females. Such aphids never have soldiers. They would make no sense: why sacrifice yourself for someone who does not carry your DNA?

Self-sacrifice does not always demand absolute identity. Plenty of animals give up their future to help brothers, sisters, parents or more distant relatives. In a closed community, common ancestry becomes the rule. Any animal living in isolation – in an Alpine village, or a fallen log – has no choice but to mate with its kin, because there is nobody else around. If (as in inbred lines of mice) this goes on for long enough the residents become almost as similar to each other as are members of an aphid clone. Selection measures the value of its subjects in the currency of the genes they carry and estimates their worth in relation to that of others.

The naked mole-rat has a society as unattractive as its name. It lives in underground colonies of up to three hundred in arid parts of Southern Africa. The soil is so hard and the roots upon which the rats feed so scattered that no solitary rat could survive. Within a colony just one female reproduces. She is the largest inhabitant of all and has sex with a few of the biggest males. Her fellows, both male and female, raise the young, dig tunnels and, with a special class of fighters, attack (often at the cost of their own lives) marauding snakes and foreign mole-rats. In a good year, when the soil softens after rain, the queen rat has twenty young at a time and the workers can dig for half a mile a month.

In spite of the dangers of childbirth, the queen (and her favoured males) live for many years. So do the workers, if they are kept alone, in a zoo. In the colony, most die young. Their lives are full of stress – not just because of the snakes and the digging, but because they are bullied by the queen and her partners. So intense is the social pressure that the hormones needed for sexual development are shut down. An animal removed from the community at once becomes mature, with a puberty that lasts a week. In the same way, a dead empress is soon replaced by the largest of her inferiors. However, for most of the time the queen keeps her fellows in a perpetual and bitter adolescence. DNA is the key. It shows that every inhabitant of a mole-rat colony is a close relative. The society has become a great extended family. By helping their relatives the workers and fighters promote their own genes.

Charity can go a long way. Most of the time, animals have the same number of sons and daughters, as each passes on the genes with equal efficiency. Under some circumstances, the balance can change:

it may be more expensive to have sons, daughters may survive better, or the offspring of a particular female might be particularly good at finding a mate.

One way to ensure the right mix is quite straightforward. Mothers do not hesitate to kill the sex less able to pass on genes. The marsupial mouse, a small and fierce Australian carnivore, has an interesting sex-life. Males live for only a year and die exhausted after copulation. Some females, too, last for just twelve months. Those fortunate enough to survive another season are old and tired and make worse mothers than they were when young. In their first year, mothers kill all their daughters but one, to give their well-brought-up and aggressive sons a chance. In their second try at reproduction, though, mothers destroy not daughters but sons, who, feeble though they may be, do a worse job at passing on her genes than do her new-born and relatively healthy female offspring.

Other creatures are more subtle in planning their families. Parrots in cages have runs of twenty sons or twenty daughters in a row, while lemurs housed in groups prefer to have sons. How – or why – they do it nobody knows. Rank may play a part, for mothers high in society can choose what sex best passes on their genes. The female wasps who lay eggs in caterpillars prefer daughters if they are lucky enough to find an unusually large grub, because a large daughter does better than a large son; but high-ranking red deer mothers have more sons, because their offspring's large antlers impress potential mates.

The Seychelles are an island paradise that, like Paradise itself, has been abused by those who live there. Among the survivors is a small bird called – with some lack of imagination – the Seychelles warbler. By 1988 fewer than four hundred were left, all on the barren island of Cousin. Then, a small group was moved to a more luxuriant islet, Aride.

On Cousin, males outnumber females. The new colonies, in contrast, have many more female birds. On the new and fertile island parents have more daughters than sons, while those who still starve in their native home prefer male progeny. Again, it is a matter of passing on genes. Young females do not look for mates of their own but stay close to their parents and help them to raise their young (who are, needless to say, their own brothers and sisters). When food is abundant, this is helpful. The parents lay more eggs and the genes of the generous daughters gain as a result. When food is short, the balance changes and the helpful spinsters become a nuisance. They eat more than they can give – and their parents retaliate. Plenty of food makes it better to bear daughters who can lend a hand. A shortage favours investment in sons who fly off and find mates of their own. In favourable habitats nine out of ten chicks are female, in starvation territory just one in five. Birds moved from poor habitats on Cousin to the luxury of Aride at once switch their sexual preference towards daughters.

A wasp's nest, a beehive or an anthill has many more females than males. Most, like the warblers, seem to commit evolutionary suicide. They have no offspring of their own, but help others to reproduce. If evolution depends on inherited differences in the ability to copy genes, how can sterile forms arise?

Again, the balance of cost and kinship is all, and a mother invests in the sex best able to copy her genes. The fate of the childless matters less than what they can do for those who have children. Natural selection, sometimes helped by odd patterns of relatedness, solves the paradox of the infertile. It can lead to sterility, to evolution in animals that have no offspring and to individuals that put their efforts into helping others.

Bees, together with ants and wasps, have an unusual way of deciding who is born male and who female. It gives mothers control over their progeny and leads to unexpected patterns of relatedness. They can make it worthwhile to abandon sex, to slave on behalf of others of one's own kind, or – now and again – to murder them.

Unlike mole-rats (or humans), male bees have but a single copy of all their chromosomes whereas females have two. As a result, a female has twice as many versions of every gene as does a male. A simple rule determines bee sex. Any individual with two different copies of a special sex gene is female while any with but a single form is male. Any animal with a solitary set of chromosomes is hence doomed to masculinity.

The queen can choose the sex of her progeny. She stores sperm and can decide whether or not to use it. Unfertilized eggs lead to sons (with just her own genes), but those that are fertilized, with genes from her mate, make daughters. In this way the mother controls her relatedness to her descendants. A female shares fewer genes with her daughters than she does with her sisters.

The logic is simple. Human brothers and sisters have half their genes in common. A pair of bee sisters, in contrast, shares every gene passed on by their father but have a 50 per cent chance that any gene comes from their mother. Sisters share not half their genes, but three-quarters (all their father's genes plus half their mother's). Any female bee is hence more related to a sister than to any daughter she might have. Given the iron rules of kinship, it then pays her to help raise those sisters, rather than to have daughters of her own. The apparent reproductive suicide of a female worker is, as a result, a matter of genetic self-interest.

Such strange patterns of relatedness may alter the economic balance, but do not themselves cause sterility. Termites make that point. They look like ants (although they are not much related to them) and live in huge colonies, most of whose members are sterile workers. Some are ruled by single females, in a society even more rigid than that of a beehive. A termite queen may live for fifty years and lay thirty thousand eggs a day.

However, termites determine the sex of their offspring in a way not much different from our own, with males and females differing only in a single chromosome. In consequence, a termite mound has both male and female workers. Only when blood ties are distorted, as in ants or bees, does the burden of sterility fall on females alone. Unlike bees, in which such behaviour has evolved many times, all termites are social, so that the habit started long ago, and stuck. Perhaps the strange pattern of relatedness between mother and offspring in bees does make it easier for a solitary animal to take up socialism when necessary.

Within a beehive all is not, as Aesop had it, sweetness and light. Not only is there sterility but murder and cannibalism. Deplorable though that is for those who make judgements from Nature, such acts help to understand the evolution of behaviour. As any stockbroker knows, kindness is hard to analyse. How does a donor lose

and a recipient gain? Greed is simpler. Often, losers forfeit everything and winner takes all – which makes it easier to work out the
balance sheet.

Cannibalism puts paid to the comfortable idea that Nature is not
really red in tooth and claw. Thousands of animals go in for it and
death by fellows can be the main cause of mortality. Some fish eat 90
per cent of their peers. Wall-eye (a sport fish of North American
lakes) eat each other tail first, with whole chains of fish busy at a
shared meal. The name itself comes from a Carib tribe, the
Canibales. In some of the sixty cultures known to have practised
anthropophagy, corpses were the second most important source of
meat. The pastime has many excuses, and kinship can be one. For
most cannibals, blood is thicker than water. Given the choice, they
keep themselves, their relatives and their genes alive at the expense
of others. The losers, although they may not realize it, show the logic
of martyrdom, of death so that others might live.

Many animals assess the genetic cost of a slaughter of the
innocents. Often, males kill and eat the young of any unprotected
female. A pregnant female mouse exposed to a new male makes the
best of a bad job – the death of her pups – with a pre-emptive strike.
She reabsorbs her foetuses, in a prenatal feast that gives the meal to
her, rather than to the hopeful male. Slugs show how costs and
benefits depend on kinship. Hermaphrodites though they are, sex
often proceeds in a mitigated way, as each animal fertilizes the other.
A box of outbred slugs turns into a loathsome soup as the survivors
feed on the corpses of the rest. Inbred animals, in contrast, the offspring of self-fertilization, respect their own genes held within their
comrades and refrain from that macabre feast.

Such rules can lead to murder in the beehive. Both queens and
workers invest in the next generation, as they lay eggs or help to raise
larvae. Because of the uneven relationship between the sexes,
attitudes to the young by the ruling and the working class are different. Two sisters share three-quarters of their genes; a brother only a
quarter of his sister's heritage, through their maternal connection.
For a worker (a sterile female), the best strategy is hence to give three
times as much to sisters as to brothers. Then, her genes are
transmitted with equal efficiency through either sex. Queen bees, in

contrast, are related to the same degree to sons and to daughters. For them it makes sense to put the same investment into each, as all the young pass on a mother's genes with equal efficiency.

In other words, within a colony, sons are worth more than brothers. It is in the queen's interests for the hive to contain equal numbers of males and females, but from the workers' point of view the proportion of males should never rise above one in four. Most of the time, the workers have their way, because there are lots of them and they hold most of the weapons. They kill and eat what they see as the excess among their brothers. This increases their own genetic chances and gives them a free meal into the bargain. Life in a hive can become quite gothic. Unhappy with attacks on their investment some queens fight back against worker control. They lay all-male broods, or – with a certain ingenuity – give birth to sons who are saved from murder because they masquerade as females.

The problem of sterility is solved when seen through the eyes of kinship. But another great difficulty for the idea of gradual change through natural selection lies in the workers' differing widely from both the males and the fertile females in structure, yet they are absolutely sterile so that they could never have transmitted successively acquired modifications of structure or instinct to their progeny. The neuters of several ants differ, not only from the fertile females and males, but from each other, sometimes to an almost incredible degree, and are thus divided into two or even three castes. The castes, moreover, do not generally graduate into each other, but are perfectly well defined. The honeypot ant of American deserts has workers whose simple job is to hang like great barrels of sugared water from the ceiling, to be tapped when needed, while others have classes with huge heads to block the nest entrance against intruders. It may seem odd that sterile animals can evolve a variety of forms, but this too comes from indirect natural selection.

The problem is no harder (or easier) to solve than the origin of soldiers in aphids, or teeth in tigers. With all the genes of an organism the same (albeit confined within a gall rather than a cloak of flesh) the focus shifts from evolution to development: cells with the same genes can make organs as different as teeth and toenails, or plant-suckers and soldiers. It is all a matter of division of labour.

In bees or ants, the sterile castes do not share all their genes with their mother. However, they have enough in common to enable the colony to control – with as little conscious effort as the embryo of a tiger – the growth of its members, using cues of place, time and expedience. Certain ant castes have risky lives. In some, a twentieth die each hour as others attack them; but the same is true of tiger skin cells as they are worn away. Some castes vary in size, just like the teeth of tigers. Among the leaf-cutter ants of South America are individuals that differ by three hundred times in weight, from tiny fungus-gardeners to massive soldiers. The soldiers, expensive as they are, do not appear until the colony is at least a hundred thousand strong. Like those of a young cub, the first days of a colony, with its shared genes, must be devoted to food rather than defence. That is risky, but unavoidable. Not until her extended family is well established will the daughters of the queen be enabled to grow into her teeth and claws.

The beehive or anthill, a society that appears to co-operate, is in fact full of compromise and deceit; a place patrolled at all times by the genetic police, who punish those who break its laws. They are microcosms of evolution. The worth of every animal is scrutinized, the economy analysed, and natural selection comes up with whatever best transmits genes. Some individuals give up their own future because it is contained within someone else, while others kill to ensure that their heritage persists. Selfishness and charity are closer than they seem. They follow biological rules which are, in the end, no different from those that determined the evolution of the bees, ants and mole-rats themselves.

The ethical dilemmas posed by the ant or the cuckoo – so useful to savants who read human behaviour into Nature – began long ago. The Shakespearian fashion for society as a beehive has been succeeded by the fad for seeing it as a converted monkey-house. George Romanes, a friend of Darwin, made a fifty-point scale of intellect. Worms were at step eighteen as they could 'feel surprise or fear', and dogs and apes at twenty-eight because of their 'indefinite morality, along with the capacity to experience shame, remorse, deceit and the ludicrous'. *Homo sapiens*, in all its moral variety, occupied steps twenty-nine to fifty.

The moral lessons of Nature, if such they are, began long ago. Dinosaurs had societies as eccentric as those of birds today. Some were cannibals: a two-hundred-million-year-old fossil has the skeleton of a young member of its own kind within its ribcage. Others were as generous as the Seychelles warbler. Great herds of eighty or more giant herbivores – brontosaurs and ornithopods – left tracks that reveal a life based, like that of elephants, on a few adults and many young. Parental care started long ago. Dinosaur eggs were first found in France in 1859 and have now been unearthed all over the world. Some young were fed by their parents, in great rookeries that lasted for millennia, while others fled at once, to live a solitary existence. A few species laid eggs beneath mounds of rotten vegetation that warmed as it decayed. The modern birds – the megapodes – with the same habits have not changed much from their dinosaur ancestors. Some make hillocks thirty feet across, while others are happy with mounds of grass-clippings. A few live in huge colonies on volcanoes, their eggs hatched by the earth's heat. For some megapodes, two brothers co-operate to help a female build her mound. Dinosaurs might have been just as generous.

Although those great animals have gone, they followed the same rules of instinct and habit as did their descendants. No doubt they could feel surprise or fear and had, perhaps, some capacity for shame, remorse or even deceit. We will never know: nobody will ever peer into the mind of a dinosaur (or, for that matter, of a megapode). Because the laws of descent with modification act on behaviour as much as on bones, the instincts, if not the souls, of dinosaurs and all animals make as much biological sense as does everything else in their lives.

Summary. I have endeavoured briefly in this chapter to show that the mental qualities of our domestic animals vary, and that the variations are inherited. Still more briefly I have attempted to show that instincts vary slightly in a state of nature. No one will dispute that instincts are of the highest importance to each animal. Therefore I can see no difficulty, under changing conditions of life, in natural selection accumulating slight modifications of instinct to any extent, in any useful direction. In some cases habit or use and disuse have probably come into play. I do not pretend that the facts given in this chapter strengthen in any great degree my theory; but none of the cases of difficulty, to the best of my judgment, annihilate it. On the other hand, the fact that instincts are not always absolutely perfect and are liable to mistakes; – that no instinct has been produced for the exclusive good of other animals, but that each animal takes advantage of the instincts of others; – that the canon in natural history, of 'natura non facit saltum' is applicable to instincts as well as to corporeal structure, and is plainly explicable on the foregoing views, but is otherwise inexplicable – all tend to corroborate the theory of natural selection.

This theory is, also, strengthened by some few other facts in regard to instincts; as by that common case of closely allied, but certainly distinct, species, when inhabiting distant parts of the world and living under considerably different conditions of life, yet often retaining nearly the same instincts. For instance, we can understand on the principle of inheritance, how it is that the thrush of South America lines its nest with mud, in the same peculiar manner as does our British thrush: how it is that the male wrens (Troglodytes) of North America, build 'cock-nests,' to roost in, like the males of our distinct Kitty-wrens, a habit wholly unlike that of any other known bird. Finally, it may not be a logical deduction, but to my imagination it is far more satisfactory to look at such instincts as the young cuckoo ejecting its foster-brothers, ants making slaves, – the larvae of ichneumonidae feeding within the live bodies of caterpillars, not as specially endowed or created instincts, but as small consequences of one general law, leading to the advancement of all organic beings, namely, multiply, vary, let the strongest live and the weakest die.

CHAPTER VIII

HYBRIDISM

Sex and the marketplace — Laws of sterility — What keeps species apart — Natural selection against hybridism — The sterility of crosses, within and between species — Cues of identity — Rules of estrangement, from brain to gene — Sex and species: male ardour and female discretion — Fertility of varieties when crossed — Laws of sterility and the genetics of species barriers — Promiscuity and new kinds of plant — Breaking the frontiers — The future of species

ADAM AND EVE fell into temptation when they ate the tomato of the tree of knowledge. They did, that is, if one takes the European view of celestial botany. For most of its nations, the tomato was such an exotic fruit that it deserved a noble title. The French call it the *pomme d'amour*, the Italians the *pomo d'oro* – the golden apple – and the Croats the *paradis* (a name that explains itself). Ridiculous, perhaps, but whatever Eden's fruit may have been it was not an apple (which is rare in the Levant) and at one time was painted as a tomato, an orange or a fig, fruits succeeded in sacred art for a time by the banana.

Tomatoes, figs, oranges and bananas have different names because they have different genes. They often grow near each other and have at least the chance to mate, to exchange their DNA, and to merge into some kind of amalgamated plant. They do not – but why not?

Sex, the trap into which the inhabitants of Eden fell, is a biological marketplace. Every species is, more or less, a sexual republic. Within each, open exchange is the rule and every inhabitant has a chance to barter its DNA with any other. The world as a whole is broken into fiefdoms – species – separated by barriers to genetic free trade. To understand the origin of species we need to know what these frontiers are and how they arise.

The differences between apples and tomatoes (or anything else) can be measured from their DNA. Most of the changes so revealed have appeared by simple descent with modification since an ancient break in sexual relations or are relics of a shared and distant ancestor. They have nothing to do with the barriers themselves. The real impediments to the marriage of tomato and apple are genetic, but are impossible to test because the two never mate. That is Darwin's dilemma. It denies to biologists the main tool of their trade. Rather like nations who find it impossible even to discuss their differences, a lack of intercourse between species makes it hard to establish what keeps them apart.

The failure of species to cross is not some magical property but comes from descent with modification. Somehow, a continuous process – natural selection – has given rise to discontinuous entities called species. How it does so was once a mystery; but genetics is beginning to reveal the truth about what species are and how they arise.

Sterility itself cannot be selected for, as success would spell the automatic end of the line involved. It can, none the less, emerge from other evolved changes (as it does in bees, in which reproductive failure pays because it helps others to pass on genes). The sterility of species when crossed also comes from indirect selection, in this case as a result of the workings of evolution as it builds creatures able to cope with what the environment throws at them.

Many genes are involved in that vital task. Within any species, in different places selection may call for (or be offered) different mixtures of mutations to do its job and may come up with subtly different products. Quite often, one local blend does not combine well with others. Sometimes – as in the two genes that cause the smoky grey fur of the Persian cat – the nature of the interaction is known, but more often it is not. If the failure of adapted mixtures to work together becomes complete, the populations find it impossible to exchange genes when they meet, and each becomes, in effect, a new species.

DNA's inability to copy itself without mistakes – mutation – means that evolution is inevitable. Natural selection does no more than capitalize on that fact. Species, too, are by-products of the Mendelian machinery. They emerge from the apparatus of inheritance, from the ways that genes co-operate to do their job. If genes for size, shape or behaviour work together only in the right combinations, then, as different mixtures build up in different places, the origin of species becomes inevitable. Once established they may evolve further, but, of their nature, species happen by accident.

Laws governing the sterility of first Crosses and of Hybrids. The first line of the *United States Army Mule Training Manual* is, it is said, 'First, catch the animal's attention by striking it smartly between the ears with a stout stick'. Mules, the offspring of a cross between a donkey and a horse, are famous for their obstinacy, but are sterile and are, as a result, no threat to the integrity of their parental species. In other wild creatures, though, distinct species can mate and may have fertile offspring when enticed to do so in captivity. Ducks do it, geese are happy to exchange genes and pheasant species have been hybridized to make a great variety of colours and forms. The fact that the barriers among life's divisions can so easily be reversed by man shows that they are not irrevocable. We must hence look at sterility not as an indelible characteristic, but as one capable of being removed by domestication.

Biological frontiers, plastic though they sometimes are, have an identity that extends beyond simple difference. When outwitted by human ingenuity, strange alloys can be made with ingredients from different creatures. Shoots of certain plants can be grafted on to the roots of another – apricot on to oak, for example – and a pig's heart-valve can be transferred, with some medical wizardry, to a man. These mixtures of genes survive quite happily, but are made without benefit of sex. Their success or failure does not much depend on the relatedness of those involved; a pig valve does better than one from a rabbit, although rabbits are more akin to ourselves.

The ease with which genes can be exchanged by mating, in contrast, fits – like evolution itself – a hierarchy of biological change. Consensual sex yields to grudging acceptance of a less than desirable

mate, to rare liaisons with unlikely partners and at last to complete
reproductive failure. Only a small proportion of all species – perhaps
one in a hundred – will accept genes from another. There are,
though, plenty of groups in which sterility when crossed is so different in degree and graduates away so insensibly, and the fertility of
pure species is so easily affected by various circumstances, that for all
practical purposes it is most difficult to say where perfect fertility
ends and sterility begins.

The dusky salamander lives in the southern Appalachian
Mountains, and likes to stay at home. It has been in its fastnesses for
many years, each population evolving to its own rules. Animals from
within ten miles or so mate promiscuously, but those offered the
choice of a local partner or one from far away almost never choose
the stranger. Plants, too, may prefer a neighbour to a distant individual who is adapted to other things (although for them the choice
involves the growth of pollen towards egg rather than the rejection
of an ardent swain).

Sometimes, simple choice fails, and hybrids appear. Some are
sterile, some so feeble that they die almost at once, while others do
well for most of the time. If they do so for long enough, one species
may be absorbed into another. Human interference is often to blame.
John Constable, when he painted the meadows running down to the
River Stour, saw plenty of poppies and thistles in his hayfields. The
cows and sheep that ate the fodder in winter had to cope with that
less than nutritious part of their diet. Nowadays, many of his fields
are filled not with scarlet or blue flowers, but with yellow: the blooms
of poisonous plants. They descend, in part, from plants introduced to
the Oxford Botanic Garden from Sicily in 1794. Britain already had
a native relative, the groundsel; and the Oxford ragwort as it spread
along the cinders of the new railways – a place not unlike its native
volcanoes – found a local partner with which to mate. Two centuries
later, the mongrels have found an identity of their own, and have
become a great pest. Many of Constable's great English oaks, too,
have been succeeded by hybrids between native plants and those
brought in from the Continent. Bastards as they might be, they seem
perfectly able to cope with their new home.

However well such hybrids may survive, and however stable they might appear, they can fail Nature's test when life gets too hard. The island of Daphne Major in the Galapagos has four kinds of ground finches, the small, the medium and the large, together with the cactus finch. It contains a hundred or so of each of the first two, a mere dozen or so large ground finches and even fewer cactus finches. Their lives have been followed for twenty years. These birds, the icon of all evolution textbooks, show why species adapted to different ways of life stay separate. It is because their hybrids do not survive the scrutiny of natural selection.

The frontiers between finches can leak. About one union in fifty is between birds of different kinds (and, on some islands, they mate with birds not counted as Darwin's finches at all). Such illegitimate behaviour is born of desperation. With a mere half dozen cactus finches on Daphne Major, some are forced to turn to another species for any chance of sex. Finches recognize each other by the shape of their beaks and by song. Rather like cows and sheep, they learn who is who at their parents' knee. This Freudian relationship can go wrong, as some males fail to pick up their father's song (perhaps because their natal nest is close to that of a noisy foreigner) and court females of the wrong sort. Their startled mates accept the alien and produce intermediate young.

At first sight, their progeny are no worse off than those of legitimate unions and, in most years, survive and reproduce as well as anyone else. If the four kinds of finch on Daphne were to carry on in this way, before long they would fuse into an average bird. Fossils, however, show the finches to have been distinct for thousands of years. Natural selection against the crossbreeds has put a stop to the urge for union. It destroys animals that do not fit into their evolved place in the economy and ensures that life stays in separate compartments.

Every few years, the Pacific winds change direction. No longer do they blow warm water west towards Australia. Instead, tropical seawater moves east towards South America and as far north as California. The current – El Niño, the Christ Child – arrives around Christmas. It brings rain and storms to the Americas and drought to Australia. The warm waters rise by six inches along American coasts, and because they contain far less oxygen and food than does the usual cold current, the numbers of fish and of the birds that feed upon them collapse.

All this leads to inconvenience and more. The El Niño of 1998 was the strongest of the twentieth century. It caused ten billion pounds' worth of damage. Rain in East Africa swept away villages and farms, while Indonesian drought led to huge forest fires. Even Microsoft had to cancel its annual party because of storms along the California coast.

The Galapagos in an El Niño has a huge increase in rainfall – by up to seven times that of a normal year. The sea warms and loses much of its nutrition. On the land, nature explodes. The plants are thick and lush, the finches breed several times in succession, and hybridism flourishes. As a result, many birds halfway between species appear.

As El Niño ebbs away, drought follows the torrential rain. Most of the finches die and, at once, selection sets to work. It reserves its greatest punishment for the hybrids. For large ground finches the main item of diet after a drought is a seed called the caltrop, after the spiked iron ball used to trip horses in medieval battles. In such difficult times, finches that have shifted down in size because they accepted genes from a lesser relative find no food soft enough as it has already been taken by an even smaller form. The offspring of matings with a larger species do just as badly, as their beaks are not tough enough to manage what is left by their more adapted relatives.

Excavations in Peru show that fishermen were washed out of their campsites by an El Niño twelve thousand years ago, with major disasters every few centuries since then. One struck in the 1780s, leading to failed harvests in Europe and to Marie Antoinette's dismissive comment: 'Let them eat cake.' The hybrid finches' lesson about the baleful force of evolution has – like that of the aristocracy – been repeated many times. Each species has evolved to deal with life in separate compartments. The sexual frontiers may leak when times are good, but, eventually, the wandering genes are stopped by the death of their carriers.

Causes of the sterility of first Crosses and of Hybrids. Sex and species have a lot in common. Asexual plants and animals are a classifier's nightmare, because they exist as a range of forms, each grading into the next. Dandelions are divided into thousands of distinct kinds, fitted to where they live and blended into an almost continuous series. Botanists often give up and refer to such groups as 'aggregates' or 'complexes'.

Sexual species are rather easier to define. They can be marked out in genetic terms, as reproductive units, isolated from each other. For them, there is courtship, a series of challenges and responses. Within a species, a male has at least a chance of persuading his mate to accept his donation. However, a gene that tries to pass between two species faces an obstacle race in which one barrier – on the first lap or at the finish – is impassable. The obstructions are those facing any uxorious male; but when a partner is of the wrong kind they cannot be penetrated. They range from distance, to learned preferences, to genetic barriers embedded within the DNA. Each one involves, in one way or another, a breakdown in the genetic teamwork needed to build a relationship, fertilize an egg, or produce a healthy offspring.

Much of natural history is the scientific study of mating barriers and, like birdwatching, soon degenerates into somewhat of a list. Species are divided from each other in many ingenious ways – by space, by time, by mating preference, by the inability to fertilize an egg or produce healthy young, or by the sterility of offspring. The hurdles at which the sexual athletes fall are as various as life itself. Those involved may never meet, or may mate at their own special time or season. Males and females of different kinds may choose not to pair, or may – with more or less enthusiasm – mate but fail to make a fertile egg. The geographical checks can be as narrow as the few inches between different orchids upon which certain bees feed or as wide as the Atlantic. When it comes to time, some flies mate in the morning and some in the evening and some crickets in the spring and others in the autumn; but two kinds of cicadas in North America emerge and mate every thirteen or every seventeen years. The difference ensures that they almost never get together (in spite of a certain confusion every couple of centuries).

Colour, song, scent and more all play a part in settling who is, and who is not, acceptable. The differences may be obvious to our own eyes or ears (which is helpful to birdwatchers) but quite often they are not. Some crickets make chirps attractive to their own kind and to ourselves, but others retain their identity with sounds too low to be heard by humans.

More barriers await any hopeful sperm that has penetrated too far. Some females mount an attack on sperm recognized as foreign, while others, if mated both by a male of their own kind and an alien, much prefer the local product. Plants do the same, with a refusal by a female to allow foreign pollen-tubes to grow through the wall of tissue that must be breached before the egg is reached.

The cues may be on the sex cells themselves. The eggs of many marine animals, from abalones to sea-urchins, have a protective membrane into which a hole is punched by the first successful sperm. Every kind has its own receptor. Each female change is matched by a male equivalent. In abalones, the lock is a repeated sequence of DNA, which changes at great speed. It binds many copies of the sperm's identity molecule. Although the seven abalones of Californian waters look much the same, their egg receptors are quite different. The females evolve, the males are forced to follow them, and the chances of inappropriate sex are much reduced.

The signals that separate species – colour, scent or song – are often those used as statements of excellence by males in their struggle for females. For many plants, the process involves a pollinator – one of three hundred thousand different kinds – who must be persuaded to act as their brain and mobile sex-organ. Bees, flies, birds, lemurs and tree-kangaroos are all pressed into service. The plant strikes a bargain with its emissary. Pollinators are rewarded with food, a place to sit (as in the lotus, whose metabolic heat keeps its flower warm while beetles copulate within), or with a sexual signal (as in the bees that pick up their own sex cues from orchids). Plants want their pollinators to be busy, hungry and faithful to their own species, while the other party prefers to get as much as it can from its host with the least effort and to spread its favours as widely as possible.

Any failure to hybridize involves a loss of communication some-where in the biological conversation that is sex. As in any dialogue, two parties are involved (with, in the case of plants, an added go-between); and if they speak different languages things will not get far.

Sometimes, it is a matter of simple recognition. Many birds follow the first object they see, be it parent, clockwork toy, or biologist, and take it as a role model. Such creatures – ducks, doves and finches – tend to be divided into many different kinds. For gull chicks, the colour of the spot on the parental bill is the cue. When herring gull eggs are brought up by lesser black-backed gulls – birds with red rather than yellow bill spots – the young see themselves as members of the latter group and, when adult, prefer to mate with black-backeds. Such boundaries turn more on the ability to see who is who than on any deeper incompatibility. That may be why birds differ so little in their genes. The average genetic divergence between related birds, from eagles to tits, is half that of mammals and even less than that of lizards.

Fragile though they seem, boundaries held in the mind can be quite secure. Some African lakes contain hundreds of kinds of perch-like fish called cichlids. The males are bright in colour, the females dull. Each species has its own signal. The transparent waters of some of the great African lakes allow the messages to flash from fish to fish and to keep them separate. Although the frontiers are guarded by preference alone, they allow dozens of different kinds to live together. Clear lakes have more species than do those with muddy water. Each trademark diverges most when in places with a danger of confusion, and a cichlid with red males tends to share a section of shore with others whose males are blue.

Reds and blues put together in a tank stay apart. Females are faithful to their own – as long as they can tell who they are. In red light, which makes it hard to separate the colours, a female will accept whatever male courts her and confusion reigns. When the two parts of the incompatibility system – male signal and female choice – break down, what were once species become, at best, varieties of the same one. That has been the fate of many of the cichlids of Lake Victoria. As forests around the lake were felled and farmers moved in, soil was washed into the lake. Victoria has lost her clarity and her fish have become confused. Species that have been separate for thousands of generations have, as a result, amalgamated into a federation.

If a female cichlid (or anything else) chooses an inappropriate mate – a feeble member of her own kind, or the wrong species – she pays a price as her progeny do not have the correct genes. Males are less worried about sex with an alien, as each act is so cheap that it is worth wasting seed upon barren ground on the chance that, now and again, one succeeds.

The number of species within any group is, as a result, related to sexual habits. Plants and birds with plenty of sexual choice tend to be divided into many species. More faithful beings have fewer forms. Seven species of flowering land plant exist for every one without flowers, because each flower attracts a limited range of visitors; and the more precise the fit, the easier it becomes for a plant to retain its identity. Ducks, with males and females very different because of sexual selection, have two dozen kinds in Britain alone. Swans, in which pairs may bond for life (and in which the sexes resemble each other), have far fewer.

When it comes to the origin of such species, females (with their evolved ability to choose a mate) are in the driving seat. When males take the wheel, the urge to split is reduced. A male let loose on those who have not gained the ability to fight him off can often breach the boundary. Take, as an example, the cuckoo. The common cuckoo – the bird with the repetitive note, the 'word of fear, Unpleasing to a married ear' – is found from Japan to Ireland, from Africa to Norway. The bird, widespread as it is, has resisted the temptation to diverge into numerous kinds found among many nest parasites. That is odd, given that cuckoo females are divided into distinct races, each able to mimic the eggs of its own host.

The answer lies in cuckoo cuckoldry; in its male's insistence on sex with any female, whether or not she belongs to the race that brought him up. Female cuckoos bear more allegiance to a particular host, be it redstart or warbler, than do their males. As a result, a cuckoo who was himself brought up by warblers may father an egg found in a redstart nest. Egg colour itself is inherited down the female line, so that females stay with the bird by whom they were fostered. As males are so much less faithful, their promiscuity explains why the cuckoo has not split into many species, each true to its dupe.

Africa, in contrast, has dozens of species of egg-dumping finches, each tied to a single host, and quite unalike in appearance. The parasites are all close relatives, with an evolutionary tree quite different in shape from that of their prey. Each specializes on a single victim, and the chicks mimic the bright mouth markings, cheeps and movements of those of its unwitting step-parent. The barriers are in the finch's early experience. The male finch learns its foster-mother's song and this cue of identity is used by its own females, who will mate only with birds who sing the right one. Such forced fidelity means that the finches have broken up into many species rather than remaining as an entity bound together by masculine ardour.

When males are unleashed, there may be evolutionary mayhem. More than a million mallard ducks are released in the United States each year to allow every hunter the kill that is an American's right. The male, with its green head and red breast, is a forward bird. Some female ducks, such as its close relative the pintail (whose own male is quite distinct, with a long tail and chestnut head), have long lived with it and can cope with its ardour. Although they can be prevailed upon to interbreed in captivity, they never do so in the wild. Wherever it goes (and it has travelled all over the world), the male mallard makes its intentions clear. The Australian black duck, the New Zealand grey, the Florida mottled and the Hawaiian duck have all suffered from its advances. Each – once a separate species more distinct in its genes from the mallard than is the pintail – has begun to hybridize with the invader. Soon, they will be absorbed into it and disappear from the birdbooks.

This has led to calls for mass slaughter. Although activists have denounced what they call 'avian eugenics', the death of hybrids is backed by the Royal Society for the Protection of Birds. Bird-watchers have an austere view of existence: that which cannot be pigeonholed should be shot. They ignore the evolutionary message hidden in the duck soup: that the origin of species can result from the struggle between the sexes as much as against the elements.

The joint action of genes as they respond to the demands of selection means that those who accept an incompatible partner pay a penalty. As discord grows, it begins to pay each party to stick to its own kind as, if negotiations are bound to fail, it is cheaper to break them off at once rather than to go through the whole painful business of courtship. Hybridism itself becomes a liability and the barriers begin to rise with an impetus of their own.

Mines have grasses that – like the copper flower in Africa – grow in places where their neighbours would shrivel. The grass responsible for the scent of new-mown hay is among them. Its internal economy changes in many ways as a result of metal-tolerance and the mine populations cannot compete except on the mines themselves. There they prevail as their neighbours are poisoned. Grasses exchange genes with the help of the wind. In a wild Welsh valley, an ancient lead-mine is surrounded by unpolluted pasture. As the wind howls over the workings it blows pollen from the non-tolerant population on to the mine, and a great cone of male sex cells from the mine plants into the downwind meadow. Neither set of genes does well except on its own patch. Selection has acted to reduce the amount of exchange. The mine plants flower a week earlier than their neighbours, which limits the chance of either sort of pollen finding a partner whose offspring would be unfit. The plants have taken a first step towards an individual identity. Should such discrimination continue, in time they might become quite unable to accept genes from outside and would be, in one sense at least, a new species.

Monkey flowers, with their showy yellow blossoms, have gone further. They have split into tolerant and non-tolerant kinds. A cross between the two gives young that sicken and die even on ordinary soil. Although what is selected for is an ability to grow on poisoned ground, sterility emerges as an incidental. When tested together the genes give each plant an identity of its own.

Fertility of Varieties when crossed, and of their Mongrel offspring. In Berlin, Friedrichstrasse meets Zimmerstrasse at a very ordinary road junction across which traffic flows freely. A decade ago this was Checkpoint Charlie, one of the few gaps in an otherwise impermeable barrier a hundred miles long. However, even a boundary as firmly defined as the Berlin Wall can, with suitable manipulation, leak or collapse; and the same can happen to the frontiers between species. As a result, biologists are now beginning to understand not just the displays that keep species separate, but the genes responsible for their very existence. Again and again, it seems, new forms have arisen because genetic systems that have, like the German Federal and Democratic Republics, faced different challenges and evolved different ways of dealing with them are unhappy when brought together.

For species, as for nations, descent with modification builds obstacles, almost as an afterthought, as populations diverge. As selection goes on there emerge genetic principalities that are reluctant to trade with their former partners. Each contains its own set of genes that have never been tested except in their own surroundings. The longer two societies have been apart, the more they become fitted to local conditions and the greater the chance of a mismatch. The greater the divisions forced by time and circumstance, the less the chance of a new gene feeling at home in the company of others. The process feeds on itself and, soon, local entities evolve towards the full statehood, which, when achieved, stops all exchange between their citizens.

Fruit-flies are evidence of how past distaste helps to guard a species' virtue. For them, familiarity has bred contempt. How long ago two species of fly last shared an ancestor can be worked out from how different they have become in their DNA. Some pairs split not long ago, while others have lived their own lives for millions of generations. When tested in the laboratory, those with a recently established identity are more reluctant to mate with their relative if the two species live in the same place than if they have never experienced each other's company except in a bottle. For flies that live as neighbours natural selection against those who made sexual mistakes meant that mild reproductive failure soon ripened into real aversion. Only the blocks to mating itself – and not to development or to the fertility of hybrids – are involved. Those first barriers in the obstacle course were, it seems, reinforced because they were so often tested by hopeful males forcing their unwanted attentions on a related species.

The best place to track down the genes responsible for the origin of species is in fruit-flies. They are happy to mate in vials and breed in a couple of weeks. Their chromosomes can be seen down the microscope, and thousands of genes have been tracked down. Even better, the group has many species, some of which will cross in the laboratory. Their willingness to compromise their reproductive purity is a great help in understanding what keeps species apart.

The rules of genetics hint at the truth. Male flies have a single X chromosome, matched by a small Y. All genes on the X manifest their presence in males, whether or not they are recessive in females. The same is true in mammals, in which the male has an X and a separate Y chromosome and the female two Xs. In butterflies and birds, in contrast, the pattern is the other way round, as females are (in effect) XY and males XX. DNA structure shows that the actual chromosomes involved in the various groups have different origins. Even so, sex determination works in much the same way, but in reverse.

There is a consistent pattern when two species are crossed. If, among the offspring, one sex is absent or sterile, it is nearly always that with two different kinds of sex chromosomes — males in mice or fruit-flies, females in birds or butterflies. This pattern is known, after its discoverer, as Haldane's Rule. Because different sexes are affected in each case, the rule is not there only because males are more sensitive about sex, but must reflect some deeper genetic truth. In pairs which (to judge from genetic divergence) have been separate for just a short time, the first sign of any sexual barrier is that predicted by the Rule. It is a universal step on the way to a new identity.

Haldane's Rule gives more support to the idea that a gene useful at home can be harmful when mixed with others to whom it is a stranger. The wrong blend leads to disaster, most of all when genes of one of the partners are exposed in all their nakedness, as happens, in one sex, for genes on the X. The exposure of hidden genes to an unfamiliar set of companions is what counts. Fruit-flies can be manipulated to put foreign X chromosomes — one from each of two species — in a single female. She is fertile, although the lack of harmony of her alien X with her other chromosomes is no greater than that of a hybrid (and sterile) male. However, in the experimental female, the mismatched genes are masked, while in the males they are exposed to do their worst. Simple dominance hides most of the genes that might be dangerous out of context; but in male mammals (or female birds) their failure to get on with the neighbours is revealed.

Many related species of fly have been hybridized in the laboratory in the hope of chasing down the genes that keep them apart. The general picture is clear: many genes are involved, each with a small effect on its own, but each liable to a violent interaction when put into combination with genes from its relative – just as expected if species are well-suited communities of genes, easily disrupted by an intruder not used to their ways. Occasionally, hybrid sterility seems to need rather few changes in the DNA, as the mongrel flies can be saved with a single gene introduced into one of the parents. Quite how this snatches the progeny from the jaws of death or childlessness is not known, but the ability of one gene to do the job suggests that the block may be rather simple.

The two mice in Europe, *Mus musculus* and *Mus domesticus*, can also be persuaded to cross. Almost all laboratory stocks come from *domesticus* ancestors. Some bear sterile young when they are mated to wild *musculus*, but some are able to mate with it and to produce healthy offspring. Perhaps, as in flies, only a few genes are involved in the sexual barrier and, by chance, certain strains have inherited them while others have not. When it comes to a cross with a more distant species, a gene involved in early development finds it impossible to work in the wrong context. The mismatch means that sperm and egg are no longer made.

Although many speciation genes have been identified, almost none give any clue about how they work. Evolution, though, makes unlikely bedfellows and can give birth to unexpected offspring. Part of the truth about speciation comes from research into cancer.

Cancer is a genetic disease of body cells. It arises from a failure of the machinery that controls cell division. In the absence of brakes, the system runs wild. It is not a single illness but many, each the result of a change somewhere in a cascade of instructions that rules the life and death of cells. Cell division, like the United States Constitution, has many checks and balances. Some genes promote the process, others slow it down. They exchange signals about when and where to act. An equilibrium between suppressors and promoters allows the body to grow and to sustain itself, but if, for some reason, an enthusiast for division refuses to respond to a command to stop, the result is uncontrolled growth.

Two tropical fish, the swordtail and the platyfish, are each decorated with a series of attractive black spots. When put together in a tank, they can hybridize (although they never do so in the wild). In the progeny of a hybrid and its parent, the elegant spots turn into large and fatal black lumps called melanomas. The hybrids inherit from one parent a gene for a molecule promoting growth, and from the other a partner that keeps it in check. In hybrids, the second fails to suppress the first, too much of the growth molecule is made, and cancer sets in. The two genes work in harmony when they have long lived together, but are doomed to conflict when inherited in the wrong combination. The battle kills hybrids and keeps the species apart.

The genes for cell division did not evolve to separate different kinds of fish. Instead, they are an incidental to evolution within each one. The crucial genes have equivalents in humans and are of much interest to medicine (which is why so much is known about them). They also tell a tale about frontiers. Since swordtail and platyfish parted, each has diverged in the machinery of cell control. Two genes – each perfectly at home in its own milieu, but lethal in combination – are enough to keep them pure.

Plants work to rules rather different from those of animals. They are more ready to accept a foreign mate; and, quite often, the offspring of such a liaison find themselves with combinations of genes that fit so well together that the hybrids flourish, even at the expense of their parents. Hybridization becomes a fast track to a new existence, rather than (as in animals) a crack through which genes leak to dilute the prospects of a hopeful species. A glance at chromosomes shows that many higher plants – wheat and rice included – arose in this way, with complete sets of genes from two ancestors that, long ago, got together, found that they were able to work in harmony and flourished. Other plants arise by the crossing of two kinds which then reshuffle their genes to make a new one. A wild sunflower of the American West is, its genes show, the bastard progeny of two others. It looks somewhat different and lives in drier places, but is in every other way a mixture of its parents. The ancestral genes stay in large blocks in the daughter form, as it arose too fast for them to be broken up by sex before the new plant gained an identity. Crosses between the parents in a greenhouse can make plants resembling the new hybrid in a mere five generations. Plant species can, it seems, originate at some speed, with no need for the long probation involved in new forms of animal life. Their genes are more ready to cooperate than are those of animals.

Hybrids and Mongrels compared, independently of their fertility. Mongrels – crosses between mere varieties – are uncertain beasts, mixtures of the attributes of their parents. They attract little attention. Hybrids, the results of matings between distinct species, have always been special, with enigmatic opportunities ready to be explored by evolution. Long before the hybrid nature of certain plants was understood, Aristotle believed that a camel could mate with a panther to produce a giraffe, while Oppian, a more radical thinker, argued that to cross that animal with a sparrow made an ostrich. That is fantasy, but the progress from mongrel to hybrid and beyond now transcends the imagination of even botanists and Greeks. It throws new light on the nature of species themselves.

Take the tomato. The fruit was once shunned because it was thought to be poisonous, and the leaves are indeed filled with noxious stuff called solanine. The Latin name *Lycopersicon* means 'wolf-apple' and the plants were once grown for ornament alone (although Thomas Jefferson is known to have eaten one). The crop has now increased to eight million tons a year. The tomato seems a simple, not to say banal, item of diet. It is more than that, as artificial selection has made a splendid variety of sizes, shapes, colours and tastes. Farm tomatoes differ among each other more than do their wild relatives and, like farm animals, they tell a tale of how species arise.

The tomatoes (and there are many kinds) are natives of South America. They belong to a diverse and talented group of plants. Their relatives include the deadly nightshade and the mandrake (known for its aphrodisiac properties, its forked root of human form, and its lethal shriek when uprooted). Tobacco and the potato are not far away. Wild tomatoes are sexual, but will accept genes only from a separate individual of their own kind. The farm version is self-fertile, which places it behind a barrier. If its own pollen is there, why accept an outsider?

The readiness of the tomato and its relatives to hybridize has long been known. The great Kolreuter, whose accuracy has been confirmed by every subsequent observer, proved the remarkable fact, that one variety of the common tobacco is more fertile, when crossed with a widely distinct species, than are the other varieties. Crosses such as his have tracked down some of the genes that prevent sex in the tomato family. They show that the crucial shift to self-fertilization in the cultivated plant involves but a single gene. It also controls the number, shape and size of the flowers. As pollen success and flower form keep the plants apart, the genetic barriers between tomatoes may result from a simple interaction between this gene and others.

The dozens of kinds of wild tomato can be divided, on the basis of flower and fruit, into three large groups. Many flourish in places where their domestic relative would die. Botanists have travelled from Peru to the Galapagos and Chile in the search for new kinds able to resist their problems of wilt, blight, mildew, rot, spot, speck, aphids, nematodes, cold, drought and salt. On the farm, their genes would be useful indeed.

The United States Department of Agriculture has mounted an assault on the tomato and (as an incidental) on the laws of hybridism. In glasshouse or field, pollen and egg from wild tomatoes were tested for the ability to cross with cultivated plants. Almost none of the thousands of attempts succeeded. Most of the time, no plants emerged and in the few that did, the progeny were sterile. In a very few, there appeared a fertile hybrid. These rare individuals had, by chance, inherited an unusual mix from each parent that, instead of the conflict that killed most of their hybrid kin, had combinations of genes that were able to co-operate. A few of the genes, detached from their jealous fellows, are willing to work in an alien background. They are useful indeed.

Each fertile hybrid is a Checkpoint Charlie between sexual republics. Through it, by means of a cross with one of its parents, DNA can pass. Often the combinations fail, but dozens of genes for disease resistance, for fruit colour and for sugar content have now been moved from the wild on to the farm. Many of the new genes mix and reshuffle quite happily with those of their domestic host, to produce even better mixtures in later generations. Such limited free trade has already been a spectacular success. Selective breeding improves the yield of farm tomatoes by a few per cent a year, but wild genes can push it up by half. Even when a wild plant seems feeble in comparison with its improved relative, it may contain genes that work wonders in their new home. Wild tomatoes are green rather than red, but some have a powerful enzyme early in the pigment pathway that, once placed in conjunction with domestic genes further down the production line, turns the fruits a brilliant scarlet. Genes from different species, put together, transcend what either one can do alone; given the chance they do not conflict to produce sterility, but co-operate.

Botanical diplomacy has moved genes a long way. One group of wild tomatoes is so distinct from the domestic kind that it refuses to cross with it, however many attempts are made. It will, though, allow its genes to enter another wild tomato. The intermediary can be induced to mate with the farm variety. Its DNA moves down a chain of allies to a destination far from its native home. To an evolutionist, that is no great surprise. All tomatoes share an ancestor and, by retracing their pedigree, breeders do no more than reverse the history of their species.

Now, though, a real Common Market of biology has arrived. Frontiers can be penetrated not with sex, but with technology. The nature of mongrels, hybrids and even species has been called into question. To fuse cells of potatoes and tomatoes together in a bottle moves whole chromosomes at a single step. The bastard cells give a new plant: a tomato-potato, never seen before. Genetic engineers can do much more. With technical tricks – DNA attached to viruses or blasted into an alien home on tiny bullets made of gold – they can separate genes from their native background and move them from anywhere to anywhere else. Sometimes the deported DNA makes a protein, but more often – as evidence that genes can work with others from their own, but not from alien, populations – it does not. Even so, hybrids of the most unexpected kind are becoming commonplace. They may be sterile but, in this new and open world, who cares?

Tomatoes were pioneers in this libertarianism. Supermarket

tomatoes look good, but have no taste. Fruits ripened on the vine are tasty but soft and difficult to transport. The Calgene company altered the gene responsible for maturation to allow them to remain on the bush and to soak up sugar for a few crucial extra days. Now these Flavr-Savr fruits can be picked when they are red (rather than green) and taste as tomatoes are supposed to taste. In 1994, the United States Food and Drug Administration defined them to be as safe as tomatoes bred by conventional means, in spite of their acceptance of an alien piece of DNA. They did not even need a special label, as the Flavr-Savr 'maintains the essential characteristics of traditionally developed tomatoes'. But is that true? Is this new fruit, with its new genes, even the same species as before?

One of the persistent myths of the Middle Ages was that of the Lamb of Tartary, the Borametz. It was both an animal and a plant. The fruit of a tree rather like a melon burst open to reveal a little lamb within, with a fleece of surpassing whiteness. It was attached by its navel to a root in the earth, around which it ranged until it had eaten all the grass, whereupon it died. Erasmus Darwin himself hailed the creature that 'Eyes with mute tenderness her distant dam,/ And seems to bleat – a vegetable lamb.'

Not even in medieval Tartary did sheep mate with melons (and the Borametz was probably the 'wool-bearing tree' of Plato, now known as cotton). Consider, though, today's tomato and its many relatives. Genetic engineers have, they claim, inserted into them not just genes for ripening, but one for an antifreeze, obtained from an Antarctic fish. The plant should delight the most ardent free-marketeer, as it marks the removal of what seemed an impenetrable barrier to trade. Oscar Wilde was fond of green carnations, but he had to depend on dye to achieve the right effect. Now the flower comes in a blue version, made by putting a petunia gene into that normally pink, red or yellow plant. Other plants now contain genes from bacteria that give them resistance to insects, and scorpion toxins have been put into viruses to kill caterpillars. There may not yet be vegetable lambs, but scientists have bred whole flocks of sheep bearing human genes. Technology has broken the sex barrier; the Borametz is no longer the fantasy it was, and the distinction between mongrels and hybrids has gone for ever. The engineers have globalized the genes and have moved DNA to places it could once never have reached.

However, biotechnology, with its twenty-first-century powers, has an eighteenth-century view of what species are. So confident are the technicians of the safety of their products that each one is seen as no more than an arbitrary mix of independent lengths of DNA. A protein that has been proved to be harmless to man – an antifreeze, even – moved to a foreign plant known to be nutritious does not even have to be tested in its new home: the new plant is, say the authorities, 'substantially equivalent' to the old. Their view takes no account of the notion of species as interacting groups of genes, the properties of one – as shown so starkly in the cancers of hybrid fish – depending upon the others with which it is placed. With even greater disregard for Darwinism, those who market unnaturally tasty tomatoes or herbicide-resistant crops, claim their new commodity to be distinct, stable and free of exchange with others, so that genes moved in can never leak out. Public concerns (and they are real, most of all in Germany where genetics has, for good historical reasons, a sinister image) are, they say, unreasonable: their modified plants are safe.

Genetic engineers deny the central facts of evolution: that the action of a gene can depend on the species in which it finds itself and that all species were once varieties (which means that the boundaries between many can still be breached). Already, genes for herbicide resistance put into oilseed rape have seeped into wild mustards and radishes. Many other candidates for gene manipulation – beets and rice included – are certain to mate with their undomestic relatives as soon as they meet them and to pass alien genes to unexpected places.

The experience of the tomato suggests that industry's optimism about the nature and stability of species is unfounded. Those who cast down the barriers to hybridism will soon be reminded of what evolution can do.

Summary of Chapter. First crosses between forms sufficiently distinct to be ranked as species, and their hybrids, are very generally, but not universally, sterile. The sterility is of all degrees, and is often so slight that the two most careful experimentalists who have ever lived, have come to diametrically opposite conclusions in ranking forms by this test. The sterility is innately variable in individuals of the same species, and is eminently susceptible of favourable and unfavourable conditions. The degree of sterility does not strictly follow systematic affinity, but is governed by several curious and complex laws. It is generally different, and sometimes widely different, in reciprocal crosses between the same two species. It is not always equal in degree in a first cross and in the hybrid produced from this cross.

In the same manner as in grafting trees, the capacity of one species or variety to take on another, is incidental on generally unknown differences in their vegetative systems, so in crossing, the greater or less facility of one species to unite with another, is incidental on unknown differences in their reproductive systems. There is no more reason to think that species have been specially endowed with various degrees of sterility to prevent them crossing and blending in nature, than to think that trees have been specially endowed with various and somewhat analogous degrees of difficulty in being grafted together in order to prevent them becoming inarched in our forests.

The sterility of first crosses between pure species, which have their reproductive systems perfect, seems to depend on several circumstances; in some cases largely on the early death of the embryo. The sterility of hybrids, which have their reproductive systems imperfect, and which have had this system and their whole organisation disturbed by being compounded of two distinct species, seems closely allied to that sterility which so frequently affects pure species, when their natural conditions of life have been disturbed. This view is supported by a parallelism of another kind; – namely, that the crossing of forms only slightly different is favourable to the vigour and fertility of their offspring; and that slight changes in the conditions of life are apparently favourable to the vigour and fertility of all organic beings. It is not surprising that the degree of difficulty in uniting two species, and the degree of sterility of their hybrid-offspring should generally correspond, though due to distinct causes; for both depend on the amount of difference of some kind between the species which are crossed. Nor is it surprising that the facility of effecting a first cross, the fertility of the hybrids produced, and the capacity of being grafted together – though this latter capacity evidently depends on widely different circumstances – should all run, to a certain extent, parallel with the systematic affinity of the forms which are subjected to experiment; for systematic affinity attempts to express all kinds of resemblance between all species.

First crosses between forms known to be varieties, or sufficiently alike to be considered as varieties, and their mongrel offspring, are very generally, but not quite universally, fertile. Nor is this nearly general and perfect fertility surprising, when we remember how liable we are to argue in a circle with respect to varieties in a state of nature; and when we remember that the greater number of varieties have been produced under domestication by the selection of mere external differences, and not of differences in the reproductive system. In all other respects, excluding fertility, there is a close general resemblance between hybrids and mongrels. Finally, then, the facts briefly given in this chapter do not seem to me opposed to, but even rather to support the view, that there is no fundamental distinction between species and varieties.

CHAPTER IX

ON THE IMPERFECTION OF THE GEOLOGICAL RECORD

On the absence of intermediates, today and as fossils — On decay — On the vast lapse of time, as inferred from deposition and denudation, and in other ways — On the power of water — On means of preservation — On the sudden appearance of whole groups of species — On gaps in the record as revealed by genes — On the Cambrian explosion and before — On the lost history of ancient times — On the fate of the dead

A MURDER VICTIM becomes an ecosystem in his own right. Within hours after a body is dumped in an English woodland, blow-flies lay eggs around its eyes and mouth. Soon, they are joined by flesh-flies that drop larvae on to the skin. As maggots burrow in, gut bacteria work their way out. The intestine bursts and its contents attack other organs. The cadaver liquefies. Soon it begins to ferment, with a strong smell of cheese. This, in turn, attracts carnivorous beetles and a fly called the cheese-skipper (named from the larva's odd habit of jumping into the air when disturbed) joins the fray. Three months after death, five hundred kinds of insect may feast upon the corpse.

The sordid details depend on the weather and where the body is hidden. They are of interest to the police, as they make it possible to work out the date of death. If blow-fly larvae are still around, then the murder was within the past week. Cheese-skippers are evidence of an older crime.

Forensic entomology has solved many murders. A thirteenth-century Chinese work, *The Washing Away of Wrongs*, describes how flies settled on a certain sickle and proved it to be the weapon. Now, science has joined the study of decay. At the Anthropological Research Facility in Tennessee, cadavers are put in wire coffins above a concrete slab, or buried at different depths, to relive the history of death. Corruption has five stages – fresh, bloat, active rot, post-decay and dry remains. An infant is reduced to a skeleton in a mere six summer days, while an adult takes three weeks. The bones themselves can last much longer but, a few years later, the chances are that not many will be left. The gravedigger in *Hamlet* has it about right: 'How long will a man lie i'th'earth ere he rot? . . . He will last you some eight or nine year'. For murder victims the return to dust is swift. The same is true for all corpses, human or not. As a result, most of what evolution has made has mouldered away.

Death, decay and dissolution all help to solve its greatest puzzle. The fossil record – in defiance of Darwin's whole idea of gradual change – often makes great leaps from one form to the next. Far from the display of intermediates to be expected from slow advance through natural selection, many species appear without warning, persist in fixed form, and disappear, leaving no descendants. Geology assuredly does not reveal any finely graduated organic chain, and this is the most obvious and gravest objection which can be urged against the theory of evolution.

The unexpected chronicle of ancient life is not proof that the idea of slow change is of its nature flawed. Much of the reason lies in the record itself. Few bones rest in peace and most of those who die decay almost at once. Even the few that are preserved are soon washed away. As a result, great periods leave no token and dramatic events stay unrecorded. The archive of the rocks is a series of snap-shots, taken at long intervals with a badly focused camera. As in a Victorian photograph album, the past appears as a series of apparently unrelated images. Worse, these are themselves faded, torn, stained, lost and muddled by the passage of time. Even a figure as prominent – and long-lived – as Charles Darwin himself has left, a century or so after his passing, a mere dozen or so likenesses. The lives of most of his contemporaries are quite unrecorded.

This dismal picture has a few exceptions. Geologists have unearthed whole pages of the past. Everyone knows about the dinosaurs of the American West and the fossil bird *Archaeopteryx*, in which even the details of feathers are retained. Two kinds of elephant exist today, but a hundred and sixty-five species of their extinct relatives are preserved, many as complete skeletons, to make a seamless set of links between past and present. The remains of our own ancestors – unknown at the time of *The Origin* – are firm proof that we descend from apes. Any of those bones is conclusive evidence of evolution and is in itself enough to demolish the creationist case. The smugness shown by students of living creatures when they decry the gaps in our knowledge of those long gone also skates over our ignorance of the modern world, with, some say, only one species in a hundred yet described.

However, so persuasive is the *fait accompli* revealed by the relics of prehistoric beings that it is easy to forget how little the record can say about the machinery – rather than the pattern – of prehistoric change. That leads to noisy (and much-publicized) disputes between students of the ancient world and that of today. Most arise because each party asks different questions of the past and is satisfied with different kinds of answers. Not much is left of the great tree of life which fills with its dead and broken branches the crust of the Earth. Today's creatures are its twigs; and so few remnants of its limbs have been preserved that to work out what controlled their growth when they were themselves still young may be impossible.

Fossils have long been accepted as messages from history. The ancients saw shark teeth as the petrified tongues of snakes fallen from the sky, and amber as congealed wildcat urine. The curled fossils called ammonites (named after Ammon, an Egyptian god with ram's horns) were thought to be snakes decapitated by St Hilda and turned to stone. They were sold in Whitby with carved heads attached.

Ammonites (relatives of today's squids and nautiluses) are not snakes turned to stone, but those who found them could make no other tie to the animals of their own day. The traders of Whitby recognized that to link the past with the present the details must be filled in. That is still true. The number of intermediate and transitional links, between all living and extinct species, must have been inconceivably great. Almost none are found (although, as in the elephants, or in the ancient snakes with legs, those silent few are eloquent indeed). Where have they all gone?

Their fate is a matter of time and chance. In the Old Testament, the Song of Deborah tells that 'the mountains melted from before the Lord'. In Biblical days her vivid image was a statement that God is indestructible and that to Him even the landscape does not last. On evolution's span, the ground is a turbulent sea and the dead are subjects of its uneasy movements. Their remains disappear at once, or are borne far from where they fell. They may sink below the surface to rise again, and are churned up with the relics of animals who expired far away and long ago. Most of the evidence has gone for ever. Those who hope to use the fragments of the past as the key to the workings of the present will, as a result, almost always fail.

On the lapse of Time. The passage of time, so great as to be quite inappreciable by the human intellect, disturbs all graves. Time once seemed a domestic thing. Deborah lived in the twelfth century before Christ, quite soon after what the writers of the Old Testament saw as the start of the universe. The passing of the ages could be measured out in just one way, in lifetimes. To count the scriptural generations (as listed in the margin of many Bibles, Darwin's included) would date the Earth. Archbishop Ussher's estimate for its origin was, famously, 4004 BC (which means that, for real believers, the sixth millennium passed in 1997). When John Burgon described Petra as a city 'half as old as time', he meant it.

Even those who dismissed such notions could not imagine how far they would have to go in search of yesterday. The great French naturalist Georges Cuvier noted that the five-thousand-year-old ibis mummies brought back by Napoleon from Egypt were identical to their fellows alive today. That, he felt, showed that animals could not change, as five millennia seemed to him to encompass most of the earth's history.

Copernicus and Galileo put an end to the idea that space could be measured in human terms. The geologist James Hutton did the same for time. In 1788, on a visit to Siccar Point near Edinburgh, he realized that the rocks were far more ancient than anything contemplated by theology. A layer of sandstone rested on the ground-down and tilted edge of a still older bed of slate and grit – which must have been deposited and worn away long before. 'The mind', he wrote, 'began to grow giddy by looking so far into the abyss of time'. The earth, it became evident, was old: it had 'no vestige of a beginning – no prospect of an end'. Since his day, geology has made it clear how incomprehensibly vast has been the span of the years.

The notion of ancient landscapes shaped by unhurried change was at first denied. Henry de la Bèche, founder of the Geological Survey of Great Britain, circulated a sketch called 'Cause and Effect' in which a baby urinates at the head of a gorge and the products of his labour flow down the middle. His nurse comments: 'Bless the baby! What a walley he have a-made!' The belief that species were immutable productions was almost unavoidable as long as the history of the world was thought to be of short duration. Only if there has been enough time for slow change can Nature mould its products into novel forms by gradual selection, without the need to make leaps.

How old are the Earth's dead inhabitants? Because their rest has been so much disturbed, it is difficult even to sort out the sequence in which they perished, let alone the date of their demise.

The first job is to put the layers in order. The task seems easy: those on top are young, and those buried deep are the oldest. However, our planet is stoked by radioactive fires and from the start has suffered from indigestion. What were its guts are on its surface and what was once its skin has been driven into its depths. Because the Earth was built from within, its rocks were lifted or buried, its beds turned over and their order reversed. The Lizard Peninsula, in the south of Cornwall, is a piece of its molten interior squirted to the surface. It has no match anywhere in Europe. Even worse, modern material is often worn away to show the ancient strata below. The first geological map of Britain, made in 1815, managed to trace only one thin seam of limestone called the Cornbrash across southern England. Now, after years of exploration, the logic of the whole world has been laid bare. Layers of material can be linked with others far away, to rebuild the ancient Earth.

A glance at any geological map shows how much has been lost. Certain strata can be followed for thousands of miles. In a few places, the beds are almost as thick as when they first settled, but in others they have been worn thin or have gone altogether. If the maximum depth of each separate layer across the globe is added together, the

Earth should be cloaked in sedimentary rock a hundred miles thick. Most of its coat is a thousand times thinner because a history written in sand has been torn to pieces, its fossils with it.

Time can be measured only by change, in the flesh of a decaying corpse, in the tension in a watch-spring, or in the landscape itself. The first attempt to age our planet (rather than just to order the pages of its biography) compared the saltiness of the sea with that of the rivers feeding it. If the sea was once fresh, and its salt came from the land, it should be possible to work out how long the ocean had taken to become saline. The estimate – of a hundred million years – was a death-blow to those who measured time in human terms. It was the first hint of the unimagined antiquity of life.

It was far too low, because salt is laid down in great deposits on the sea bed and is returned to the land as it rises. We live not a hundred million but four thousand five hundred million years since the Earth was spun from dust and rock around the sun. The evidence comes not from our own planet, but from its satellite. The Moon flew off its parent after a giant impact. Because it stayed small, cold and undisturbed it gives a better picture of the past than does its parent. A quick trip by the Apollo XI mission was enough to date it. The Earth's turmoil makes it harder to trace its own origin. Its oldest rocks, found in Greenland and Western Australia, are just under four billion years old. The most ancient in Britain, on the Isle of Lewis, are a billion years younger.

For living things, the answer to time lies not with deposition, but with decay, and not of murder victims, but of atoms. Radioactive materials throw off parts of their structure and, as they do, change into other forms of matter. How long it takes is measured by the 'half-life' – the period needed to convert half the original into its derivative. This varies from the four and a half billion years needed to change uranium-238 (the radioactive version of the element) to lead, to the twelve years or so for the shift of tritium (a form of hydrogen) to helium.

Some chemicals are useful in dating the past. Potassium has a half-life of rather more than a billion years as it breaks down to argon. A comparison of daughter with parent says when the decay began. As the element breaks down, the inactive gas produced shows when the rocks were made (or at least when they were last melted and the gas driven off). Carbon-14 – a radioactive form of the atom upon which life is based – has, in contrast, a half-life of a mere five thousand seven hundred and thirty years. It is made when nitrogen is bombarded by cosmic rays, high in the atmosphere. Plants and animals all pick up traces of carbon-14, which begin at once to decay. When the animals themselves die, the rare form of the element is no longer replaced and the proportion left is a key to the time of their demise.

To explore the past it helps to have benchmarks. Volcanoes help because their ash blankets the Earth to give layers that can be read as a series of pages upon which history is written. The explosion of Mount St Helens in 1980 generated so much debris that the Columbia River was blocked and ships were trapped upstream. The cloud of particles circled the Earth. Fossils are mixed into the ash itself, or are in the layers of rock between each band. Some of the first evidence of our own ancestors is in a line of two-million-year-old footprints left as two upright primates strolled across an African plain coated with soft volcanic dust, stopped, and looked to the east. Lake-beds, too, are blank sheets upon which the past is recorded. Each spring, they accumulate a layer of silt. The Green River beds of Wyoming contain twenty million of these annual bands. They allow its fossils to be ordered with great accuracy.

Such evidence has aged the Earth. Its many layers have been named (some – the Cambrian, the Ordovician, and the Silurian – after Wales and its ancient tribes) and dated. Eighty per cent of history is in the Pre-Cambrian, an era once thought to be barren. The sixth or so of the past into which most fossils are crowded itself represents a lapse of time impossible to imagine in human terms.

A scheme is under way to build a clock – the Clock of the Long Now – in a remote desert, driven by a twisting pendulum powered by the expansion and contraction of a metal bar as night follows day. The hands will move once every year, it will chime once a century, and every thousand years a cuckoo will pop out. To geology, that clock is a mere stopwatch, and the Long Now but an instant in the narrative of the earth.

On the poorness of our Palaeontological collections. The Natural History Museum in London owns six million fossils of marine snails. It seems an impressive collection, but is not. A square yard of sea-floor can accumulate a hundred of those unpretentious animals a year, and within any geological period – some tens of millions of years – could make enough shells to swamp all the world's museums. The crust of the Earth is itself a gigantic depository; but its collections have been made only at intervals of time immensely remote. They are a reminder of how many of the products of evolution have been lost. Any single place retains mere fragments of the past, either a large part of a small time or, more often, a few glimpses of a much longer period.

The greatest destructive force is water. American builders shift eight billion tons of earth a year. The rain does much more. It consumes the countryside and planes great mountains smooth. Water destroys, but it also builds. As it does, it edits the record of the past.

At the time of *The Origin* the California Gold Rush was in full swing. In its first days, miners were content to use technology not very different from that of the Greeks and the Golden Fleece to trap small particles of gold from rivers; not with animal skins suspended in the stream to filter out the metal but by running a trickle of water over the gravel in a pan. A good miner could work through a cubic yard of pebbles a day and become modestly rich from his efforts.

In 1853, Edward E. Matteson had the idea of using high-pressure water from the mountains to wash out the gold. It was a spectacular success. Two men could break up fifteen hundred tons of ore a day. The *Nevada City Transcript* declared that the miners had 'broken into the innermost caves of the gnomes, snatched their imprisoned treasures, and poured them, in golden showers, into the lap of civilized humanity'. Matteson's nozzles, sixteen feet long, were called the Dictator, the Monitor or the Giant. In five years the miners built five thousand miles of ditches and flumes to feed them. By the end of the boom (when Matteson died bankrupt), twenty-five million ounces of gold had been hosed from the gravels of California.

As the dirt ran downstream, it covered thousands of acres of farmland and blocked San Francisco Bay. Within five years, a century's worth of sediment was washed from the mountains and laid down in the plains. Even today, mile-wide gullies in the Sierra

Nevada are monuments to a single decade in which man used a tiny part of Nature's energies. Now, the miners have moved to Alaska, and geologists comb through the detritus blasted from the rocks by today's Monitors and Giants in a search for the bones of extinct mammals.

The land is also demolished as the sea pounds it. The Japanese are desperate to save the remains of their distant and decayed island of Okinotorishima. At high water it consists of two lumps of coral, each fifteen feet across. Those tiny fragments allow Japan to claim a coastal economic zone of a hundred and fifty thousand square miles, with all its fish and minerals. They have spent millions of dollars to build a steel-reinforced concrete wall around the rocks (not in touch with them, as it would then be classed as artificial and could not qualify as a territorial claim). The force of the waves means that the task is hopeless. Soon, Japan will lose another great swathe of what was once its Empire; and another page of fossil history will be lost from view.

Water shows its real force when it turns to ice. The California miners soon learned to pour water into cracks high in the mountains in daytime and to let it freeze at night. The stress on the rock as it expanded was that of a sledgehammer swung with the force of ten men. Great blocks of ore could be shattered with a bucket dipped into a stream.

The rain can melt rock as well as smash it. The granite columns of Egyptian temples lie in the desert, their inscriptions almost as sharp as when they were cut. Roll a column over and it is blank. Damp in the soil has reacted with the stone, turned it to clay, and wiped off its message. In 1879, two obelisks that had stood for three thousand years in the ruins of Thebes were shipped to New York and to London to allow the public to admire their hieroglyphs. A century later, the New York stone is unadorned (although its London twin, Cleopatra's Needle, has survived rather better).

The Pyramids were time machines, designed to project the ego of a Pharaoh into the future. They are a better monument to how fragile even stone can be. Their place in geological history is secure, as Herodotus identified the ancient shells found in the rock from which they are made as the remains of lentils discarded by their builders (Voltaire, in turn, thought that fossil fish in the Alps were the remnants of pilgrims' packed lunches). Their structure is less safe. The earliest tombs, those at Memphis, were built of brick and have long gone. They were succeeded by the Pyramids themselves, the first built about four and a half thousand years ago. Until the time of William the Conqueror, many were covered with polished limestone that shed the rain as it fell. Their covers were removed to build mosques in Cairo, and water began to soak into their structure. Many are now mere stumps, covered by the debris of their own destruction. The Great Pyramid has lost sixty thousand cubic yards of material – and its innumerable fossils – since its shield was removed.

The Pyramids are young compared to the hills from which they were quarried. Those seem almost timeless; but that, too, is an illusion. Most mountains arise as the plates of the Earth's crust strike each other. The Himalayas began a mere fifty million years ago – after the death of the dinosaurs – as India crashed, at a foot a year, into Asia. Mountains determine their own fate because they make their own climate. Because it faces the wind and the rain, the Indian slope of the Himalayas is gashed by huge valleys like the Khumbu Ice Fall, up which the first climbers struggled, while Tibet – half the size of the United States and most of it higher than America's highest point – is dry, flat and cold. With no rain, it has lost less of its substance.

Mountains float, like icebergs, upon the mantle of the earth, with most of their mass miles below the surface. They are buoyed up in a sea of lighter material, their summits renewed from below. This makes them older but less sturdy than they seem. As glaciers grind millions of tons from their path, the peaks thrust even higher. The Appalachians are a foot taller than when first seen by Europeans because so much soil has been lost from the valleys; the mountains are more or less unchanged in shape, but their material has been washed away and replaced from within.

In the end, the rain always wins. Great mountains are popular, but rare. The mean height of the land across the world is two and a half thousand feet (the height of the Brecon Beacons) and most of it is flat. The turmoil of the peaks means that those who die there will soon be gone. Australia is the most exhausted country of all, its low hills in the last stage of decay. Its fossils include those of alpine plants, remnants of a more elevated past. The rest of Antipodean history has been swept into the sea. Sooner or later, the Himalayas, too, will be gone. Their inhabitants will follow.

The remnants move downstream to the ocean. A slow river carries more earth than it does water. Every stream is an engineer. Its silt settles into sediment and forms the rocks that cover three-quarters of the globe. In their turn, these are disturbed by new downpours. As it falls, the mud builds new soil. It was explained to Darwin in South America that the bones of giant sloths deep below the surface proved how, in an earlier age, they lived in holes. The idea is seductive, but the truth is simpler. Given long enough, mountains – and plains – flow, to use the Biblical metaphor, like rivers. They take their contents, vandalize them, and bury them in the refuse. That is stirred up and reburied as gain follows loss. As the remains of a shattered landscape pour downhill, the earth's surface is reshuffled.

The Aswan Dam was completed in 1970, and Lake Nasser began to fill behind it. The Nile slowed down and stopped. As it did, it dumped its load of sand, silt and corpses on Nasser's floor. Below the dam, the pure water ran faster than before and picked up more silt. It scoured the stream bed, with the loss of five hundred bridges and of many miles of the Delta. To build the dam disturbed the balance of deposit and withdrawal in the mud bank.

Cairo sits on a huge rift – bigger than the Grand Canyon – which was cut and then filled with sediment by an ancient Nile. The Grand Canyon itself began at about the same time, as the Colorado River sawed through the Earth's crust. Around it the land was uplifted to give the mile-high cliffs of today as the river sawed through their soft rocks. Like those of the Nile, the Colorado's waters are valuable. Thirty years ago, the Glen Canyon Dam domesticated the river to feed Los Angeles. Its lake, like all others, has begun to fill with mud. Below it, where once were sandbanks and rapids, are thickets of mesquite and a channel clogged with boulders. Concerned by the changes, ecologists opened the dam for a week to feign a flood. Millions of gallons of water poured out and the river rose by twenty feet. The effects were dramatic. Huge boulders shifted for hundreds of yards as fans of sand were destroyed and re-formed as fresh beaches. The Canyon was born again after decades of decline; and now the Sierra Club is calling for its two-hundred-mile-long lake to be drained to allow nature a permanent return.

Among the debris were thousands of bones, any of which might in time have become a fossil. They were churned by the torrent into a soup, to settle as a new and arbitrary mix. A week of flood muddled the evidence of thirty years. In the Grand Canyon's five-million-year history, such deluges must have happened uncounted times.

Science can be magnificent in its simplicity. How better to test the movements of the uneasy earth than to simulate the fate of a giant sloth, dead fifty thousand years? Take the bones of slaughtered cows, mark, and scatter around the landscape. Leave for a decade or so, and return to see how many remain, where they might be and what company they keep. The experiment was carried out around the famous fossil sites of the East Fork River in Wyoming. Its results were clear. Most of the bones have gone for ever. Many of the survivors have moved and are buried deep in sediment. The remains of cows that expired in the 1970s mingle with those of animals last seen on Earth ten thousand years before. A geologist yet unborn would have a strange image of today's America, with herds of cattle pastured among the mammoths.

Sometimes, life adds to the confusion of the landscape. In a few places the shells of sea-snails are mixed with the bones of land mammals and birds. The snails were picked up by ancient seabirds who dived for food and left the fragments around their nests. Even the famous human fossil, the Taung Child, is so mixed with the bones of other animals that its remains may have been taken by a bird of prey to feed its young.

The turmoil of the rocks means that fossils are not laid down in neat sequence in an ordered world. Instead, they fall into a universe of change. The passenger pigeon flourished in North America as late as the Civil War. At the time of the *Mayflower*, nine billion were alive – more than all the birds of America today. A single flock was said to have been a mile wide and two hundred long. So stupid was the bird and so ruthless its pursuers that the last, a female called Martha, died in Cincinnati Zoo in 1914. Nobody has ever seen the fossilized bones of a passenger pigeon. Without a written record we would never know of its existence. Untold numbers of other beings have accompanied it into oblivion. Like the products of past labour, too much has been made for it all to be preserved.

The hope of survival for any corpse depends on what its owner was, and how and where it died. Some places give the dead a chance but others are less kind. Almost all fossils are mere replicas of animals and not their flesh. Their bones or shells are transformed after death as they become saturated by minerals and turn to stone. The image, not the real thing, is saved, and must be petrified at some speed if it is to have a chance. In Victorian times stone bowler hats were a popular souvenir of a visit to a cave. All that is needed is to drip limestone-rich water on to a hat for a few months and its shape is preserved. Already, half a century after it went out of fashion, almost as many survivors of that strange headwear are preserved in stone as in felt. A thousand years hence there will be many more.

An animal stifled in mud will last better than its cousin destroyed by fire, and a cadaver has more hope in a chalky sea than an acid marsh. One of the oldest of all fossil beds, the Burgess Shale (first found in the Canadian Rockies), was buried half a billion years ago. The animals were infiltrated by soft clay, which made perfect images of parts of their bodies. Whole communities were embalmed in a fall of mud on to the sea bottom after a flood. *Archaeopteryx*, in contrast, died on land. It was swept into an anoxic sea, where it was conserved in exquisite detail, along with many of the plants and animals living alongside it.

Sometimes a triumph of the frail allows the animal itself to survive. Human remains in peat bogs retain the hair, clothes and pained expression of their original owners. As the flesh was tanned by acids in the soil, the bones were dissolved to leave mere bags of skin. Animals can topple into a lake of pitch such as the La Brea Tar Pits of Los Angeles or stick to tree-gum that turns into amber. All the household means against decay – deep-frozen mammoths, Spanish woolly rhinos marinated in a salty swamp, ground sloths in South American caves reduced to beef jerky by slow evaporation – help to conserve these Mona Lisas of the fossil record.

Some animals have more hope of immortality than do others. A crab has a better chance of a monument than does the sea-anemone that grows upon it. Worms and jellyfish rot at once, while snails, corals and the like have better prospects. Two-thirds of marine animals today – and throughout history – have soft bodies. Almost none are known as fossils. A solitary ancient specimen of krill (the food of great whales) has been found, although some shoals contain a hundred million tons of the tiny creatures, the weight of the world's annual fish catch. The survivor had been eaten by a fish and fossilized within it.

In Klagenfurt, in Austria, is a stone fountain carved in 1590. It depicts a beast mythical in that it has wings and breathes fire. The head looks strangely like that of the ancient European rhinoceros. The sculptor, it is said, based it on fossils found around the town. The Klagenfurt fountain is one of many reconstructions of the past in which artistic imagination takes precedence over scientific fact. The Burgess Shale remains were so decayed and scattered that single animals were at first identified as several. One made a simultaneous entry into science as a jellyfish, a sea-squirt and a beast rather like a crab. Not until a complete specimen was found was it recognized as a being quite unlike any alive today and named as *Anomalocaris*.

The La Brea Tar Pits or the German quarry where the remains of *Archaeopteryx* were found are great museums of ancestors. Both places, like the Louvre or the Hermitage, hold but a few moments of evolutionary effort. Just seven skeletons of *Archaeopteryx* (plus an isolated feather) have ever been found. As in a museum, the accidents of preservation give a biased view of the past, with too many gold ornaments and too few clay pots. The accuracy of the collection can be checked against those alive today. How many of each modern kind are found as fossils?

Sometimes, the assembly of the dead is not much different from that of those who survived. In California, where the sea floor has risen to make terraces along the coast, two-thirds of the marine snails on the modern shoreline are in the fossil beds – which give a reasonable image of how things used to be. In most places, though, the dead are not an accurate sample of the living. Three-quarters of the large mammals in the savannahs of East Africa leave their bones for at least a year, but smaller beasts disappear almost at once. Although living shrews are far commoner than their giant neighbours, the fossils do not reflect reality; the bones of elephants scattered around the African landscape are over-represented by five hundred times compared to those of shrews. Museums are short of fossils not because their curators are lazy but because they have set themselves such a difficult task. What they have is a glimpse of the unexpected worlds of the past; and too often one so fragmented that it says little about what those ancient universes were really like.

On the sudden appearance of whole groups of Allied Species. The abrupt manner in which whole groups suddenly appear in certain formations has been urged as a fatal objection to the belief in the transmutation of species. If numerous species, belonging to the same genera or families, did start into life all at once, the fact would be a severe blow to the theory of descent with slow modification.

Students of ancient barnacles, bony fish, mammals and birds have each, at one time or another, proclaimed their favourite animal's abrupt arrival in the rocks as evidence that gradual change is not enough to account for the first appearance. All have been proved wrong by finds of forms much older than the supposed pioneers. Now, genes – descending, as they do, from a past more ancient than any fossil – can help fill gaps in the record. Quite often, they show how a sudden appearance may hide years of obscurity in the wings. We continually overrate the perfection of the geological record and falsely infer, because certain genera or families have not been found beneath a certain stage, that they did not exist before that stage.

Archaeopteryx was discovered just two years after the publication of *The Origin*. As the first 'missing link' it caused a sensation and was at once seized upon as proof that birds must have arisen in a single step. Indeed, its descendants, today's birds, appear as a well-established and diverse group just after the demise of the dinosaurs. However, it is not as unique as it seems. New finds show where *Archaeopteryx* came from and where it went on the unhurried and inconspicuous path to modern birds. Its direct ancestor was a small meat-eater which, like *Velociraptor* (famous for its role in the film *Jurassic Park*), ran upright on two legs. Until the late 1990s, the record between those land-bound animals and *Archaeopteryx* itself consisted of an enormous gap. In 1997 an intermediate was found in Patagonia. It cemented the tie between birds and dinosaurs. *Unenlagia* ('half-bird', in a local language) lived ninety million years ago, long after its famous relative. As its shoulder joint pointed not downwards but to the side, *Unenlagia*, albeit earthbound for most of the time and without feathers, could move its forelimb with a full upstroke.

Soon after the discovery of *Unenlagia*, a turkey-sized, carnivorous (and emphatically flightless) dinosaur with short arms was found. It came from twenty million years after *Archaeopteryx* and had un-mistakeable feathers, barbs and all, fanning from its tail. In the absence of wings, they could not be used for flight, and may have been courtship displays or even insulation for the animal's rear end. The tail was short and fused to the pelvis, a feature that would later be useful in the air. In almost every other respect the animal is a dinosaur. Birds can no longer claim sole ownership even of feathers, let alone of an identity that sprang fully formed into the fossil record.

There still remained a great gap before the modern birds. The divers have long been suspected from their behaviour and their anatomy to be primitive; in addition, they turn up as fossils earlier than do their fellows. Most birds of modern form appear somewhat later than the divers, at almost the same time, some sixty-five million years ago – a hundred million years after *Archaeopteryx*. Cormorants, owls and geese turn up first, to be followed by the first penguins, the earliest parrots, and then by the preserved remains of a whole host of feathered beings. In spite of their noble ancestry, modern birds seem to have been sudden arrivals on the fossil scene. A few hints of a long-lost avian past remain. The sandy shores of ancient seas a hundred million years ago echoed to the cries of the waders who left only tracks as a memorial. Those birds must have looked much like modern oystercatchers or redshanks, forty million years before the first fossil of any wading bird.

The lost birds that walked the sand left biological footprints in the genes of their descendants. They can be used to track their affinities today and to estimate when they split. The new avian pedigree, based on a set of slowly evolving genes associated with cell division, shows how much has been lost. Penguins and albatrosses are close, and fall into a group with the shearwaters and the ancient divers. This puts the joint ancestor of the four at a time earlier than the first fossil divers, and long before the demise of the dinosaurs. The rhea, the ostrich and the moa (together with the domestic chicken) are so genetically distinct from other birds that their branch of the avian family must have flourished even earlier. Fossil-hunters, sceptical about such a gap, were persuaded by the discovery of a dead parrot in Wyoming from a time when dinosaurs flourished. Birds of familiar form may, it seems, go back a hundred million years before the first bird fossil of modern appearance.

They are animals of the southern hemisphere, for most of their ancestors are found on the remnants of the great broken continent of Gondwana (the progenitor of Australia, Antarctica and South America). The oldest fossils of ostriches, parrots, pigeons, songbirds, divers and penguins are all on that land mass. In the still scarcely explored southern part of the world there were many plumed and aerial descendants of dinosaurs, with diverse habits, flitting around their giant and earthbound cousins. Our knowledge of their fossils, and of those of most other creatures, is, it seems, so sketchy that it is as rash to dogmatize on the succession of organic beings throughout the world as it would be for a naturalist to land for five minutes on some one barren point in Australia and then to discuss the number and range of its productions.

On the sudden appearance of groups of Allied Species in the lowest known fossiliferous strata. For geology, life came in with a bang. The oldest fossils known in 1859 – and until the 1940s – were from the Cambrian system of rocks, which began some five hundred and forty-five million years ago. Most of the major divisions of today's animal kingdom – the phyla – appear just after the start of the Cambrian. Before then, nothing is to be found. If evolution is gradual, how could so many forms spring all at once into existence?

The nineteenth century, with its dislike of drama, appealed to a lost and ancient world in which these groups had evolved, but left no remnants. There was little evidence from before that time, and what life there was seemed quite unlike that of today. Stromatolites are limestone mounds. Some are huge – a hundred yards across. By two billion years ago, they made great reefs. Then, they disappeared, with no obvious descendants. Now, a few of their much-reduced builders have been found alive on the shores of Western Australia; each stromatolite not an individual, but an ecosystem a few inches tall, built by microbes. Not much else seems to remain from before the Cambrian and most modern groups can trace their history no further than that time.

The earth might then have been in the midst of a great evolutionary adventure, the 'Cambrian Explosion'. Within a mere five million years, there appear in the record snails, starfish and animals with jointed limbs (whose descendants include the insects, spiders and crabs). Perhaps this reflects some crucial change in DNA. After all, some mutations can at a single bound change a fruit-fly antenna into a leg. Might a great burst of genetic creativity have driven a Cambrian Genesis and given birth to the modern world?

The idea gained weight with the discovery in the 1940s in the Ediacara Hills of South Australia of some shadowy and mysterious fossils. They came from a time before that famous era and were the relics of soft-bodied beasts that look not at all like life today – further evidence, perhaps, of an eruption of novelty in the Cambrian itself. Such animals have now been found all over the world. The youngest are just below the base of the Cambrian, and the oldest go back as much as six hundred million years.

That era was, without doubt, a busy time for evolution. In its operatic landscape there appeared strange beings whose names – *Anomalocaris* or *Hallucigenia* – are testament to their odd appearance. An inhabitant of the Burgess Shale beds, *Opabinia*, had five eyes and a long nozzle; a body plan not seen since. But was it a period of unique innovation, with so many new groups leaping into existence that selection could never have made them? The answer is, with little doubt, no. First, the Burgess Shale animals are less peculiar than is sometimes made out. *Hallucigenia*, with its strange spikes on its back, had, it turned out, been drawn upside down. It has relatives among those drab animals, the velvet worms, still common in rotten logs in the Southern hemisphere and no more different from other animals with jointed limbs than are today's barnacles.

The Cambrian Explosion, so called, is a failure of the geological record rather than of the Darwinian machine. Its radical new groups reflect not a set of exceptional events, but something more banal: the first appearance of animals with parts capable of preservation. Before then, there were soft creatures that decayed as soon as they died. Why shells appeared all of a sudden is not certain. Perhaps the first predators evolved and drove their prey to don expensive armour, or perhaps a surge of oxygen enabled animals to grow large enough to need a skeleton. Whatever the reason, the Cambrian marks the origin of a fossil record, rather than of modern life.

Take those evolutionary celebrities, the trilobites, the first animals to lay claim to jointed limbs. They are close to the roots of a tree that later grew branches as flamboyant as the insects and a living fossil called the horseshoe crab. If – as the record suggests – trilobites burst into existence within five million years at the base of the Cambrian, one brief event changed the whole direction of evolution. In fact, a closer look shows that among the earliest to be preserved were many distinct kinds. Such diversity shows that trilobites had a past dating to long before that famous era. What made them seem new was no more than their skeletons. Their predecessors had died and decomposed, but their more solid descendants were preserved in millions. The Cambrian was a busy time for trilobites, but it marked their middle age, not their infancy.

Now, at last, the youth of the world has been revealed. Newly discovered fossils hint at a history older than anything imagined by those whose inspiration begins with the Cambrian. Bit by bit, they have pushed life further into the past; and each discovery is a new hint at the untold billions that have disappeared.

Phosphorites are rocks laid down in shallow seas. They are much mined for fertilizer, and, in some places, preserve soft-bodied fossils in fine detail. Great beds are found around the Yangtze gorges in China, where they hide remains of lives from before even the Ediacarans.

Their ancient world was full of diversity. The Chinese beds, five hundred and seventy million years old, hold the remnants of sponges, long thought of as primitive. To the amazement of those who found them, they also preserve the embryos of other animals, each cell seen in all its detail. These oldest embryos look much like those of modern insects or worms, animals with a left side and a right. This feature was, the first fossils show, established in a distant past, quite unrecorded until a crucial find was made.

Most unexpected of all are the seaweeds. Soft though they are, many of those plants are in the phosphorite beds. They are almost as varied in form as are their equivalents today. Those ancient plants hence have a hidden history that began long before the Yangtze beds. One is almost identical to a modern seaweed, so that this unlovely creature lays claim to the longest of all pedigrees. The record of past seaweeds has always seemed scanty, but this single collection shows it to be far more incomplete than anyone had feared. Unlike the Ediacara, who have no modern equivalents, such fossils show how much is missing from the early history of the life of today.

Some billion-year-old rocks in central India have sinuous grooves on their surface that look rather like those made today by worms creeping on the bottoms of muddy ponds. Perhaps they are the traces of such beasts, their bodies lost, but their spoor remaining. To burrow through mud needs muscles and an internal cavity filled with liquid, and some geologists question whether an animal so advanced could exist so far into the past. If it did, it marks a gap in the record of worms as long as the whole time from the Cambrian to today.

A few discoveries from an unknown land prove that biology's supposed preface, the Cambrian, is in fact well into its plot. So deficient was the evidence of earlier times that nobody guessed it was there. Single-celled creatures stretch back even further. The balance of isotopes in carbon trapped in Greenland rocks three thousand eight hundred million years old resembles that of the modern bacteria that live on methane. They evolved when Earth was in its infancy and show that it took several times longer to develop a cell with a nucleus than it did for life to appear. Three billion years' worth of fossils older than the Chinese seaweed are waiting to be discovered.

More familiar places are a token of how much has been lost. On London's southern fringe are low hills of chalk, now improved by suburbs, but with Darwin's own house still hidden at their foot. Those North Downs give way to a wooded and intricate landscape, the Kentish Weald, until, twenty miles further on, a second range of hills, the South Downs, takes over. At Beachy Head the Downs reach the English Channel. On a clear day, France can be seen. That, too, is white. It continues the system that built southern England. The Channel Tunnel, indeed, owes its existence to a band of solid chalk beneath the waves.

Once, long ago, the North and South Downs – and the Bas Boulonnais in France – were one. They have been worn down in what seems to us an age, but was in terms of evolution an instant. Their chronicle is that of the world, of the piling up and throwing down of rock, of times below the sea and above it, and of great sections of the earth, with their fossils, gone for ever.

It is an admirable lesson to stand on the North Downs and to look at the distant South Downs; for, remembering that no great distance to the west the northern and southern escarpments meet and close, one can safely picture to oneself the great dome of rocks which must have covered up the Weald within so limited a period as since the latter part of the Chalk formation. The history of southern England is written into that view.

Its chronicle is one of shallow seas that dried and left their remains behind. Nowadays Kent is a bowl with chalk hills on either side. Not long ago the bowl was inverted, a peak with slopes running down to the sea. It was evidence of a fold in the Kentish earth, made when the Alps began to build. The strain buckled a whole series of beds. They begin with coal from more than two hundred million years ago. Soon after the coal forests, much of Western Europe sank. A great sea covered Kent. In it settled sand and clay. The waters ebbed and rose until, a hundred million years ago, the land sagged again. Over the whole sea bottom a thick sheet of calcareous mud was laid. It became chalk.

Chalk does not just contain fossils – it is fossils, the shells of tiny animals called foraminifera, mixed with billions of tiny plates of lime secreted by marine algae. Scattered among them are the teeth of extinct sharks, the shells of nautiluses, the bones of the odd pterodactyl, and many of the sea-urchins that once lived on the ocean floor. The white expanse also has a few foot-long boulders, stones carried in the stomachs of ancient reptiles as an aid to digestion. Apart from those remnants, the rock is pure, as little was washed into it from the arid deserts that surrounded the seas over southern England.

Then, at the end of the dinosaurs (the first of which was found in the Weald), the sea fell back and the land rose. Kent became an island of chalk. Its rocks, and the sand and clay below, buckled upwards. Five million years ago, the land dried for the last time and rose still higher.

Nowadays, the chalk survives only as the North and South Downs. The central part of Kent is made of older rocks: the coals, sandstones and clays that once lay beneath the white shroud. The rest of the winding-sheet has gone. Most disappeared within a mere two million years. In the ice ages, the soft rock was shattered as water froze within it. Rain could not seep away, because the soil was so hard, but made great rivers that carved valleys, now dry, through the hills. The top of the dome was skimmed off. The denudation goes on. Sunken roads – holloways – run through steep-sided trenches twenty and more feet deep. They were formed by the horses, cows and carts that wore paths through the soft rocks laid bare when the topmost layer was lost. The cliffs still fall into the sea in great land-slips, one of which, in 1915, moved a train and its passengers fifty yards out on to the beach at Folkestone.

The Channel itself was made as the sea destroyed the earth. The waters of the Dover Straits beat at the base of cliffs and cut them back like a giant saw until the rocks collapsed under their own weight. A few months before the millennium, hundreds of thousands of tons fell off Beachy Head, and what had been a boat trip to the lighthouse became a rocky scramble. The soft chalk moves at a great rate. The Belle Tout lighthouse was built on the clifftop in 1834. The edge moved back by fifty feet a century (in spite of its owners' efforts to block rabbit burrows to save it) and the precipice moved to ten feet away from the front door. Although it has now, at great expense, been moved fifty feet inland, the tower's fate is, within a geological instant, certain.

Deep grooves on the bed of the Channel show that the first breach in the land bridge was made half a million years ago. The Thames was pushed south by glaciers until its waters filled a huge lake behind the ridge connecting England to France. The lake burst through and tore away great blocks. Soon, the sea fell back as the climate cooled and the bridge was rebuilt. Seven thousand years ago, the waters rose again and wore away the rocks to make the present Channel. It gave Britain the Straits of Dover, as a statement of national identity and a shield against invasion.

The Kentish landscape and its cousins across the Channel show how fragile is the record of death over even a moment of evolutionary time. They are the key to evolution's great anomaly, the difference in pattern between past and present, between fossils and the plants and animals of today.

From an aircraft window on an afternoon flight over the battlefields of Belgium and Northern France, great lakes of shadow fill the valleys, while the hills, low as they are, are picked out in the sunset. Eighty years ago that difference of a few feet meant life or death. It determined the results of the great battles of the First World War, Ypres, Passchendaele and the Somme. Passchendaele is notorious for the first use of poison gas, the Second Battle of the Somme as the last gasp of the War. Each was, in addition, an experiment in palaeontology.

To visit the trenches is to test the fate of the dead. The battlefields show how incomplete is the record of the rocks, even when less than a century old.

Before the Third Battle of Ypres in 1917 the Germans held the high ground and could look over the Allied lines. At the Somme, in March 1918, the order was reversed and the British line was dug through a line of low hills. The Passchendaele plain is a flat clay, its hills made of sand. Further south – where the Somme was fought – chalk takes over. The landscape is much the same as that of the Weald and the Downs.

The soldiers soon learned their geology. Clay became sodden and drowned them in thousands, while sand collapsed as they dug their trenches. Chalk was more reliable and huge dugouts could be built to escape the shells. At Passchendaele a layer of clay passed from the British lines to deep beneath those of the enemy. Miners were set to work to burrow through it. On 7 June 1917 twenty huge mines went off at once. The blast – the biggest non-nuclear explosion in history – blew a hole two hundred feet across and forty feet deep. Ten thousand Germans were killed and most of the remainder fled. Then, it rained. Tanks were bogged down and the battle became another frightful stalemate. At the Somme, the British were on the defensive and their line was broken. The troops retreated but held the enemy a few miles further back.

At Passchendaele, the slate of history has been wiped almost clean. The visitor notices little, as the soft soil has swallowed up the trenches and most of the craters. The site of the giant mine is now an unremarkable pond called the 'Pool of Peace'. Few tourists come because there is not much to see.

They prefer the Somme, where the chalk has kept a better record. Even so, just a few of the miles of trenches are still visible. Most are now ditches and mounds rather than the crisp military structures of 1918. Their fate – and that of those who died in them – is that of the chalk that once covered Kent. The erosion at Passchendaele reflects the loss of the Weald's softer and more ancient rocks. Quite soon, no remnant will be left of either field of war to remind visitors of the conflict. In less than a century the surface of the earth has been so blurred that the record of the past is almost lost.

Most of the bodies of the Somme were recovered and crowded into cemeteries. Passchendaele still spits out the dead from its dreadful mires. Its cemetery – the largest war grave in the world, Tyne Cot, named for the thousands of Northumbrians who died in the battle – contains twelve thousand graves. Three times as many are recorded only as 'A soldier of the Great War – Known unto God' or as a name on a shared monument; their bodies sunk so deep into the mud as to be unrecognizable, or to be lost altogether. The Somme cemeteries are almost as full, but more of their graves have names. Even so, to dig them up, as is sometimes done in the interests of good order, reveals, quite often, nothing that can be identified as human.

Sixty billion people have lived since man appeared in modern form. To excavate every graveyard – and every fossil site – in the world would turn up no more than a minute fraction of their remains. The lost armies of the dead have a moral for evolution.

They are a reminder that the geological record is a history of the world imperfectly kept, and written in a changing dialect; of this history we possess the last volume alone. Of this volume, only here and there a short chapter has been preserved; and of each page, only here and there a few lines. However grand the monuments, and however firm the hope of eternal life, in the depths of time a Pharaoh in a pyramid has as little chance of immortality as does a soldier stamped into a bloody swamp. The history of ancient Egypt, of the present century – and of the existence of our own species – will soon be gone for ever.

CHAPTER X

ON THE GEOLOGICAL SUCCESSION OF ORGANIC BEINGS

On the power of gradual change — Species once lost do not reappear — On disaster caused by earthquakes, deluges, and heavenly bodies — Gradual and sudden change in the fossil record — On the true rate of evolution and the sense of time — On fossil genes and the affinities of the past with the present — On the state of development of ancient forms — On the succession of the same types within the same areas — Summary of preceding and present chapters

IT TAKES THREE hours to travel the hundred and fifty miles of free-way from Thebes to Memphis (Memphis, Tennessee, that is). The journey can also be done by river-steamer, a trip that lasts a couple of days. The boat is slower, but it has to travel three times further as the mile-wide Mississippi meanders across a flat landscape. The River Meander itself, a Turkish stream a twentieth the size of its American cousin, does the same. Even rain does not take the shortest path downhill as it trickles down a road.

Heraclitus (who lived near the Meander) had evolutionary views: no man, he said, can step in the same river twice. Streams – always the same but constantly in flux – were a metaphor for life. In biology, too, slow change can have great consequences.

Streams evolve through a balance of forces. The bed shifts as it erodes one bank and dumps its remains on the other. It returns when its loops are cut off as the water finds a more direct route downhill. Complexity – meandering – is opposed by simplicity, the shortest path to the sea. Raindrop, Meander and Mississippi follow the same rules. Measurements of dozens of real rivers, and computer simulations of many more, show that the relationship between their shortest possible path across a plain and their actual length is always the same. It is pi, the ratio between the circumference of a circle and its diameter. Each river, whatever its size, goes a little more than three times further than it needs on its way to the sea.

As it does, it shifts uneasily. When Paul visited Miletus on his journey to the Ephesians, the city was at the mouth of the Meander. Now, the river passes it by. The efforts of the US Corps of Army Engineers are all that keep the Mississippi itself in place. Without them its mouth would be at Morgan City, Louisiana, rather than at New Orleans. Rivers are as inconstant as is their water. Stand on the bank for long enough, and you will drown, because the stream will come and get you. How soon (seconds for a trickle of rain and years for the Mississippi) is determined by a simple and general law.

Darwin's theory makes the same appeal to unity in space and time. Nature has no need to make sudden leaps. Instead, varieties, species and the larger divisions of nature emerge – like raindrops and rivers – from a process that acts without regard to scale. The very title of his greatest work – *On the Origin of Species by means of Natural Selection, or the Preservation of Favoured Races in the Struggle for Life* – joins the course of change within a species to the grander process that causes new forms to appear. DNA is a river from the past. If the trickle seen today is driven by selection, why should its majestic passage through time not obey the same rules?

It is possible to argue that all the great leading facts in palaeontology seem simply to follow on the theory of descent with modification. Old forms are supplanted by new and improved forms, produced by the laws of variation that still act around us, and preserved by natural selection. Like the bed of a great river, the course of evolution is in its substance simple.

As streams move on, they abandon their earlier path. Why should not natural selection, too, have over the years led the inhabitants of each successive period of the world's history to beat their predecessors in the race for life? If some have improved, others must pay the price. On this view, new forms are destroyed by the process that forges them. A species once lost can never reappear, even if the very same conditions of life, organic and inorganic, should recur.

The rocks are filled with evidence that forms which do not become in some degree modified and improved, will be liable to be exterminated. Until a few years ago we knew little of our own roots, as they, like most others, have suffered that very fate. Once, the vertebrates – the group to which mammals, birds and fish belong – seemed to have had a timorous childhood. Today's versions of their predecessors – the lancelet and the hagfish – feed on soft tissues. The lancelet and the young hagfish filter particles from the water, while the hagfish adult is more interested in the flesh of dead or crippled fish. They are not, perhaps, a noble set of ancestors for the lords of creation.

Now, the image of our past has changed and the lost world from which we descend has at last been revealed. An abundant but enigmatic fossil was the key. Many chunks of limestone contain thousands of tiny pointed objects. Conodonts, as they are called, were discovered in Russia in the nineteenth century and have since turned up all over the world. Because they are so widespread, and because their shape changes over geological time, they are much used to identify from which layer a particular rock might come. The animal who made them was quite unknown. Like the tooth of the Buddha at Kandy, what remained was not enough to reconstruct what might have been. The many guesses as to its appearance included a version that looked rather like a Swiss roll studded with razor blades. All were, at best, implausible.

In 1982, in some rocks from the shore near Edinburgh, were found the preserved remains of the animal itself. It had a soft eel-like body a few inches long, with large paired eyes, a stiff rod down the back, and tail fins. The conodonts themselves were not separate beings, but its teeth.

Before these dozen or so specimens – the first examples of the animals that made the tens of millions of conodonts seen by geologists over a century and more   the vertebrate skeleton was thought to have started as a set of defensive plates on the body of a primitive fish. The first vertebrates were, it seemed, victims, prey rather than predators.

The complete conodonts changed all that. The first sign of the skeleton was, it seems, in the mouth. Sets of conodont teeth, when pieced together, look as if they were used to shear flesh. The conodonts flourished and diverged before they were driven out. More – and larger – conodont animals have now been found, from Wisconsin to South Africa. One, the size of a small fish, even preserves a pair of eyes (themselves at first classified as the remains of a plant). The conodonts prove that our predecessors were not grazers, sifters or suckers, but carnivores.

Now, the conodonts have gone, and can never return; but their descendants can do all that they did, and much more, with grazing geese, sifting whales, sucking lampreys and dogs that eat meat. Such animals are more diverse and have more complex relations to their organic and inorganic conditions of life, in land, sea and air, than did their predecessors. Conodonts had a long and varied history, but were at last driven out. Extinction, as environmentalists so often remind us, is for ever, and we will not see their like again even if the vertebrates themselves should some day disappear.

The idea of a dignified passage of origin, divergence and departure has, for conodonts and many other vanished creatures, much in its favour. In the real world, however, rivers or raindrops – or species – are often deflected from mathematical virtue. The unexpected intervenes, and the beautiful simplicity that links their actual story to that set by theory disappears.

In December 1811 a gigantic earthquake struck between Memphis and Thebes. It was equal in power to the combined energy of all those in North America since. The shock was enough to ring bells in Boston. Columns of coal dust rose from the ground, and few houses within two hundred miles were left upright. The crew of the first steamboat on the Mississippi had moored to an island. When they awoke, it was gone: 'The numerous large ponds ... were elevated above their former banks ... A lake was formed on the opposite side of the river, in the Indian country ... It is conjectured that it will not be many years before the principal part, if not the whole of the Mississippi, will pass that way'.

Although a ten-mile lake – today's Reelfoot State Park – appeared that day, the river disdained it as a new course. Even so, it shifted its path in many places as islands sank and bluffs collapsed. One cataclysm did more to determine the Mississippi's fate than had centuries of gradual change. Now, the river faces another catastrophe: control by man. Although the State Line between Arkansas and Mississippi still meanders along the bed of an abandoned stream, the waters themselves flow down a straight and narrow channel.

Life, too, has had its disasters. Why should they not, at least in part, determine patterns in the fossil record? No doubt, gradual selection was important in the rise and fall of many beings. But could not ancient tragedies of geology or genetics sometimes have done the job? The record has many sudden deaths and just as many unexpected entrances. Some, no doubt, are there only because the fossil record is incomplete, but some might be the biological equivalent of the Mississippi earthquake. Are disasters needed and, if so, how often do they happen? Are they a merely destructive force, or might they be a help to evolution? Such questions are at the heart of the disagreements between palaeontologists and those who study the plants and animals of today.

On Extinction. Many creatures disappear because they are replaced by something better. That process, busy all over the world, is itself evidence of natural selection. The nineteenth century preferred to see the past swept away by acts of God. Disasters were an excuse for not thinking. Charles Lyell was scornful: 'We are told of general catastrophes and a succession of deluges, of the alternation of periods of repose and disorder, of the refrigeration of the globe, of the sudden annihilation of whole races of animals and plants, and other hypotheses, in which we see the ancient spirit of speculation revived, and a desire manifested to cut, rather than patiently to untie, the Gordian knot.'

As any visitor to the Galapagos can see, plenty of extinction takes place in the Darwinian way, as an old order gives way to a new. However, lots of life has gone out with a bang. General catastrophes there have certainly been; disorders greater than the Mississippi Earthquake that have wiped out the well-crafted products of evolution over much of the Earth. Any animal caught up in such an event, however well adapted it may have been, has every chance of annihilation. In geological terms, even the great glaciations (slow as they appear to us) were sudden, and the obliteration of much of the northern hemisphere over a few thousand years is, on the scale of the rocks, abrupt enough to count as a rude interruption of a gradual climatic trend.

When it comes to deluges, the Black Sea tells part of the tale. It was once a small freshwater lake, the Bosphorus no more than a dry ridge between two basins. Then, to the north, a river carved a valley. To the south, a slip in the earth's surface made a deep fjord. Five and a half thousand years ago, as the sea rose, the Mediterranean broke through and the Black Sea filled. It did so over a cascade the size of two hundred Niagara Falls. The roar could be heard three hundred miles away and in some places the shore moved back by a mile a day.

The salt waters were an evolutionary tragedy for the animals of the Black Sea plains, however well they were adapted to their humble lives. Five hundred years later, the Mediterranean itself died, leaving a thin layer of rot on its floor as evidence of a holocaust. Soon after the Pharaohs arrived there was a time of heavy rain. It ran off the denuded land to cover – and smother – the ocean with fresh water. No oxygen could get through as the layers did not mix. Like today's Black Sea (in which a layer of oxygen-free salt water sits below a fresher cap), the Mediterranean was dead in its depths. The whole process took sixty years and swept away thousands of species. Given today's pollution, the sea could die again just as quickly.

Flood legends are common in the Middle East. The earliest, that of the *Epic of Gilgamesh*, dates from about the time of the inundation of the Black Sea Basin. Noah's Flood, much derided by evolutionists, may be one of many sudden reverses that changed the course of their science.

On Friday 9 October 1992 a meteorite damaged the right rear fender of a 1980 Chevrolet Malibu, the property of Michelle Knapp of Peekskill, New York. The rock was sold for fifty-nine thousand dollars and the car, damage and all, for ten thousand.

The chance of real harm from a celestial body seems tiny. In spite of the threats of Apocalypse ('a star from heaven, burning as it were a lamp'), the sole recorded case of a direct hit was in Alabama in the

1950s when a lady asleep on a couch was bruised by a meteorite that bounced off her hip.

In the end, though, disaster is inevitable. A trillion comets orbit the sun. Each trails a stream of gravel that, when it hits the atmosphere, appears as a shower of meteors. The Earth gains a ton in weight every hour from their dust. Two thousand asteroids big enough to destroy civilization orbit nearby. Given the likelihood of a hit in the course of a lifetime, the safety of aircraft, and the few who fly, the chance of an average person meeting death from a comet is higher than that of death in an air crash. Apocalypse is, sooner or later, guaranteed. The United States Government even established a committee (chaired by the then Vice-President, T. Danforth Quayle) to find out how to deflect or destroy any celestial body audacious enough to threaten to land on the United States. There are plenty of reminders of the danger. In 1996 an asteroid missed Earth by a mere quarter of a million miles. A direct hit would have been equivalent to an explosion of all the globe's atomic weapons at once. As Voltaire said of such an event: 'What a Disaster would it be for our Earth ... The Idea of two Bombs, which burst on clashing together in the Air, is infinitely below what we ought to have of such an encounter as this'.

Meteor Crater in Arizona was not recognized as such until 1929, giving rise to a wave of cosmic fears to go with the Wall Street crash. The greatest crater of all is further south, near the port of Progreso in the Yucatán Peninsula of Mexico. The surface shows little sign of the object six miles across that struck with the force of five billion Hiroshimas sixty-five million years ago. Its testimony is buried under a mile of limestone. An ancient seabed in New Jersey reveals, at just that time, a three-inch layer of glassy spheres, the remains of the liquefied rock ejected from the crater. All over the world, the rocks of those days have high levels of the element iridium (brought, perhaps, by a meteor), together with much soot.

The seventh-century Chinese philosopher Li Ch'un Feng had it that 'Comets are vile stars. Each time they appear ... something happens to wipe out the old and establish the new'. The Yucatán disaster wiped out an old regime for ever and helped to establish the modern world. Below the layer marking its arrival are the remains of tiny dinoflagellates and foraminifera – hard-shelled animals present in billions in the oceans – while above it, all are gone. No new marine animals appeared for thousands of years and, after the bang, the seas were almost dead. The dinosaurs, too, disappeared at about that time.

The mass extinction is itself smaller than the cataclysm marking the end of the Palaeozoic period two hundred and forty million years ago, when ninety-six per cent of all marine species disappeared. It did for the trilobites, until then the lords of the oceans, and almost destroyed the ammonites (who made a modest recovery). There have been more recent cataclysms. In central Siberia is the fifth largest impact crater on earth. The rock that splashed out is thirty-six million years old. At almost the same time another comet struck on what became the eastern coast of the United States to make the Chesapeake Bay. The two big bangs marked the appearance of ice sheets in Antarctica and the greatest extinction since the loss of the dinosaurs. It marks the boundary between two geological periods, when the first simple horses, deer and whales gave place to mammals close to those of today.

Three extinctions at the time of a comet must be more than a co-incidence. But how sudden were the bangs – or were they roars, or even distant rumbles? In some senses, they were abrupt indeed. At the famous collision of Yucatán, everything within a mile of the impact was extinct within a second. Within a minute the Americas were shaken by earthquakes and nine minutes later most of the continent's forests burst into flames. Later the same day, huge areas across the world were destroyed by tidal waves, and at sunset a nine-month night of smoke and dust began. Voltaire under-rated the disagreeable effects of even a small meteor.

Even so, for most of the Earth's inhabitants the comet meant dis-comfort rather than death. It took a hundred thousand years before all the dinosaurs were gone, and the last ammonite lingered on for as long again. Many bivalves (mussels and the like) disappeared at about the same time – but their holocaust happened thousands of years before the impact. Indeed, the dinosaurs themselves had begun a decline before their fate was (perhaps) sealed by an agent from outer space. Land plants and fish seem unperturbed by the blow and their history continues more or less as normal through those trouble-some times. What is more, some marine animals rose, like Lazarus, from their graves, long after the comet. They were dead to all appearances, but resurfaced as a statement of the record's deficiencies rather than of divine assistance.

It is, nevertheless, certain that an event out of the ordinary hit the world sixty-five million years before the present. It took two million years to get back to normal; and the new normality was noticeably different from the old, with many novel kinds in the place of those who had gone. Evolution's progress (if such it is) is not uninterrupted. Catastrophe on a scale unknown to history has played a part. Whether it had a constructive, rather than a merely lethal, effect is another issue. Some claim that mass destruction led to biological explosions as the survivors evolved to fill the gaps. If they are right, cataclysms drive change as much as does slow modification. Certainly, some patterns in fossils suggest that gradualism is not enough and that the river of life has suffered many earthquakes, of several kinds.

On the Forms of Life changing almost simultaneously throughout the World. A world designed by a physicist (or a god) might start simple and grow more complicated and stop when it reached perfection. Life has never been like that. A year after *The Origin*, John Phillips recognized three peaks of diversity among fossils, each followed by a sudden extinction and a gradual climb back. The record has long periods in which not much happens, followed by episodes of rapid change.

In spite of arguments about just how flat were the plains and how tall the peaks, most geologists accept that there were indeed three high points of variety. They came at the height of the Cambrian around half a billion years ago; in the Palaeozoic from around four hundred to two hundred million years ago; and in the modern period, which marks the richest mixture of species ever seen. What caused such deviations from divine (or at least arithmetic) perfection?

The idea that birds and mammals had to wait for the demise of the dinosaurs before they could come into their own is wrong. The genes show that, for both, there were tens of millions of years of profitable change before the comet struck. All the modern groups were well on their way before that vexatious event. Even so, most lineages appear in the record quite suddenly. One section of marine fossils in upstate New York is typical: it stretches over a hundred million years and all its many kinds stay the same for millions of years before most alter over just a few hundred thousand. For marine fossils with a long enough history – corals, snails and small bottom-living creatures – nine-tenths of all records are like that: a sudden appearance followed by millions of years of tedium until the line disappears or emerges unchanged today. A few alter slowly, in true Darwinian style, as proof that the patterns found in the others are not due to some error of sampling, but the general picture is that life is calm for most of the time. One tadpole shrimp is so static that it has kept the Latin name used for its ancestors of a hundred and eighty million years ago; but it copes so well with the modern world that it is a pest of irrigated rice-fields.

Proponents of gradual change, of stability and of revolution can each call on convincing evidence to support their views. No universal pattern unfolds within the rocks. Now that the fossil record is so much better understood than once it was, evolutionists can no longer point to imperfection as the sole explanation for its structure. Some lineages alter in what seems a non-Darwinian way, not gradually but with episodes of calm punctuated by change. Whether such conservatism lays bare some inherent reluctance to adapt, or whether it shows merely that not much evolution is needed because Nature's challenges tend to stay the same, we do not know. However much Darwinians may protest (and they do), millions of generations of inertia scarcely fit his image of life as poised for an instant response to any challenge. The argument between supporters of evolution as unhurried Victorian progress and those with the modern view of history as boredom mitigated by panic is unresolved.

In part that is because the question asks too much of a fossil record so battered by the accidents of time; but it also turns on a disagreement about the meaning of words. What is gradual, and what instantaneous? Chiang Kai-shek, when asked his opinion of the French Revolution, said that 'It is too early to tell.' His was a long view of the past, which to Western eyes seems quite out of proportion. The Nationalist regime in China had by then occupied just a couple of decades in a political continuum from the Shang dynasty, four millennia before. As its leader saw, the events since the Bastille, dramatic as they appeared to those involved, were an instant in history, and perhaps an unremarkable one. It is all a matter of scale. Much of the argument about whether the fossil record shows slow or rapid change depends on what those terms signify.

The Turkana Basin in East Africa holds the remains of many ancient primates, together with those of our own tool-using predecessors. All are preserved in strata separated by well-dated layers of volcanic ash. Spectacular as these bones might be, they are rare and say little about gradual or interrupted evolution.

In the same Koobi Fora beds are preserved millions of shells of twenty or so different kinds of freshwater snail. Their fossils have remained almost undisturbed since they were laid down. At first sight, what they reveal is not at all the pattern expected of slow and successive modification. Instead – as is true for so many marine fossils – long periods of stability are interrupted by sudden bursts of change. A new variety appears, persists unaltered, and disappears as quickly as it came. This is not just a matter of the inadequacy of the record, as huge numbers of the inhabitants of these lakes are preserved. The intermediates are around for just a tiny part of each lineage's history. How can this be, if nature does not make leaps?

The problem comes, as Chiang Kai-shek saw, from the sense of time. Those who peer into its depths find it hard to see things in proportion. When one referee in Nature's race is used to a stopwatch and the other to Big Ben, disputes are to be expected. An instant to a palaeontologist may appear an infinity to those who study life today. In the Turkana fossils, the 'intermediates' last for just a tiny part of the duration of their ancestor or descendant forms. As a result, they seem a classic case of an evolutionary leap over a moment in the snail pedigree; a pattern quite unlike anything expected from slow and successive modification.

However, that moment represents, for different lineages, between five thousand and fifty thousand years. The snails have one or two generations each year. A blink of a palaeontologist's eye hence covers about twenty thousand mollusc generations. For dogs, that is equivalent to forty thousand years, a period longer than it took to evolve chihuahua from wolf by conventional selection. Why appeal to anything else in the case of the snails?

Natural selection can certainly do the job of changing one snail into another. But why, if it can, are its energies confined to such a short part of the history of each kind? And why are the snails immovable for so long when they have the capacity to evolve? Is something more needed to explain such unexpected patterns? That is the argument of 'punctuated equilibrium' – the notion of evolution as stasis interrupted by sudden change: a pattern that might, perhaps, result from some intrinsic property of the organism, rather than of the environment in which it lives. It points at a great disparity between what life can do and what, over evolutionary history, it does.

A universal measure of the rate of evolution that can be used on fruit-flies over a few weeks in the laboratory or on dinosaurs over the millennia points up the contrast. A unit of evolution per million years is called a 'Darwin'. It is based on how much the average size of any feature (corrected for the absolute size of the structure itself) alters with the years.

Laboratory experiments on flies can generate values of over a hundred thousand Darwins, and selection on the farm often gives rates of several tens of thousands. Transplant experiments (such as those of the guppies moved from a dangerous stream to a calmer place) give rates of change of up to fifty thousand Darwins. English sparrows were introduced into the United States a century ago and have filled the Americas. As they spread, they evolved a pattern similar to that in the Old World, with heavy birds with short limbs in the cold north and lighter and more graceful animals in hot places. The legs of the birds from the warm south have lengthened at around a hundred Darwins (which is about 5 per cent in a century). Although it is not known to what extent the change in sparrow shape is due to genes rather than to a direct response to the environment, evolution at that speed would bring forth a sparrow with the legs of an ostrich in just ten thousand years.

Fossil rates of evolution are far smaller. Horses are a classic of rapid modification, as their teeth altered to keep up with a shift from soft leaves to hard grasses. Over several million years, they changed at a tenth of a Darwin – which means that it took millions of generations for the average height of the tooth to shift from that of the smallest to the largest present in the population at any time. In the deeper record – the hundred and fifty million years of the dinosaurs included – evolution of shape or size is thousands of times slower still. Life, it seems, has done much less evolving than it could. Why, we are by no means sure.

On the affinities of extinct species to each other, and to living forms. The genes of today link every plant and animal together in an unbroken chain of ancestors that descends from the ancient past. Now, they can do more, for some of the molecules of life are themselves preserved as fossils.

A coalmine is a hecatomb of past existence, full of the remnants of ancient trunks and leaves. Oil is a degraded form of chlorophyll, the green material of plants. As about a fiftieth of all the rocks – limestone, chalk and so on – that fell as sediment from oceans is made up of biological remains, such rocks contain ten thousand times more organic material than does the whole of life today. Some of the ancient material preserves its structure. Chitin, the solid material of the insect skeleton, has been found in twenty-five-million-year-old remains from a lake-bed, and the amino acids in scallop-shells may last for a hundred and fifty million years.

Snail shells, too, retain a genetic record of the past. They keep their marks for thousands of years. Those buried, unnoticed, by the ancient farmers who erected the great monuments of Silbury Hill and Avebury Trusloe in Southern England were dug up, just a few feet beneath their descendants, when the sites were excavated. The genes for shell pattern had changed. Ancient populations from colder times are darker in colour, because dark shells are favoured at times when it pays to soak up the sun's heat.

The idea that ancient DNA could be preserved gained fame from the film *Jurassic Park*, in which dinosaurs were recreated with the help of fossil DNA. The molecule certainly can persist after death. It was first found in a nineteenth-century museum specimen of a quagga, an extinct zebra. More turned up in a two-thousand-year-old mummy dried in Egyptian sands (exciting the Copts, who hoped to test their claim to Pharaonic descent). There were claims of its presence in twenty-million-year-old magnolia leaves from an Iowa bog, in a hundred-and-twenty-million-year-old weevil preserved in amber and in dinosaur bones themselves.

That was every evolutionist's dream. What better than to have the genes of ages past to compare with those of today? Reality made an unwelcome appearance when it emerged that almost all so-called ancient DNA results from contamination with modern material (a real problem, given the sensitivity of the methods used to search for it). The molecule cannot last for long. The measure of its frailty lies in proteins. After death, they break down into individual building blocks called amino acids. These come in two mirror-image forms, which in life are biased to the left. The leftward inclination begins to disappear after death, with a slow chemical switch back to a mixture of the two. The smaller a fossil's chemical twist, the older it must be. DNA decays at about the same rate as proteins. Any fossil whose amino acids have lost their natural bias must hence, for simple reasons of chemistry, have lost all its DNA.

Almost all supposedly ancient DNA fails the test. As Hamlet's gravedigger points out, 'water is a sore decayer of your whore-son dead body'. Whatever its parenthood, any DNA in an animal whose body lies in a wet place is at once destroyed. Ice, in contrast, can be a positive help (which is why pathologists have searched for the genes of the 1918 flu virus in the frozen remains of Alaskan fishermen). Genuine (albeit fragmented) DNA is in the bones of the giant sloths that led Darwin to call the Pampas 'one wide sepulchre of these extinct gigantic quadrupeds'. A North American relative from twenty thousand years ago left faeces filled with DNA from its last meal of grapes, flavoured with mint. Frozen Siberian mammoths from forty thousand years earlier still seem, from their genes, to have been more related to African than to Asian elephants.

Rare though its remains may be, DNA has left plenty of less direct evidence of its history; for although the molecule itself may disappear, its footprints remain in the bones themselves. Lungfish are living fossils: animals with an agile and creative past that nowadays have sunk into deep conservatism. Long ago, they slowed down, and have stayed unchanged for hundreds of millions of years, while their relatives moved on.

Bone contains many cells, all of them with a nucleus. The hard material squeezes each one so that its size is a measure of how much DNA it once contained. Early in lungfish history, the size of the cell nuclei – and the amount of genetic material – began to creep up. Soon, the animals had hundreds of times as much as do most mammals. As it did, evolution slowed. Now, the lungfish are stuffed with DNA (most of it with no apparent function) and their evolution has stalled altogether. The fit between DNA content, a lethargic lifestyle and evolutionary sloth is widespread. To copy that chemical takes energy. Bacteria are speedy and have no excess genetic material, while salamanders, torpid as they are, are filled with genes. Plants, too, have a close fit between habit and nucleic acid content. All weeds have small genomes, while more established plants, packed with DNA as they are, grow slowly and can take a month to make a single egg cell. Whether an indolent life allows the amount of genetic material to build up, or whether the extra dose itself slows down evolution, nobody knows.

Fossils show that the same happened to the lungfish. Whatever slowed it down, the imprint of its lost genes left in its bones shows the affinity of an ancient species to its descendants. Although life, as usual, does not live up to what Hollywood can do, the tragedy of the lungfish links today's nucleic acids to those of long ago.

On the State of Development of Ancient Forms. The idea of a present in decline from the past has been around since history began. Homer and Virgil deplored the decrease in stature of the human race, and Greek actors in the role of mythical heroes played them on stilts. The notion was supported by fossils. The bones of the monster Polyphemus, found in Sicily in the fourteenth century, joined many others hung in churches as proof of the Genesis claim of 'giants in the earth in those days'. Polyphemus's bones were in fact those of fossil elephants. His single giant eye (as leader of the Cyclops he was blinded by Odysseus when cast ashore on Sicily) came from the large nostril in the centre of an elephant skull, which resembles a huge eye-socket. Other ancient elephant bones were paraded around Europe reassembled into saints. They led a French anatomist to calculate that, given the rate of decline since the Creation, Adam must have been a hundred and thirty feet high.

When it comes to evolution, size still counts. Buffon had claimed that in America dogs lost their bark and 'all animals are smaller'. Even men lacked virility: the American Indian was 'feeble . . . He has small organs of generation . . . and no ardour whatever for his female . . . Nature, by refusing him the power of love, has treated him worse and lowered him deeper than any animal'. Jefferson sent Buffon measurements of bears and beavers and, as final proof, a stuffed moose, but not until the first all-American mastodon was exhibited in Paris was the reputation of the New World saved.

The notion of decline was succeeded by the idea of advance, by the assumption that today's plants and animals are more developed than were their ancestors. Nature, it seemed, moves inexorably from microbe to man, in Pope's 'Vast chain of being! Which from God began,/Natures aethereal, human, angel, man,/ beast, bird, fish, insect, what no eye can see,/ No glass can reach; from Infinite to thee.' But has life improved, and, if it has, are its latest models better than what went before?

The theory of evolution by natural selection differs from other models of the past because it has no inbuilt need for progress. Richard Owen, inventor of the word 'dinosaur', founder of the Natural History Museum in London and polymath to the British government, was unabashed about what history was for. When it came to fish: 'those species, such as the nutritious cod, the savoury herring, the rich-flavoured salmon, and the succulent turbot, have greatly predominated at the period immediately preceding and accompanying the advent of man; and that they have superseded species which were much less fitted to afford mankind a sapient and wholesome food.' His view that life changes for man's convenience contrasts with Darwin's stark – and accurate – warning that no animal exists for the good of another. The evolution machine can do no more than adapt its products to whatever they face at any instant. It has no inbuilt direction, be it forwards, backwards or sideways.

In Italy, such was the force of habit that it was until 1923 customary to drive on the left-hand side of the road in town, and on the right in the country. The Darwinian machine, too, once started, is hard to force into reverse. Life might be short on progress but has plenty of inertia. The European ancestor of domestic cattle, the aurochs, was much hunted before it had the sense to form an alliance with its enemy. The last one died in Poland in 1627. It was black, six feet high, and had fearsome horns. There have been many attempts to recreate it with crosses between fighting bulls and other breeds. Although these animals resemble their noble antecedent, they are a cheap copy: like it in shape, but quite different in their genes. To reverse through the maze of descent is impossible.

The vehicles of evolution often become trapped in a one-way system. Most organisms are sexual. Some of those who take up the pastime are able to return to the innocence of Eden when the chance arises but many others have taken an irrevocable step along the sexual road. Pines stick with it because the chloroplasts (the green factories that use sunlight as a source of energy) are passed through pollen, while fish and amphibians are forced to hold on to their males because the sperm provides the machinery of cell division. Virgin birth is unknown among mammals because males put a stamp upon genes that pass through sperm. Some creatures go to great lengths in their attempts to escape their reproductive fate. Certain fish use sperm of another kind to activate an egg, but they, too, are imprisoned by sex as although they do not use his genes, they need the male to make the crucial sperm. A state of development descending from the most ancient forms is, it seems, hard to escape.

Even so, in one particular sense the more recent forms must be higher; for each new species is formed by having had some advantage in the struggle for life over other and preceding forms. From the extraordinary manner in which European productions have recently spread over New Zealand we may believe that in the course of time a multitude of British forms would exterminate many of the natives. Under this point of view, the productions of Great Britain may be said to be higher than those of New Zealand. Yet the most skilful naturalist from an examination of the species of the two countries could not have foreseen this result.

Some natural trends do suggest progress, but on a limited scale. The tiny worms who left evidence of their ancient hunts as tunnels in mud that later became stone started off searching at random, but

– four hundred million years later – were tracking back and forth across the bottom in a more purposeful way. In mammals, most ancient species are smaller than those alive today. At the end of the dinosaurs, the largest mammal was the size of a cat, but now we have the elephant and the whale. The dinosaurs themselves grew from cats to eight-ton giants within a few tens of millions of years. However, the leviathans did not last. Not just the dinosaurs lost out. The giant sloths and mastodons of the Americas have disappeared (as have most of the ocean's whales); modern horsetails and mosses are tiny compared to those of the coal-forests and in spite of signs of progress in fossil behaviour, plenty of worms today blunder at hazard across lake bottoms. Evolution has no escalator of increase, be it in smartness or in size.

The idea of evolution as a ladder is (or ought to be) dead, but life has certainly got more complicated since it began. Today's bacteria are not much different from those of two billion years ago; but now we have animals and plants as well. Some organs have become more intricate than before. Insect limbs have evolved from simple jointed tubes into pincers, paddles and more, while the mammalian brain is more folded than it was. Other structures (such as the skull, with fewer bones in humans than in ancient fish) have stalled. Plenty of animals go backwards, with males as parasites inside their females. In parasites themselves, all is lost but the organs of digestion and sex.

Complexity sends mixed messages about progress along the evolutionary road. The earliest sponges had a mere half-dozen cell types, but humans have hundreds. There has, even so, not been much increase in that measure of complication over millions of years, since fish have about the same number as ourselves. Sponges have become more colonial since they began, but the corals, almost as old, are nowadays more solitary than they were. In economics, in the long term, as Keynes said, we are all dead. Biology, too, never thinks of the future and what matters happens now, in an ecological instant, with no grand plan.

Too often, the notion of progress is used as a code-word for perfection; the chain of being in a different guise. The term should be employed with caution. Some see an arrow of time in biology, as in physics, but in the opposite direction – a relentless tendency to improve, just as the universe has a built-in trend towards chaos and disorder. That is too optimistic. Some lineages get more complicated, some simpler, and much of life has to struggle to stay in the same place. If everyone is evolving, nobody can afford to stop, and there may be constant change with no overall advance at all. Although living things have become more complicated in the past four billion years, the issue of which form is higher and which lower usually depends on who asks the question. Evolution does not need progress. After all, transfixed by time's arrow, all its products will soon be dead.

On the succession of the same Types within the same area, during the later tertiary periods. Science fiction sees the ancient world as much like today, with added dinosaurs. For biology, though, the past is another country, a story of different worlds as much as of different beings.

Evolution is rather like history itself, a drama in which different actors succeed to the same role. The world has fifty or so large carnivores – lions, wolves, jackals, bears and more. Wherever they are found, they fall into four groups with different habits. Some, big cats included, eat only meat. Others, hyenas, eat meat and crush bone. Jackals and foxes have wider tastes, and take half their food as roots and vegetables, while bears have shifted further towards vegetarianism. Each faction has its own skull structure, and any skull can at once be ascribed to the group to which it belongs.

There is more than one way to skin a cat. Five million years ago, in what is now Yellowstone National Park, all the groups were present, but different animals played each part. Hyenas had not yet appeared. Instead, a large dog went in for bone-crushing. Bear-like animals roamed the land. They were not real bears, but huge raccoons with a taste for berries and roots as well as flesh. In spite of much extinction and the appearance of new forms, the four elements, made up at various times of quite dissimilar beasts – bear-dogs, bear-raccoons and a lost set of cat-like meat-eaters called nimravids – lasted for millions of years. Birds show the same consistency. Today's vultures, from the Americas or from Africa, fall into three guilds, each with distinct heads and beaks. They are rippers, gulpers and scrapers; and scrapers, gulpers and rippers from quite a different set of birds are preserved in the La Brea Tar Pits.

The play has had such a long run that there have been inevitable changes in its plot. Five hundred million years ago the air had twenty times as much carbon dioxide as it contains now. This led to a natural 'greenhouse effect', which was reversed two hundred million years later when the level of the gas dropped. Oxygen, too, has swung between extremes. Twice as much of the gas as today allowed the growth of enormous plants, of spiders the size of a hardbacked book, and of scorpions a foot long. A later burst led to the development of aerial reptiles such as *Quetzalcoatlus*, with wings forty feet across. Its abundance allowed animals to burn energy at a rate great enough to persuade them into the air. In today's attenuated atmosphere, nothing so large could carry the burden of gravity. Even the days of the Earth have changed. As the moon saps its neighbour's rotational energy, the globe slows its spin. Corals have daily and annual surges of growth and growth rings from four hundred million years ago show that there were then four hundred days a year. Whatever its state of development, the pace of life in those short and energetic days was faster than it is today.

The geological record is the court of last appeal for all theories of evolution. Although biologists still argue about how the process works, fossils make it impossible for anyone, biologist or not, to deny that it happened. Cuvier, faced with a pile of bones of extinct animals from the Paris Basin, said, 'We will take what we have learned of the comparative anatomy of the living and will use it as a ladder to descend into the past'. Now, the fossils themselves provide, if not a ladder, a set of hesitant steps through time and a proof of a past more dramatic than anything he imagined.

The days when fossil-hunters were, like the College of Heralds, engaged in a futile search for missing links have gone. The record of ancient beings is the most forceful statement of what evolution can do. A century ago, geology was at the centre of the conflict between science and belief. For a time, its patterns seemed so different from those of today as to lead to scientific arguments that were almost theological in their intensity. The disagreements are still there; but now that those who study ancient and modern forms of life speak the same language they may, some day, be solved.

Summary of the preceding and present Chapters. I have attempted to show that the geological record is extremely imperfect; that only a small portion of the globe has been geologically explored with care; that only certain classes of organic beings have been largely preserved in a fossil state; that the number both of specimens and of species, preserved in our museums, is absolutely as nothing compared with the incalculable number of generations which must have passed away even during a single formation; that, owing to subsidence being necessary for the accumulation of fossiliferous deposits thick enough to resist future degradation, enormous intervals of time have elapsed between the successive formations; that there has probably been more extinction during the periods of subsidence, and more variation during the periods of elevation, and during the latter the record will have been least perfectly kept; that each single formation has not been continuously deposited; that the duration of each formation is, perhaps, short compared with the average duration of specific forms; that migration has played an important part in the first appearance of new forms in any one area and formation; that widely ranging species are those which have varied most, and have oftenest given rise to new species; and that varieties have at first often been local. All these causes taken conjointly, must have tended to make the geological record extremely imperfect, and will to a large extent explain why we do not find interminable varieties, connecting together all the extinct and existing forms of life by the finest graduated steps.

He who rejects these views on the nature of the geological record, will rightly reject my whole theory. For he may ask in vain where are the numberless transitional links which must formerly have connected the closely allied or representative species, found in the several stages of the same great formation. He may disbelieve in the enormous intervals of time which have elapsed between our consecutive formations; he may overlook how important a part migration must have played, when the formations of any one great region alone, as that of Europe, are considered; he may urge the apparent, but often falsely apparent, sudden coming in of whole groups of species. He may ask where are the remains of those infinitely numerous organisms which must have existed long before the first bed of the Silurian system was deposited: I can answer this latter question only hypothetically, by saying that as far as we can see, where our oceans now extend they have for an enormous period extended, and where our oscillating continents now stand they have stood ever since the Silurian epoch; but that long before that period, the world may have presented a wholly different aspect; and that the older continents, formed of formations older than any known to us, may now all be in a metamorphosed condition, or may lie buried under the ocean.

Passing from these difficulties, all the other great leading facts in palaeontology seem to me simply to follow on the theory of descent with modification through natural selection. We can thus understand how it is that new species come in slowly and successively; how species of different classes do not necessarily change together, or at the same rate, or in the same degree; yet in the long run that all undergo modification to some extent. The extinction of old forms is the almost inevitable consequence of the production of new forms. We can understand why when a species has once disappeared it never reappears. Groups of species increase in numbers slowly, and endure for unequal periods of time; for the process of modification is necessarily slow, and depends on many complex contingencies. The dominant species of the larger dominant groups tend to leave many modified descendants, and thus new sub-groups and groups are formed. As these are formed, the species of the less vigorous groups, from their inferiority inherited from a common progenitor, tend to become extinct together, and to leave no modified offspring on the face of the earth. But the utter extinction of a whole group of species may often be a very slow process, from the survival of a few descendants, lingering in protected and isolated situations. When a group has once wholly disappeared, it does not reappear; for the link of generation has been broken.

We can understand how the spreading of the dominant forms of life, which are those that oftenest vary, will in the long run tend to people the world with allied, but modified, descendants; and these will generally succeed in taking the places of those groups of species which are their inferiors in the struggle for existence. Hence, after long intervals of time, the productions of the world will appear to have changed simultaneously.

We can understand how it is that all the forms of life, ancient and recent, make together one grand system; for all are connected by generation. We can understand, from the continued tendency to divergence of character, why the more ancient a form is, the more it generally differs from those now living. Why ancient and extinct forms often tend to fill up gaps between existing forms, sometimes blending two groups previously classed as distinct into one; but more commonly only bringing them a little closer together. The more ancient a form is, the more often, apparently, it displays characters in some degree intermediate between groups now distinct; for the more ancient a form is, the more nearly it will be related to, and consequently resemble, the common progenitor of groups, since become widely divergent. Extinct forms are seldom directly intermediate between existing forms; but are intermediate only by a long and circuitous course through many extinct and very different forms. We can clearly see why the organic remains of closely consecutive formations are more closely allied to each other, than are those of remote formations; for the forms are more closely linked together by generation: we can clearly see why the remains of an intermediate formation are intermediate in character.

The inhabitants of each successive period in the world's history have beaten their predecessors in the race for life, and are, in so far, higher in the scale of nature; and this may account for that vague yet ill-defined sentiment, felt by many palaeontologists, that organisation on the whole has progressed. If it should hereafter be proved that ancient animals resemble to a certain extent the embryos of more recent animals of the same class, the fact will be intelligible. The succession of the same types of structure within the same areas during the later geological periods ceases to be mysterious, and is simply explained by inheritance.

If then the geological record be as imperfect as I believe it to be, and it may at least be asserted that the record cannot be proved to be much more perfect, the main objections to the theory of natural selection are greatly diminished or disappear. On the other hand, all the chief laws of palaeontology plainly proclaim, as it seems to me, that species have been produced by ordinary generation: old forms having been supplanted by new and improved forms of life, produced by the laws of variation still acting round us, and preserved by Natural Selection.

CHAPTER XI

GEOGRAPHICAL DISTRIBUTION

On geography and change — Present distribution cannot be accounted for by
differences in physical conditions — The five mediterraneans: their invaders
and their residents — Importance of barriers — Means of dispersal by land,
sea and air; and by changes of climate — Centres of Evolution — Dispersal
during the Glacial period — The uneasy Earth and the geography of life

IN THE ARIZONA desert in the early 1990s an island was built.
Biosphere Two, as it was called (the Earth itself was Biosphere One),
tried to isolate itself from the pollution and vice around it. The plan
was to create a world that never was – an unadulterated place in
which man and Nature could live in harmony. Eight Biospherians
set up a community sufficient unto itself, an ecosystem in the balance
that had once, they claimed, ruled Nature. The immense greenhouse
was sealed off from the air as a two-hundred-million-dollar microcosm of diversity, from desert to rainforest to million-gallon ocean.

Within a year, its inhabitants faced reality. Microbes in the soil
caused the amount of carbon dioxide to shoot up and the level of
oxygen to fall to that on the summit of Mont Blanc. Vines strangled
whole sections of the Biosphere as other plants died out. The animals
had even less success. Nineteen of twenty-five kinds of vertebrate
perished, as did all the insect pollinators (which meant that most of
the plants were doomed). The 'desert' grew grass and the water
could be kept clean only by cutting great mats of algae. In 1994 the
Biosphere was abandoned.

The United States itself was established for the same reason as that great glasshouse. It was a new world, isolated from the evils of the old; a chance to start again in harmony with Nature. Much of what its pioneers saw was familiar. Laurel, walnut and ivy; robins, black-birds and larks, all were there.

Although the new England seemed much like that left behind, it was not. The American laurel is a poisonous heath plant, unrelated to European plants of the same name, and the robin is a thrush. The settlers' names were based on nostalgia rather than biology. Each colony – under glass, and under God – faced the unpalatable truth that when life is isolated, it changes.

The Origin begins with travel: 'When aboard HMS *Beagle*, as naturalist, I was much struck with certain facts in the distribution of the inhabitants of South America . . .' That sentence is full of mean-ing. It was the door to what became a science of its own. Darwin realized that to prove the fact of evolution all that is needed is to go somewhere else. As he saw, some of its best evidence comes from maps, with the geography of life 'a grand game of chess with the world for a board'.

There is a striking parallelism in the laws of life throughout time and space: the laws governing the succession of past times being nearly the same as those governing at the present time the differences in different areas. Why should Australia be the only place with kangaroos? Do we need St Patrick to explain the absence of snakes from Ireland? And why is coal, the remnant of tropical forests, found in the Antarctic? All this makes sense if existence altered as it moved. Geography is an escape from the sad truth that (except through fossils) we cannot visit the past.

In considering the distribution of organic beings over the face of the globe, the first great fact which strikes us is that neither the similarity nor the dissimilarity of the inhabitants of various regions can be accounted for by their climatal or other physical conditions. There is hardly a climate or condition in the Old World which can-not be paralleled in the New. Notwithstanding this parallelism in the conditions of the Old and New Worlds, how widely different are their living productions!

The wine counter of any supermarket (except in France) shows how different places can, given the chance, support identical forms of life. It may sell twenty different Chardonnays. They taste much the same, although some are smoky and others redolent of butter, peach or passionfruit – subtle contrasts, but important to those old and rich enough to disguise their favourite drug. Some bottles are expensive – Chablis can cost fifty pounds – others as cheap as South African or Chilean labour can make them. Those who buy them are out of fashion, and Britain has an Anything but Chardonnay club, founded to stamp out the ubiquitous invader.

Distant parts of the world have millions of vines that grow the grape. Only an expert can separate the best French from, say, the best Australian product of that name. Even the bottles tend to look the same. Whatever the price, and from wherever they come, their labels often show pantiles and vineyards set among scrub-covered hills. From France (the home of Chardonnay) to Western Australia (where it was first grown only twenty years ago), via Chile, California and the Western Cape, the weather and the landscape are similar. The wine trade makes, although the oenophile might deny it, an identical product in each location.

The Chardonnay grape flourishes in these scattered places because each has the same weather and soil. They are the five great mediter-ranean ecosystems of the globe. All are about the same distance from the Equator, and all have wet winters and hot dry summers. Vines, first domesticated around the Mediterranean itself, now grow in every region lucky enough to share its climate.

Although much has been displaced by grapes, each of the five has its own native vegetation – maquis or garrigue in the Mediterranean, fynbos in South Africa, chaparral in California, mattoral in Chile, kwongan in Australia. Whatever the local flora might be called, it is made up of hard, spiny and fire-resistant shrubs. To wander through maquis, kwongan or fynbos is to sniff the heady scents of nature to the sound of birds while being torn apart by thorns. Such ecosystems, wherever they might be, represent some of the most diverse com-munities on earth. The Mediterranean itself has more than twenty thousand kinds of higher plant (a tenth of all those in the world). The five communities taken together contain a greater variety of plants than the whole of tropical Africa and Asia combined. The birds, too, are marvellous. Three hundred and fifty kinds breed around the

Mediterranean, almost as many as in the whole of the rest of Europe. A wet winter and a hot dry summer are, for some reason, a great promoter of biological variety.

Similar as they seem, the native plants and animals of each place (unlike the plant introduced by man) are in reality quite unalike. Each faces much the same conditions – that, after all, is why vines do well – but their resemblance, in thorns or in song, is superficial. If we compare large tracts of land in Australia, South Africa and western South America, we shall find parts extremely similar in all their conditions, yet it would not be possible to point out three faunas and floras more utterly dissimilar. Those natural communities, and their twins in the Mediterranean and California, have almost no biological affinity. Instead, all have their own residents, descended from local ancestors and adapted to a regime of fire and hard grazing. In southern Europe, dwarf oaks do the job, in South Africa heathers dominate, and in Australia most of the trees are wattles. Chilean spines are as likely to be those of cacti as of shrubs. The plants descend from separate sets of rainforest ancestors that flourished before the last glacial epoch.

Nature's work is in great contrast to our own. The natives look similar but are in fact distinct, while the wine tastes the same because it is the same. Faced with the same conditions, ready for exploitation by the narrow and commercial mind of man, biology has moulded different sets of inhabitants to do the same job. Their resemblance is on the surface, and a glance at the actors in each ecological play, similar though their plots might be, shows that all have their own history. The likeness of fynbos, kwongan, mattoral, chapparal and maquis reflects five independent responses to the same force of natural selection. The locals – unlike the vineyards – make the case for evolution.

Except in gardens, there are no cactuses around the Mediterranean, or wattles in Chile. Such plants are confined to their native land because they are confined behind barriers – the Sahara or the Andes – that prevent them from spreading further. All plants or animals face obstacles that exclude them from places where they might otherwise do well. No two marine faunas are more distinct, with hardly a fish, shell or crab in common, than those of the eastern and western shores of South and Central America; yet these great faunas are separated only by the narrow, but impassable, Isthmus of Panama. Any plant or animal will migrate to wherever it can, given the chance. As soon as an impediment is removed, life pours through and destroys any less adapted forms in the way. To breach a barrier is to experiment with evolution.

In the 1980s, Israelis were forced to give up bathing. Their beaches were clogged with immigrants: poisonous jellyfish, twenty in every cubic yard of water. They choked power-station inlets and forced fishermen to stay in port because their nets were filled with decayed flesh. The culprit was a native of the Red Sea that had broken through a five-million-year-old barrier.

The Mediterranean has an unexpected history. It has been not one, but several, seas (and, now and again, deserts). Five million years ago, it was dry, because the last of many great evaporations had left a layer of salt a mile thick across its floor. Some of its inundations were from its eastern end and brought warm-water plants and animals that were well adapted to the tepid waters of what is, of its nature, a sub-tropical ocean. The Mediterranean's modern waters, though, came from the Atlantic, over the great Falls of Gibraltar. Because it last filled from the west, the Mediterranean is now a warm sea filled with the descendants of plants and animals from a cold ocean. At the end of the last ice age, just eighteen thousand years ago, there were penguins in the South of France, and even today the Mediterranean has more than three thousand whales (together with a tiny remnant of its once abundant seals).

The Suez Canal was opened in 1869. It was the successor to several earlier links with the Red Sea, the first made three thousand years ago by Rameses II of Egypt (the builder of the Abu Simbel temples, themselves inundated by today's engineers). The present Canal is half a mile across in parts of its hundred-mile length. It acts like a giant new Mississippi as its waters flow downhill from the Red Sea to its younger cousin. The connection between the seas allowed their animals, confined for millennia, to move.

The traffic was one way. Three hundred Red Sea natives made it to the Mediterranean, almost none the other way. The migration continues at a rate of ten new forms a year, with the newcomers flourishing at the expense of the locals. A third of the Israeli catch now consists of Red Sea fish and the wave of aliens has reached Sicily. The native prawns have almost gone and many fish are in decline. The survivors manage only because they leave space for the invaders. The local mullets, for example, live in deeper waters than do their newly arrived relatives.

The immigrants succeeded because they are more adapted to today's Mediterranean than are its natives. The locals have had no time – nor, in the absence of competition, much need – to respond to the challenges presented by its warm and salty waters since the sea last filled. As a result, they were soon driven out by the subtropical outsiders. The breach in the barrier allowed animals to swarm into the new space. In time they will fill it – as far as they are able, and until they are stopped by the cold or by others more suited to local conditions. Wherever it arises, and however it travels, a species will move on until something restrains it. As it does, it must evolve or die.

As a result, the continents generate their own mixtures of in- habitants. On each one, successive groups of beings, specifically distinct yet clearly related, replace each other. On the plains of La Plata we see the agouti and bizcacha, animals with nearly the same habits as our hares and rabbits, but they plainly display an American type of structure. We look to the waters and we do not find the beaver or musk-rat, but the coypu and capybara, rodents of the American type. We see in these facts some deep organic bond, prevailing throughout space and time. The naturalist must feel little curiosity who is not led to enquire what this bond is.

The bond is inheritance, modified by natural selection. Such community of descent casts its net wide. The world can be divided into great provinces – the New World Arctic and tropics, their equivalents in the Old World, Africa, the Far East and Australia, with more domains beneath the sea. Each has its own identity, shared not just by mammals, but by insects, snails, worms and trees. The differences are not absolute, and many creatures range over more than one province, but the existence of such huge areas of affinity is evidence of a shared past.

Means of Dispersal. Once evolved, plants and animals face a constant struggle against the pressure of their own numbers. They move as far as they can, by their own efforts or with the help of others. Some can travel for huge distances by land, sea or air, while others are confined to the place where they were born.

Often, a hopeful migrant faces an impassable barrier. The capacity of migrating across the sea is more distinctly limited in terrestrial mammals than perhaps in any other organic beings; and, accordingly, we find no inexplicable cases of the same mammal inhabiting distant points of the world. For other animals the sea is a highway. Pytheas of Massalia was the first sailor to venture beyond the Pillars of Hercules (and the first Greek to visit the British Isles). As he explored the Atlantic shore of Spain he noticed that the ocean flowed south, like an immense river, an *okeanos*. His river runs on as the Canary Current, part of a great girdle of water flowing around the North Atlantic. The Gulf Stream has the force of three hundred Amazons; a river forty miles across that takes a month to cross the Atlantic and is but one of many great conveyor belts bearing flotsam across the world.

All oceans have their currents. About a tenth of their waters are always on the move. Most flow on the surface (although slow streams in the deeps take a thousand years to carry icy water from the Antarctic to the Galapagos). The top ten feet of the ocean store as much heat as does the entire atmosphere. The movements of water are started by heat, by winds and by the rotation of the Earth. Because water at the Equator is warm, it expands. As a result, the sea in the Caribbean is three inches higher than at Newfoundland. The warm water flows downhill towards the Poles, and is twisted in a clockwise direction in the northern hemisphere (and its reverse on the other side of the Equator) as the world turns. The wind obeys the same rules, and its storms help generate the sea's drift as they spin round the globe.

Some animals float to a new home. The surface of the water between the Galapagos Islands carries dozens of insects – aphids, cicadas, ants and mosquitoes. Most will drown, but a few will make the journey. Others hitch lifts on the many vehicles that pass by. Every day, ten million pieces of garbage – bottles, bags and plastic sheets – are dumped from ships. From the land, the sea receives much more. The island of Pitcairn is three thousand miles from the nearest mainland. Its best-known detritus was the *Bounty* mutineers, who landed in 1789. The island was so remote that their refuge was not discovered until all but one had died. Nowadays, its beaches are as filthy as any in Europe, with a piece of rubbish every yard. Pitcairn has the European mix of buoys, bottles and bags, but a relative shortage of disposable nappies. The whisky bottles suggest that many of the migrants come from South America. Not all the flotsam is useless, as the local land crabs are fond of shoes as shelters, but it is a dismal reminder of how the most remote places have been forced to join the modern world.

Some marine debris is still mysterious: why are twice as many left shoes as right washed ashore in The Netherlands, while the opposite is true for Scottish beaches? All has a message for evolution. Nowhere is isolated. Given the chance, plants and animals will float, fly or drift through the air to reach the most remote parts of the earth.

Plenty of animals travel on rafts (shoes and bottles included). Off the coast of Cuba float substantial islands of vegetation that may bear mature trees. On Christmas Island, three thousand miles from North America, American redwood, fir and walnut are used as firewood because so many come ashore. Canoe-builders on coral atolls in the deep Pacific once depended on logs that had floated halfway across the world. A complex etiquette determined who got the biggest. The natives procured stones for their tools, solely from the roots of drifted trees, these stones being a valuable royal tax. Many of the unsinkable vagrants bore not just stones, but animals and plants. Great clumps of seaweed full of shore animals have been found a hundred miles from land. Glaciers, too, deliver their contents as they float from the ice caps. Antarctic boulders on the sea floor off Cape Town were brought within the past ten thousand years. Among them are the remains of penguins. However, the sea can be a formidable barrier. Few freshwater fish can manage to cross a strait more than a couple of miles across, elephants stop swimming at thirty miles, and tortoises and snakes cannot manage more than five hundred.

To find a raft is not enough, because the travellers must survive their journey. Most land plants find it hard to deal with seawater. Their seeds do rather better. Some simple experiments prove how tough they are. Often, dried seeds do best. To dry stems and branches of many different plants and to place them on sea water shows that some when dried floated; for instance, ripe hazel-nuts when dried floated for ninety days and then germinated. As many currents run at sixty miles a day, plants might be floated across miles of sea; and, when stranded, if blown to a favourable spot by an inland gale, they would germinate.

Seeds are the genetic memory of the plants that bore them. Many survive not for weeks, but for years, with germination sparked off by a change in the environment.

An unplanned test of the power of the seed began when the wheatfields of northern France were abandoned after the economic collapse that followed the Franco-Prussian War of 1870. The crop was in those days full of weeds. Poppies were everywhere. The plant can generate thirty thousand seeds in a square yard of soil. After the French collapse the farms of Flanders stayed grazed and flowerless until 1914. Then, the land was cultivated again – not with ploughs, but with swords, shells and blood. Once disturbed, the poppies bloomed at once, from long-buried seeds. A quarter of a century later, in the next round of human folly, the Natural History Museum in London was bombed. The fire-hoses caused many seeds to germinate, among them a mimosa, collected in China in 1713, and revived in a sudden flood two centuries on and five thousand miles away. The poppy is now much used as a symbol of war's destruction; but it bears a more hopeful message about how well life can survive in the face of adversity.

With a third of the world's plants in danger, there is a new interest in conservation. The toughness of seeds is a great help – and is a reminder of their vital role as containers for genes. A Millennium Seed Bank at Kew aims to store dried and frozen seeds of a tenth of the world's kinds of plant (together with the whole of the British flora). Most should last for centuries. Simply to dry the seeds of beet, rice or elm allows them to survive for a decade and more. Some weed genes allow the plants to lie low even in good times. They sprout over weeks or months rather than all at once – which is useful when it comes to long and risky journeys.

Plenty of travellers fly, rather than float, across the globe. Every few years, after an Atlantic storm, dazed North American birds reach Europe. Most die, but the survivors carry a cargo. The crops of birds do not secrete gastric juice and so do not in the least injure the germination of seeds. All the grains do not pass into the gizzard for twelve or even eighteen hours. A bird in this interval might easily be blown five hundred miles, and hawks are known to look out for tired birds. Some bolt their prey whole, and after an interval disgorge pellets, which, as seen in experiments made in the Zoological Gardens, include seeds capable of germination.

Locusts soar upwards as the sun falls, and are caught in the winds of the upper air, to move fifty miles a night. They concentrate on the edge of weather systems to form plagues as they descend to earth. Less conspicuous things also move – as anyone who suffers from hay-fever knows. Every summer afternoon, on the Costa del Sol, a great cloud of marijuana pollen descends from the illicit fields of Morocco. A constant rain of pollen and spores from South America, mixed with the odd seed and insect, falls on Signy Island, on the edge of the Antarctic. Few places on the Antarctic continent itself are warm enough for mosses and liverworts to grow, but where they can – around hot springs and the like – they do, evidence that nowhere, remote though it may seem, is safe from migrants. The presence of the same form in distant places is not evidence that it was created twice, but that it can move.

Dispersal during the Glacial Period. The identity of many plants and animals on mountain-summits, separated from each other by hundreds of miles of lowlands, where the Alpine species could not possibly exist, is one of the most striking cases known of the same species living at distant points, without the apparent possibility of their having migrated from one to the other. The pattern arises not because each peak is a separate factory for the same product, but because a brutal landlord has broken up great estates of the living world.

Their inhabitants are a relic of ancient ice-sheets, now retreated. Glaciers were the first hint that science and the Bible do not coincide. How could even a believer in Noah's Flood account for the Scottish boulders, too big for any conceivable deluge, that turned up in Wales, where geologists found great chunks of the island of Ailsa Craig? At a conference in Glasgow in 1840 the young Swiss geologist Louis Agassiz persuaded the assembled sceptics that these 'erratic blocks' (later the subject of a celebrated Spoonerism) were moved not by water, but by ice. Now, glaciers are known to explain many enigmatic patterns of distribution.

Most of the familiar landscapes of today are a statement of the power of frozen water. The evidence of past ice ages is everywhere around us. The ruins of a house destroyed by fire do not tell their tale more plainly than do the mountains of Scotland and Wales, with their scored flanks, polished surfaces and perched boulders, of the icy streams with which their valleys were lately filled.

Ice is the commonest rock in the solar system. Even the moon has plenty beneath its surface, but Earth is the sole planet upon which it exists in consort with water itself. The balance of the two has waxed and waned, with rapid swings between ice ages over the past three million years. Great waves of cold are not common. Bouts of glaciation hit seven hundred million years ago, and five hundred million years later, but for most of the time the world has been warmer than today. It has cooled for a hundred million years. When the last long winter began, Northern Alaska had ferns and lush forests of gingko, and the banks of the Thames Estuary were covered by a subtropical forest, with crocodiles and turtles in its waters.

The Pleistocene has had sixteen cycles of cold and warmth. At times, glaciers covered more than a quarter of the land. Each glaciation lasted about a hundred thousand years, with brief incursions of warmth, in one of which we live. In the most recent cold period, ice reached to Missouri and central England and the tropics cooled and dried. As the sea fell, it exposed thousands of square miles of land. Water is heavy stuff. The lake behind the Hoover Dam depressed the deserts around it by six feet. The cloak of ice, too, forced the land to sink and, after its burden was shed, it rose again. Arctic Canada has elevated itself by a thousand feet – and it still bounds upward at an inch a year. Scotland's rise is less precipitate, but its release from a frigid past means that these islands are tipping over, with Scotland on an upward path and the south of England falling into the sea.

When the last glaciation ended, a mere dozen millennia ago, the summer temperature rose by eight degrees Celsius – the difference between London and Lagos. Great changes in temperature over shorter times are recorded in tree rings, in ice cores, and in the sediment of Arctic lakes.

Cycles of warmth and cold continue, and animals and plants still travel in their wake. The 'little ice age', which began in the sixteenth century and lasted to Victoria's day, explains both the Dickensian Christmas and the failure of explorers from Cabot to Franklin to find a sea route around North America. It has switched into today's global warming. A peak of heat in 1945 has been succeeded by another. Three of the five warmest years in England (and the records go back to 1659) have been since 1988, with 1998 the warmest year ever recorded for the entire planet.

Satellite photographs show that spring in the temperate parts of the Northern Hemisphere now starts a week earlier than it did in the 1960s. A network of schoolchildren tracks the key events – the first snowdrops, the earliest oaks to flower and the first cuckoo. They follow a noble tradition. The descendants of Robert Marsham, of Norfolk, pinned up a chart of the arrival of spring in 1736 and kept it for two hundred years. It recorded how often – and how late in the year – the chamber pot had a skim of ice. Frozen urine was more of a feature of an eighteenth-century Easter than of that festival today.

Why is the climate so unstable? In part it turns on the Earth's orbit, an ellipse whose shape changes with a rhythm of about a hundred thousand years. At its most extreme it causes winters to be colder and summers hotter as the planet moves further from and closer to its source of heat. A series of cold winters is enough to tip the globe into frost. The sequence is modulated by a shorter cycle in the Earth's tilt and wobble. As forests bloom – and later rot – in warm periods, carbon dioxide and methane escape into the air. These allow energy from the sun to pour in, but trap the ground's own long-wave radiation to give a 'greenhouse effect' that pushes the temperature up still further. As ice can pass through a glacier at five miles a year, it does not take much of a drop in temperature for it to race across the landscape, or much of an increase for an ice cap to shrink.

For much of the time the climate is poised on edge. The end of the last European ice age was marked by two sudden spurts of cold. In each, for a few centuries, glaciers returned to the Pyrenees and the Alps. They retreated ten thousand seven hundred and twenty years ago. The short interval of warmth between the ice's last breaths lasted a mere moment, but was enough to allow Mediterranean beetles to flourish in Britain before they were again wiped out. Greenland, too, has seen many increases of eight degrees over a few decades, followed by sudden drops in temperature. Today's concern about heat will be succeeded by anxiety about a new ice age if the warm and salty Gulf Stream is displaced by cold fresh water from what used to be the ice cap – which, in parts of Greenland, has lost three feet of its depth in only a decade. The last great interglacial, ten thousand years long, ended in just four centuries as the Atlantic conveyor belt was stopped by melted ice.

As ice retreated at the end of its most recent advance, it forced the animals and plants of cold climates to withdraw with it. Most made it back to the far north or south, but some were marooned in mountain ranges. They moved up the valleys as the weather warmed. Now, each range has its glacial relicts, identical or almost so to those of distant mountains. For them, the evidence of shared descent remains, but as fragments. They serve as a record, full of interest to us, of the former inhabitants of the surrounding lowlands. Such refugees show that the difficulties in believing that all the individuals of the same species, wherever located, have descended from the same parents, are not insuperable.

The evidence of climatic change is in the inhabitants of the mountains as much as on their scarred flanks. The varying hare (so called because it goes white in winter) is found in Scandinavia and in Northern Canada – and in the Alps, a thousand miles away. Insects, too, follow the ice. Two thousand different kinds are known from Britain's glacial times. Almost all have gone from these islands, but fewer than twenty are altogether extinct. Instead, they survive – not in today's subtropical England but in Siberia, Arctic Canada or the high mountains of Europe. What was once the commonest dung beetle in England now lives mainly in Tibet. Genes show that many other forms now widespread (and with no obvious sign of past disaster) were fractured. In Europe, oaks and grasshoppers have deep divisions, with a great split between the DNA of west and east which reflects their isolation and advance from refuges in Spain and in Turkey. For the same reason, the plants and animals of the north are less variable than are their southern relatives. As they moved step by step to colonize Europe, they went through one bottleneck after another. As a result, the genes of a few formed the fate of their many successors.

Forests are always poised to move. In a churchyard near Edinburgh live a group of trees known as 'the walking yews of Ormiston'. Legend has it that, with some supernatural help, they have shifted from place to place in response to crisis. And, of course, they have. Yews put out new shoots from old branches where they touch the ground. When the old trunk dies, its branch replaces it. In time it grows into a new plant, yards away from its previous incarnation. No miracle is required.

As glaciers fell back, trees pressed on behind. In North America, spruce was the first to migrate, followed by pine. Each travelled at a great rate – a mile every three years. Behind came the chestnuts. Their progress was more dignified, at a hundred yards a year, but for a large tree with heavy seeds that is almost a sprint. Birnam Wood, with its pines, could have marched on Dunsinane (ten miles away) well within Macbeth's lifetime.

The forest journeys continue. The river of ice at Glacier Bay in Alaska has withdrawn by sixty miles since first seen by Europeans in 1750. It took just a century for mature woodland to cover land that was once frozen water. As it did, it lost diversity, and – as in Europe – the pines and the birds of Canada are less variable than are their ancestors to the south. The birds, mice and insects of the north arrived almost at once. The climate changed too fast for them to evolve out of trouble, but instead they moved.

Life's geography is like that of nations. It seems natural that a state's frontiers should be determined by the wishes of its inhabitants and those of its neighbours, but that ignores the facts of politics. Most Britons have the good fortune to live in a territory that defines itself, on an island. Few countries – or animals – are as lucky. History is a record of how difficult it can be to divide up the landscape.

Species and countries each vary in what they occupy. The Devil's

Hole pupfish fills a single freshwater spring in Death Valley. It has its being in a space the size of a large room (and was once reduced to a global habitat of a bucket when its home dried up and its occupants were rescued by an alert conservationist). The blue whale, in contrast, roams all the world's oceans. In much the same way, Russia, the largest political unit in the world, is a million times bigger than Nauru, the smallest.

The range of most birds, flies or plants in the tropics is smaller than that of those in the north. The same is true of states, with tropical nations a quarter the size of those overlapping the Arctic. The geography of animals and plants turns on hard times long ago. As the ice retreated those able to survive in the chill landscapes left behind could follow. The homelands of those inconspicuous birds, the warblers, vary by a hundred times. Those with the largest ranges are in the north, because they could move with the birch or pine forests marching behind the ice. Birds whose haunts (such as rhododendron thickets) are restricted to a single mountain range are perforce confined to smaller areas. Most rare plants and animals are found in the tropics, because they have not been able to expand into the space available to their hardier kin.

The Garden of Eden is always painted as a lush and sultry place. From biology's point of view, too, the tropics are a great and ancient city, with more inhabitants, more energy, more water, more production and, because the land has not been wiped clean by glaciers, more time for specialists to evolve than in the icy north and south. A quarter of all the world's kinds of bird are found on a twentieth of its land and a fifth of its plants in a two-hundredth. All these centres of origin are tropical, in Madagascar, Malaysia or Central America. Arctic Canada possesses ten kinds of ant, compared to two thousand in the same area of tropical South America; and Hong Kong, at four hundred square miles, has more kinds of bird, mammal, insect and plant than the whole of the British Isles. Such gradients from Pole to Equator are as old as the fossil record itself. They prove the claim of Adam Smith that division of labour is proportional to demand. Although a workman who made only nails would never survive in a Scottish village, in a large town such a narrow specialist would prosper.

The great tropical factories of life and the rigours of the last ice age mean that the north is a new nation for all its inhabitants. As plants and animals followed the glaciers as they retreated towards the Poles they filled vast tracts of country. Many had no real need to change, because their new homes were no colder than those they had just left. Evolutionists often see the environment as rigid and life as flexible. The story of the ice shows that, quite often, the opposite is true, and that conditions change more rapidly than those who suffer them. Life is always on the look-out for somewhere more comfortable. It is much easier to migrate than to evolve. Northern roots lie in southern parts, and the inhabitants of half the world are testimony to a catastrophe of just ten thousand years ago. Much as Scottish Nationalists might disagree, to cut themselves off from their neighbours may destroy their chance of survival in the next war with the cold.

Some journeys do seem improbable. Why are there ostriches in Africa, emus in Australia and rheas in South America? All are large, aggressive and – above all – flightless. Their genes show them to be relatives, who descend from a wingless ancestor. How did they make the trip? Trees, too, show unexpected patterns. The southern beech of South America has close kin in Australia, in South Africa and even in the Antarctic, preserved beneath the ice. Ten thousand miles separate these places but their forests are almost the same. Pheasants, partridges and quails, electric fishes and tree-frogs each have ties between their members in Australia, South America and New Guinea, with more distant links even as far as Europe; but none are much good at ocean crossings. Three thousand miles of sea are a barrier for an eagle, let alone a heavy beech seed, a tree-frog or a bird without wings.

The geography of the sea is even harder to understand. Some of its inhabitants are found both at the northern and the southern ends of the Earth but nowhere in between. Their distribution has lasted for two hundred million years so that the simple notion that they floated across the Equator on some recent iceberg does not work. Mussels, scallops and whelks all have close relatives in Arctic Canada, Siberia, Chile and the Antarctic. How can this be?

Such patterns might seem to be evidence against descent with modification. It is incredible that individuals identically the same should ever have been produced through natural selection from parents specifically distinct. To find the same plants and animals in isolation at opposite ends of the globe leads some to the conclusion that the same species must have been independently created. The truth is simpler, but almost as startling.

Scott of the Antarctic froze to death eleven miles from safety. His sledge was weighed down with thirty pounds of rock dragged by exhausted men for hundreds of miles. The rocks helped kill Scott; but held a crucial clue about history. Some contained fossils of a tropical tree, otherwise found in India and Africa.

Much later, the skull of a thirty-foot dinosaur from eighty million years ago was found in Madagascar. It is quite different from those of Africa. With its rough surface and prominent horn it looks like others found in South America and India. In today's atlas, all this makes rather little sense. Perhaps, though, today's atlas is not the one to use.

Some odd patterns of shared geology suggest that an earlier edition is called for. The folds of the Appalachian Mountains in the Eastern United States can be traced to a parallel set of rocks in Ireland and Brittany. In the same way, parts of India, South Africa and tropical South America bear scars made at about the same time by ancient glaciers.

Once there was talk of land bridges between the continents, used as highways for trees, dinosaurs and more on their global journeys. The connections rose and fell almost on demand. So many links between distant places were needed to explain the distribution of animals and plants that the seas were filled with theoretical Atlantises. Those who drew the maps bridged every ocean, and united every island to some mainland. In an attempt to explain the wide distribution of plants and animals, the remotest islands were seen as the wrecks of sunken continents.

The truth is more remarkable, for the earth itself, rather than the seas, is on the move. Continents now far apart were once part of the same mass. They have split and taken their inhabitants with them, to become great arks of land that wander the globe and, now and again, collide. The ancient movements of a fluid Earth explain much of the geography of plants and animals today.

The fragments of what were once continents travel at about the rate that fingernails grow but, in time, that is enough to shift them for great distances. The expansion of the Atlantic has been measured from satellites (leading to a rare joke by the Duke of Edinburgh: that at last we know where we stand in relation to the United States). If the Pilgrims were to repeat their transatlantic trip, they would find themselves out of place; not because Puritanism and stovepipe hats would seem eccentric in America's new conformism, but because America itself has moved on. Their landing place, Plymouth Rock, is fifty feet further west than it was in 1620. William Bradford, if he stepped off the *Mayflower* today, would get his feet wet.

Leonardo da Vinci had noticed that the continents would fit together like jigsaws. A Frenchman, Antonio Snider-Pellegrini, revived the idea in the year before the publication of *The Origin*, in his forgotten work *Creation and its Mysteries Revealed*. Unfortunately for him, he used the observation as a proof of Noah's Flood rather than the key to the Earth's structure. In 1912, the first great world continent was given the title of Pangaea, the universal land, long before evidence of its existence was found. The idea was ridiculed by one geologist as 'utter, damned rot!' because there seemed no force able to drive a land mass across the globe. Although some claimed that the continents had been pushed by the Earth's spin as they ploughed through the ocean floor, that made no sense.

Our planet is less solid than it seems. Deep inside, the core is liquid, and on the surface – rather like the lines of movement in a pot of boiling porridge – are upwellings of molten rock. Most reach the surface in long chains of submarine mountains. These mid-ocean ridges were discovered in the nineteenth century by the Atlantic Telegraph Company, whose engineers laid the first cable between Europe and North America. They assumed that the sea floor was flat, but their connection broke within weeks because it was suspended between peaks higher than the Alps.

The forty thousand miles of ridge mark the lines where the fluid contents of the Earth spew into the seas. As they congeal, the fresh rock moves outwards to make a new ocean floor. The continents float on the Earth's liquid mantle, their keels embedded deep within it. They are pushed apart as the sea floor spreads. The flow from the centre means that the edges of the oceans are cooler and thicker than the submerged ridges, and sink into the mantle below, with the younger rock dragged behind. The constant gain and loss of rock means that the ocean floor is young compared to the land. Near the ridges it dates from the past million years or so, at the edges of the continents from about two hundred million years ago. Most of the sea bottom is younger than the dinosaurs.

Iceland has the misfortune to sit astride the lava factory, the Mid-Atlantic Ridge itself. The country gets bigger by the day. Most of the island is less than twenty thousand years old. In 1783, huge rifts were formed as a hundred square miles of new rock belched from below. A cleric of the time wrote that: 'Those terrors that fell over and upon us I can hardly describe . . . In the middle of the flood of fire great cliffs and slabs of rock were swept along, tumbling about like large whales swimming, red-hot and glowing.' One Icelander in five died of suffocation by noxious gas, or of famine. The infernal stench reached Europe, causing widespread fears of imminent damnation.

When two plates meet out in the ocean, the older, cooler and heavier slips beneath its neighbour. As its mass slips back into the liquid core it pulls the ocean floor behind it. Along the line of sinking rock a trench is formed, and volcanoes burst through to give great circles of islands. They include the Aleuts, part of the 'Ring of Fire' around the western Pacific (an ocean increasing in width by seven inches a year). When two continents crash, a mountain range is thrust towards the skies. As a result, the summit of Everest is formed – like that of Snowdon – of rocks made in a shallow sea.

The atlas of the past holds many surprises. To reconstruct it, all that is needed is to subtract from the map of today the ocean floor made since the date in question. The key is in the Earth's magnetic field. Every few hundred thousand years, what was the North Pole becomes the South until the poles reverse once more in their endless dance. The switch is recorded in the rocks. Great stripes of magnetic reversal across the bottom of the ocean mark their movements. As band after band of older material is taken away, the continents reveal their ancient positions. Before about a hundred and eighty million years ago no ocean floor is left and the information must come from matched geological sequences in different places.

The land itself bears evidence of its journeys. At St Martin in western Canada, Manicougan in the east, and Rochechouart in France are three comet craters dating from two hundred and fourteen million years ago. At first, their shared age seems a coincidence. Rearrange the continents to their position at the time of the collision, and the craters lie on a straight line, proof of the break-up of a comet just before it struck and of the slower decay of the once solid earth upon which it fell.

Leonardo was right. The Americas and the Old World were at one time joined. So were many other places. Five hundred million years ago, there was but Pangaea, a single mass of land. This was later separated by the Tethys Sea into two great continents – Gondwanaland (made up of much of the present India, South America, Africa, Australia and Antarctica) and Laurasia (now North America, much of Southern and Western Europe, and Asia) and a smaller one, Baltica (Northern Europe and Scandinavia). A hundred million years ago, Gondwanaland itself broke up. Several sections declared their independence and drifted northwards to make parts of Europe, Tibet and two pieces of China. Madascar, with its unexpected dinosaurs, is a piece broken off Gondwana long after Africa gained its identity; part of a lost world stranded two hundred miles off a foreign land when its birthplace was shattered by movements of the crust.

The shape of the Earth was not much like that of today until the extinction of the dinosaurs. Not until fifty million years ago did Australia separate from Antarctica and could Europe, adrift from Greenland, at last cement its relationship with Asia. By then, the world could be navigated, more or less, with a modern chart.

As the continents drift like slow ships, they take their passengers – alive and long dead – with them. An animal that started its journey in what is now Australia may have links with others in Africa or South America. Penguins and flightless birds like the ostrich and the rhea are scattered across South America, Australia, South Africa and Antarctica because new oceans have divided their ancient home. The same is true of those of electric fish and frogs. DNA shows the rhea of South America and the Australian emu to be more similar than either is to the ostrich. Genetics and geology tell the same story, for their history is that of the continents. South America split from Africa, but stayed in contact with Australia via a junction across the Antarctic.

Evolution itself reconstructs the movements of the land. The first fossils of birds and mammals, each evolved from separate sets of reptiles, are separated by a giant gap from their modern forms. A clock based on the genes of today times their great radiations at a hundred million years ago, when Pangaea had broken up and the world was more fractured than it has ever been. Europe, the Americas and Africa were each divided into several islands. Dinosaurs, frogs and toads also split into a variety of forms at that time. The history of life and of the continents are close companions.

The restless lands acted not just as arks and cradles, but as funeral ships. The fossils of plants and animals that met their end on the same land mass have moved to form post-mortem alliances with others that expired much later. They are a reminder that the Earth – and its inhabitants – have been in turmoil since they began and a proof of dramatic changes in the atlas at a time long before the ostrich split from its flightless cousins.

In spite of today's resurgence of Celtic racial thinking, Scotland and England seem quite similar places. Their rocks show that this is not at all the case. The ancient limestone around Durness in the far north of Scotland and the matching beds in northern England each contain many fossils of trilobites. The Scottish versions are covered with lumps and warts. English rocks of the same age hold nothing like them, although they have many trilobites of their own. The Scottish fossils are in fact more similar to the trilobites found in parts of North America than they are to their relatives a few miles to the south. The rocks of Scotland tell the tale of an upheaval that formed these islands five hundred million years ago.

The Stone of Scone is an emblem of Scottish identity. The nation's kings were crowned upon it until it was removed to Westminster Abbey in 1296 to do the job for England. Seven hundred years later, faced with a slump in popularity north of the Border, the

Conservative government sent it home in a political association of object and nation more typical of Serbia than Britain. It can now be viewed in Edinburgh Castle, at £5.50 a time. Dr Johnson's comment on another piece of Celtic geology, the Giant's Causeway, comes to mind: 'Worth seeing, yes; but not worth going to see.'

According to legend, the Stone was already well travelled. It was the pillow upon which Jacob slept and dreamed his celebrated dream of a ladder of angels on their way to heaven. It passed with divine help to Egypt, Sicily, Spain and Ireland, and enabled the Scots, in the Declaration of Arbroath in 1320, to define themselves as a Lost Tribe of Israel.

What of the real movements of the Stone over that vast passage of time? Jacob's Scottish pillow is made of pale yellow sandstone. Three hundred and fifty million years ago, it began its existence in shallow tropical waters as silt and sand were washed from a range of mountains. Over many years the grains consolidated into the rock from which the Stone was made. The journey from its brief exile in London, up the slope of glacial rubble known as the Royal Mile, to the granite island in the ice upon which sits Edinburgh Castle, seems fair restitution, the righting of a geological wrong.

But geology is above politics. Two hundred million years before the Stone was formed, Scotland itself was on the move. Much of the country was on holiday abroad. A vanished ocean called Iapetus stretched between England and its northern neighbour and split Ulster from Eire. The rocks that were to become what Nationalists insist on calling the Stone of Destiny were then in North America. Soon, Iapetus – an ocean as broad as but three times older than the Atlantic – began to close.

Until it did, England was part of a separate continent, Avalonia (named for the island to which Morgan le Fay carried King Arthur). Quite where its northern shore ran all those years ago is hard to say, because most has been buried by the sediments scraped from its floor to make the Southern Uplands of Scotland as the ocean closed.

A series of cracks in the landscape – faults – can be traced across Northern England and Ireland. These, together with chemical differences in the rocks of the ancient continents, hint at where the frontier used to be. In Ireland the boundary was south of today's political line. It started north of Dublin and reached to the mouth of the Shannon. On the mainland, the northernmost remnants of Avalon are near the town of Moffat in the Scottish Borders.

Wherever their encounter, the desire for Scotland to reunite with England was such that the two halves of this island charged towards each other at the unheard-of speed of a foot a year. Scotland skated so quickly across the globe to fall into the arms of its southern neighbour because its roots are deeper within the liquid earth than are those of most continents. When their marriage was consummated, life crossed the gap, with invasions and extinctions as great as those of the Mediterranean from the Red Sea, millions of years later.

The clash of Caledonia with Avalon threw up a range of mountains as high as the Himalayas on the north side of the narrowing strait between them. Their stumps are the Scottish Highlands, and their eroded remains the source of the sand that settled, in a shallow sea full of giant sea-scorpions, to become the Stone of Scone.

Scottish Nationalists may hope to re-open Iapetus and reverse the Palaeozoic Act of Union, but geology shows that the real home of the Stone of Destiny is in Newfoundland. There, Scotland's trilobites evolved and died; their fossils a reminder of how that nation – and every other – has moved around the globe far more than its inhabitants did when they were alive. Such post-mortem journeys also explain the bands of coal and fossil coral that link Central America, southern Europe and much of Asia. They are remnants of a time when all were still part of a great tropical continent.

And what of the future? If the continents continue to move at their present rate, in fifty million years America will be close to Asia as the Atlantic broadens and the Pacific gets narrower. Australia will rush northwards to collide with Japan, and the eastern part of Africa will declare independence as the Rift Valley becomes a sea. The Straits of Gibraltar will soon close, and the Mediterranean may again dry into a salty plain before it disappears for ever.

Many lands have moved around the charts. Atlantis has come and gone, but the Island of Buss (supposed to have been discovered in mid-Atlantic by the Frobisher Expedition in the sixteenth century and big enough to take three days to sail around) shrank and shifted as successive travellers failed to find it. It was demoted first to a reef and then to the fable it had always been. Maida Island, spotted near Newfoundland at about the same time, ended up in the Caribbean on a map published as late as 1906.

Land bridges, Atlantises, and all the other myths dreamed up to explain the distribution of plants and animals are less remarkable than the truth: that the world has evolved as much as have its inhabitants.

CHAPTER XII

GEOGRAPHICAL DISTRIBUTION – CONTINUED

Babies as islands — New lands and their colonists — On the nature and inhabitants of oceanic islands and their relation to the mainland — Evolution by accident — The plants and animals of Hawaii — Fresh-water lakes as islands — The fragility of insular creatures and the grand simplification of life — Summary of the last and present chapters

A NEWBORN BABY enters the world and is at once corrupted. At birth, its body is sterile, free of the moulds, bacteria and mites that infest the rest of us. From a bacterium's point of view, the child is a large, fertile and underexploited island. Within moments, it is colonized. Some of the invaders come from its mother's guts, others from the midwife or the proud father. The pioneers are a small sample of the millions of bacteria that float through human entrails. For a time the infant bowel is volatile and unsettled. As any parent knows, that has dramatic effects on the ecology of the territory upon which the invaders make their home.

Within a few days, the child contains ten times as many bacteria as it does cells of its own. From its neighbours, clean as their household might be, comes a rain of immigrants. People, some say, grow to look like their dogs. In fact, what most unites any family (pets included) is the contents of their guts. The younger a baby, and the more isolated from a source of new bacteria, the lower – and more distinctive – its internal diversity. Because of migration, any archipelago, of entrails or anything else, begins to resemble the nearest mainland. In a large town many people have the same bowel contents, while isolated villages diverge. Cities far apart – great continents of bacteria – have a flora of their own, as soon becomes obvious to those who travel to them. The gut of a Suffolk villager at once links him to the intestines of London, while those of a farmer in Bangladesh show his affinity to the entrails of the great city of Dacca.

The eruptions of travellers and babies, messy as they are, reflect an internal struggle for existence, the colonization of a new territory from a larger expanse nearby. What arrives is a matter of chance, dictated by how far from the source the new island might be and how good each migrant is at making the journey. Some bacteria triumph by changing to deal with new conditions, while others fail in the battle and disappear. In time the digestive system settles down as its ecology changes. The process has a lesson for evolution, because the history of guts is not much different from that of islands, of mountains or of great lakes.

The further an island is from the mainland, the harder it becomes for new animals and plants to arrive. As a result, each contains just a sample of what is present on the mainland. The most isolated patches of land are the most diminished of all.

To an agent of disease, each host is an island. If the archipelago of victims is small, the infection may disappear. Iceland had too few people to support a permanent population of measles viruses until after the Second World War. If a place is too remote, a disease may never arrive – which is why Britain has no rabies. Even in Europe, the virus must pass between islands – hosts – to stay alive and to reach a new victim before the last sinks dead beneath the metaphorical waves. Where foxes are rare, each is so isolated that the disease cannot gain a hold. Rabies can be controlled with increased isolation. To shoot – or to immunize – foxes moves the survivors further from a source of infection. For a fox, a city is a continent, and the animals are so common that neither vaccines nor bullets can save them from invasion.

On the morning of 27 August 1883 a gigantic explosion shook the world. The bang was heard from Sri Lanka to central Australia. Its blast travelled four times around the globe, and the tidal wave reached Dover. Krakatau had exploded, with the force of ten thousand Hiroshima bombs.

The volcano sits in the Sunda Strait, between Sumatra and Java, on a line of strain in the Earth's surface. As a fulcrum of a gigantic geological lever, the Strait is always in turmoil. The Javanese *Book of Kings* records a huge eruption in AD 416. There have no doubt been many more. Before the bang, Krakatau had been covered with forest. The *Endeavour*, on Captain Cook's first voyage, visited in 1771: 'We saw that there were many houses and much Cultivation upon Cracatoa, so that probably a ship might meet with refreshments who chose to touch here'. After Cook's death in Hawaii, his comrades revisited the island and the expedition artist made a sketch of its rich landscape.

The eruption put paid to all that. The pressure gauge of the Jakarta gasometer is the record of its story. On the day of the explosion it rose and sank as the blast passed over. Then, the sea fell back and the sky darkened as ash rained down. Its tsunami drowned thirty thousand and the stream of red-hot fragments that boiled across the sea incinerated many more. Dust was carried around the world and led to spectacular sunsets. Fire engines were called out in New York State to fight what seemed a giant blaze in the distance. In a rare conjunction of geology and the arts, Tennyson asked: 'Had the fierce ashes of some fiery peak/Been hurled so high they ranged about the globe?/ For day by day, thro' many a blood-red eve./ . . . / The wrathful sunset glared'.

A week after the eruption two-thirds of the island was gone. The rest was covered by red-hot pumice. The first visitor wrote that 'In spite of all my searching I could find no sign of plant or animal life upon the land, except a very solitary small spider; this strange pioneer of the revival was in the process of spinning its web'. The island was almost sterile.

Life did not take long to return. By the end of the century grasses and cane to the height of a man covered Krakatau. A survey a few years later in what had become a dense fig forest revealed more than eight hundred different kinds of animals. Nowadays, to a casual glance, existence on Krakatau seems the same as that on the mainland. It is not. Although its landscape has trees filled with bats, eagles and woodpeckers, pythons and monitor lizards, it lacks the monkeys, frogs, squirrels, tree-shrews and cats abundant on Java itself, thirty miles away. Otters are its only land mammals.

Evidence of Krakatau's tenants before the cataclysm comes from a small collection of snails and plants from 1867 and the sketch made by the Cook Expedition. Although the island now has nineteen kinds of snail, they include none of the five collected before the eruption. Of the half-dozen plants, two never made it back. The Cook picture shows four recognizable plants (a grass, a fern and two trees), just three of which have returned.

The swimmers got there first. The earliest plants were followed by worms that floated across in rotten tree trunks, and by monitor lizards (themselves often seen out at sea). Grass seeds arrived by air, together with a surfeit of orchids and ferns. Then came birds, their guts loaded with the seeds of figs and other trees, and bats with other seeds stuck to their fur. In time, a new community emerged.

The history of Krakatau is an experiment in evolution. It differs from other isolated islands only in that its birth was seen by man. Its deformed mix of plants and animals makes sense because its history is known. For most islands, the evidence of the past lies only in the present; in the fact that their communities have a different shape from those elsewhere. Geography, not St Patrick, is to blame for the lack of snakes in Ireland. Frogs are absent from all oceanic islands – Hawaii, Madeira, the Galapagos – and the commonest mammals (apart from those introduced by man) are bats. In the same way, such places have grasses in abundance, but not many trees.

If frogs or snakes appeared on Earth through the agency of a beneficent architect, why should they not be brought into being on Hawaii or Ireland as much as on the mainland? They are absent not because the Creator has a devious mind, but because they cannot cross the sea. However, islands are also great manufactories of life. If New Guinea is taken as the world's largest island, then such places make up a thirtieth of the land surface. They contain, however, about one in six of all known species. Such places are proof of what Darwinism can do and those most isolated give the best evidence of all.

On the Inhabitants of Oceanic Islands. Some islands are too remote for any but the most determined migrants. The further from a source, the slower the rate of colonization, until, at last, a piece of land very distant from a continent contains far fewer forms than does an equivalent area of mainland. Although in oceanic islands the number of kinds of inhabitants is scanty, the proportion of endemics (those found nowhere else in the world) is often extremely large. If we compare the number of the endemic land-shells in Madeira, or of the endemic birds in the Galapagos archipelago, with the number found on any continent, and then compare the area of the islands with that of the continent, we shall see that this is true. The inhabitants of the most distant islands have been isolated for so long that descent with modification has done its work undisturbed. As a result, they lay claim to forms of existence – related to but distinct from those of the nearest mainland – found nowhere else.

For less mobile creatures, evolution on islands can proceed apace. Madeira is inhabited by a wonderful number of peculiar land-shells, whereas not one species of seashell is confined to its shores. Though we do not know how seashells are dispersed, yet we can see that their eggs or larvae, perhaps attached to seaweed or floating timber, or to the feet of wading birds, might be transported far more easily than land-shells, across three or four hundred miles of open sea.

Krakatau is young and close to its source of immigrants. Although its inhabitants are a biased sample of those on Java, none is unique. More distant islands, in contrast, are full of change. Evolution did not spring from Darwin's brow as soon as he saw the Galapagos finches (the best he could find to say about them was that: 'their general resemblance in character and the circumstance of their indiscriminately associating in large flocks, rendered it almost impossible to study the habits of particular species . . . They appeared to subsist on seeds'). Even so, the birds helped form his later ideas, and islands are still among the best evidence of how evolution works.

Islands come in many shapes and sizes. Most are the remains of drowned – or raised – continents. At the height of the last ice age so much water was trapped in the glaciers that there was a fifth more dry land than today. The Hebrides, the Isle of Wight, the Scillies and many more formed as the sea reclaimed its own. What once were hills became islands. Cyprus, Crete and Corsica have the same history. As sea-levels changed, the barrier at the Straits of Gibraltar allowed the Mediterranean to dry (and African apes to reach

Gibraltar). As the Atlantic rose, the Straits turned into a marine waterfall that filled it within a century. Yet other islands rose into existence as the ice was shed. The sea bottom around Finland – a country that is a dilute solution of land in water – still rises by half an inch a year and forms archipelagoes as it goes. Places such as these are mere fragments of a broken continent and do not differ much from their parent nearby.

When *The Origin* was published, Alfred Russel Wallace was in Indonesia (from whence he had sent the famous letter that caused Darwin to rush into print). He noticed a strange pattern of distribution among its inhabitants. A line – later called Wallace's Line – drawn across the archipelago, with the Philippines, Borneo, Java and Bali to its north and west, and Lombok, Sulawesi and New Guinea to the south and east, separated two biological worlds. One side had ties to Asia, with its fruit-thrushes, weaverbirds and tigers, and the other to Australia, with cockatoos, honey-eaters and kangaroos. In some places a mere fifteen miles of sea separated the two provinces.

Wallace almost got it right: he wrote of the archipelago, 'I believe the western part to be a separated portion of continental Asia, the eastern the fragmentary prolongation of a former Pacific Continent.' He appealed to the drowned land of Lemuria, now sunk in the Pacific, to explain the links with Madagascar. In fact, the history of Gondwana is to blame. The country east of his Line was part of that continent in its last days, while that to the west was part of Asia. The ancient division remains today. No lost continent is required, just one that moves around. As Wallace noticed, the narrow straits that made his Line were deep and passages of shallow water were no barrier even to animals that could not swim. That is a relic of the end of the last glaciation. The shallows are mere valleys filled with water, while the deeps mark the edges of ancient continents.

Some islands have no ties of any kind. They are isolated by vast oceans and emerged, as sterile as a newborn baby, from the sea as lava boiled from below. Hawaii is above a hot-spot, a spring of liquid rock that wells up from the Earth's core and waves like a plume of smoke as it makes its way towards the surface. As the sea floor grinds over it, volcanoes burst through. Each new expanse is borne to the west as the plate moves on and, as it moves, is worn away. The island of

Hawaii is the highest mountain on Earth, almost a mile higher than Everest when measured from its base on the ocean floor to the summit of Mauna Loa, fourteen thousand feet above the sea. It is one of fifty great volcanoes. The most distant, the Emperor Sea Mounts, stretch almost to Siberia and are eroded stumps deep below the ocean's surface. Hawaii itself is less than a million years old, the furthest sea mount more than seventy million. To the east of Hawaii, as yet unrevealed, another member of the group – to be named Loihi – is under construction. It will break the surface in thirty thousand years. As it moves on, the chain is a conveyor belt of evolution.

Distant islands, wherever they might be, have much in common. Each has a balance of species distinct from that of the mainland, and tends to be more like the polar regions than it is to the nearest point on a continent. Thus, the Canaries have a rather Mediterranean feel (although they are off the coast of Africa) while the Galapagos are deserts compared to the equatorial forests found at the same latitude in South America. Island plants become more interested in sex; with many more species having separate males and females than on the mainland. Their animals, oddly enough, tend to lean the other way, with many sexual species abandoning their males to become parthenogens. Not much of this is understood, but remote islands all share one property that gives them a special place in the case for evolution.

The most striking and important fact for us in regard to the inhabitants of islands is their affinity to those of the nearest mainland, without being actually the same species. On the Galapagos archipelago, situated under the Equator, between five hundred and six hundred miles from the shores of South America, almost every product of the land and water bears the unmistakeable stamp of the American continent. The close affinity of most of the birds to American species in every character, in their habits, gestures and tones of voice, is manifest. The naturalist, looking at the inhabitants of these volcanic islands in the Pacific, distant several hundred miles from the continent, easily feels that he is standing on American land. It is obvious that the Galapagos Islands would be likely to receive colonists from America and that such colonists would be liable to modifications, the principle of inheritance still betraying their original birthplace.

All islanders resemble the inhabitants of the nearest mainland and not those of comparable islands far away. The Cape Verdes have animals and plants resembling those of Africa and not the Galapagos. They show that isolation, not insularity, is what causes change. Distant islands – unlike, say, the zinc-coated archipelago of pylons, each of which evolves an identical response to their shared environmental challenge – are different because each received their own mix of immigrants from their mainland. Necessity – evolution – has moulded what chance provided.

All new islands – marine volcanoes, mountain-tops left empty by the ice, or lakes that open as the Earth's crust moves – are colonized by a sample of the mainland population. If the sample is small (as it often is) then a roll of the genetic dice itself may cause the new arrivals to differ from those left behind. That means immediate genetic change through the accidents of travel. What is more, the first generations in a new home will be few in number. As a result, and at random, more genes will be lost, because their bearers fail to reproduce.

Any population that goes through a bottleneck has no choice but to evolve. Cats tell the story. Cities are filled with feline diversity – tabby, black and white, orange, long-haired and all the rest. Islands are different. Every cat on the remote French Dependency of Kerguelen, in the southern Indian Ocean, is black (some mitigated by splashes of white), no doubt because the few arrivals carried only those genes. On the equally French territory of St Pierre-Miquelon, fifteen miles off Newfoundland, the cats are different again. Their genes are not like those of the nearby mainland, but a reduced sample of those of Bordeaux, two thousand miles away.

To stay alive in a distant land is not easy. As those who tried to introduce birds to New Zealand found, a hundred or more of each form were needed to survive. The much-touted idea that a whole community could descend from a single seed or pregnant female is in most cases not realistic. Any traveller able to reach a safe haven will, given time, arrive in some numbers. There are a few exceptions, like the solitary (and somewhat dazed) macaque found afloat on a log after the Krakatau explosion, but most colonists need, like the Pilgrim Fathers, a reasonable group to have a chance of success.

Genetic variation in animals that live both on the mainland and on an island – deer, wolves, mice, birds, reptiles, snails and insects – show most island populations to be less variable than their progenitor. Even so, the reduction is modest. Darwin's finches are themselves so diverse that they must descend from a solid nucleus of birds that made the journey from South America. The idea of whole species arising in such places almost by accident – a sort of supply-side evolution – has been much explored by the mathematicians who cluster around the margins of Darwinism. Most of the evidence is against it. Necessity – natural selection – moulds their inhabitants, as it does those who stay at home. An empty and remote volcano is full of opportunities. Animals and plants soon make use of them, to give new forms that might find it hard to live in the competitive world of their ancestors.

Accident does play a part in who ends up with what. On different Caribbean islands the lizards of the treetops or the trunks look much the same but are not in fact related. DNA shows that one island's climber may belong to quite a different family of lizards from the climbing lizard on another. Each has (like the partridges and hares that go white in winter) evolved the same independent solution to an ecological challenge, starting from a different place. No doubt the first to arrive in a new place filled a gap and denied it to later arrivals.

Most mainland trees (apart from those with buoyant seeds, such as the coconut) have no hope of an oceanic journey. An herbaceous plant, though it would have no chance of successfully competing in stature with a fully developed tree, when established on an island and having to compete with herbaceous plants alone, might readily gain an advantage by growing taller and taller and overtopping the other plants. If so, natural selection would often tend to add to its stature and thus convert it first into bushes and ultimately into trees. As they reach to the sky, the most improbable plants become large and woody, a habit denied to them back home. Daisies are fond of the pastime, as are many more. Tree-lettuces grow on Madeira and the Cape Verde islands; tree-cabbages, tree-asters and tree-sneeze-weeds on St Helena; tree-sunflowers and tree-fleabanes on the

Galapagos; and tree-celeries are scattered over the many islands of Macronesia. The mountain tops of East Africa – Mount Kenya and Mount Elgol – have their unique tree-lobelias. All rise above their competitors and shade out the plants below.

Because many island insects are flightless, and others – such as butterflies or bees – do not often cross the sea, the plants have to adapt to a new sexual landscape. Hawaii has a mere half-dozen kinds of moth, two butterflies and no bumblebees. Most island flowers, even in the tropics, are not the showy displays of the mainland, but tend to be small and inconspicuous. The commonest colours are white, green and yellow, in lands as far apart as New Zealand and St Helena. As wind-blown seeds – like the pollinators themselves – might blow away, many plants reduce the wings or parachutes that spread the seeds of their land-bound ancestors.

Diversity explodes in distant places, on the Hawaiian archipelago most of all. Almost a thousand flowering plants are found only on the chain. Eight hundred distinct kinds of fruit-flies, a third of the world total, live there, and the beetles have burst into even more variety.

The oldest surviving island, Kaui, is a mere five million years old. The string of drowned and more ancient lands to the west show that much of the chain's evolution happened in lands now beneath the sea. Even so, the Big Island of Hawaii, the youngest member of the group, has eighty plants of its own, each of which must have evolved in the million years since it rose above the waves. One group, the lobelioids, has a great range of forms. More than a hundred kinds are found on the archipelago, a ninth of the native flora of the Hawaiian chain. None of them is present anywhere else on Earth, and most are restricted to single islands in the group. Their DNA shows them all to be close kin, but in shape and size they vary from low rosettes, to vines and to trees twenty feet high. A small fleshy cliff-dweller whose seeds blow in the wind has as its nearest relative a large tree with seeds dispersed by the birds that eat its fruit.

Their diversity is a monument to other natives, now lost. Many of Hawaii's plants (unlike most of those on islands) have red, purple or blue flowers. Those garish blooms evolved to attract pollinators – the archipelago's own birds, the honeycreepers. Twenty-three different kinds have been seen by Europeans, fifteen of which survive. Most have long, thin curved beaks. They live on nectar and specialize on their own sweet flower. Other unique birds of the chain had a less amicable relationship with the plants. Some lobelioid leaves and stems are covered to chest height with thick spines to protect them against their enemies. Now, the grazers have gone, but not long ago Hawaii had its own giant herbivorous ducks and geese. The thorns were needed on the lower leaves alone, as the great birds were flightless. They were killed off, with many of the honeycreepers, by the first human settlers a thousand years ago.

The world has many other islands; of people, of land and of water. A third of all species of fish live in fresh water, although lakes and streams cover only one part in a hundred of the Earth. Even so, most lakes and ponds, isolated as they might seem, are in effect members of the same shared continent; only the true islands of water (as isolated as those of land) have been hotbeds of evolution, and for the same reason.

Most freshwater fish live risky lives, because a sudden drought may put paid to their home. Streams and ponds are temporary places, liable to dry up and to re-form as valleys are worn away and as rainfall comes and goes. Their inhabitants are, in most places, quite similar, living on watery islets separated by dry land though they might, because such places are less separate than they seem. Floods can move fish from stream to stream, and changes in the level of the land cause rivers to flow into each other. Most freshwater plants and animals are good at travel; as spores, attached to the legs of birds, or as live adults (the fish and frogs dropped by whirlwinds included). As a heron moves from one pond to the next, any egg stuck to its leg is guaranteed to arrive in a favourable place. Nature, like a careful gardener, has taken her seeds from a bed of a particular nature and has dropped them in another equally fitted for them.

The zebra mussel is a modest beast – a freshwater snail the size of a thumbnail. Its home is in the basins of the Black, Caspian and Aral Seas. There, mussels are everywhere. Their larvae float, or stick to flotsam or the legs of birds. The Volga Boatmen dragged plenty of the animals with them as they heaved their barges upstream. As trade increased, the mussels moved west. They reached the Danube by 1800 and the Thames soon after. In 1988, a colony was seen in Lake Saint Clair, near Detroit. It had arrived as a stowaway, in water carried by a ship as ballast (millions of gallons of which are moved from port to port each year). The zebra mussel was the most success-ful of the four hundred kinds of animal found in the tanks. In the 1970s hundreds of Russian ships came to the Great Lakes to collect the grain that the Soviet Union could not grow for itself. Some carried mussels. Within ten years the animal had spread over the Eastern United States. It took almost no time for the animals to fill the Mississippi. On the way they have driven out their native equivalents, the purple wartyback, the shiny pigtoe and the monkey-face included.

However, a few distant lakes have long been safe from such immigrants. They have managed to evolve an identity of their own. They are ancient and remote, filled with unique animals and plants, veritable volcanoes of water separated by great oceans of land. Their inhabitants have been isolated for millions of years. Like those of other islands, they have exploded into a diversity of form; and, like them, they are now threatened by the greatest traveller of all.

Lake Tanganyika was first seen by Europeans in 1858, when the British explorers Burton and Speke reached its eastern shore in their quest for the source of the Nile. It lies, with its fellows Lake Victoria and Lake Malawi, in the Great Rift Valley. Its arid shores are home to some of the most spectacular animals in the world. They live out their struggle for existence for a global audience of television viewers.

The cameras would find much more diversity beneath the water. Each of the great lakes has thousands of species found nowhere else. They are the finest evolutionary microcosm on earth. As the longest and, at almost five thousand feet, the second deepest lake in the world (after Baikal, itself the home to a strange fauna of its own), Lake Tanganyika contains a sixth of the globe's fresh water.

It was formed as rivers flowed into the basin left by two conti-nental plates as they shifted. Tanganyika is among the oldest of lakes, first filled twelve million years ago. Below about three hundred feet, the water is dead. Its sides are steep and most of the inhabitants are confined to a narrow layer around the shore. They include fifteen hundred unique kinds of animal and plant – fish, sponges, crabs, snails and even a freshwater jellyfish. Many look rather like sea creatures, and the idea that the lake was once joined to the ocean led to the first expedition. Its evolution, however, happened within its own confines.

The lake has almost as many kinds of freshwater fish as does the whole of Northern Europe. Catfish, eels and perch have evolved into an assortment of shapes, sizes and behaviours. Although just a small part of the shore has been explored, two hundred members of one group, the cichlids, are known. All except five are found in Lake Tanganyika alone. Lake Malawi, which shares much the same history, has a thousand different kinds.

Within the lakes are the fishy equivalents of the elephants, lions and hyenas that roam the nearby plains. Cichlids vary in size from an inch to two feet or so. Some scrape algae off the rocks, others browse on snails or on the animals of the bottom. Some, almost like whales, filter the water for tiny particles of food. Members of another group, with no land-based equivalent, sneak up on fish from behind and tear scales from their bodies. They have two forms, one a specialist on the left flank, the other on the right. All cichlids care – like lions or elephants – for their young. In some, females hold them in their mouths (and are open to the attack of fish cuckoos, which dump their own eggs in another's maw). Others use abandoned snail-shells as homes for their brood. Their brains as well as their bodies have changed, as the fish-eaters are better endowed with grey matter than grazers on plants or snails. All this is evidence for what natural selection can achieve when left to do its work.

Most of the fish inhabit rocky outcrops or patches of gravel close to the shore. They are sedentary beasts, and almost never venture across the sand separating each patch. They live on islets within the large lake isle itself. Even the small Lake Nagubago, separated from the main Lake Victoria by a spit of sand within the past fifteen hundred years, has several cichlids of its own. As the waters rise and fall by as much as fifteen feet over a few years, many tiny lakes have appeared and disappeared around the shores of their great parents. They were, like the drowned islands of the Hawaiian chain, part of the engine that drives diversity.

Lake Tanganyika itself almost dried up twelve thousand years ago, and split into three smaller bodies of water. This division led to an outbreak of evolution. The ancient breach in its waters is reflected in today's fish, who are separated into distinct lineages that arose within the reduced ancient lake and have diverged further in their own tiny homelands. Each lineage has its specialists – a grazer, a grinder or a filterer – adapted to the opportunities on offer. Lake Victoria dried up altogether at the same time, and its three hundred cichlids have evolved since that disaster.

Islands are the workshops of evolution, their products a microcosm of its every stage. What they make is fragile. Islanders do well in their own limited marketplace, but once exposed to the outside world are soon driven to extinction. Because they have adapted free from outsiders most are, like the flightless geese of Hawaii, doomed. Their fate comes from a simple law of economics: the need to compete. The products of islands, like those of other small factories, find it hard to survive in the global economy. More than half the mammals lost during historical times lived in such places, as did nine-tenths of the extinct birds.

Integration destroys diversity, as solitude promotes it. Islands are dangerous places, because they are small and open to accident. Any loss that might elsewhere be replaced is final. On a patch of land, things are liable to go wrong more often than on a larger tract; and there is nowhere to hide from disaster. All islanders are at constant risk from rude outsiders. The rudest of all is man. He filled the furthest points of the habitable Earth a mere thousand years ago, when boats reached New Zealand. As soon as he arrived, he had no compunction in exploiting what he found.

William Strachey wrote an account of his shipwreck on Bermuda in 1609. His book influenced Shakespeare, whose *The Tempest*, written two years later, mentions the archipelago as the 'still-vexed Bermoothes'. Strachey noticed a bird which 'for their cry and whooting, wee called the Sea Owle ... Our men found a prettie way to take them, which was by standing on the Rockes or Sands by the Sea side, and hollowing, laughing, and making the strangest out-cry that possibly they could: with the noyse whereof the Birds would come flocking to that place, and settle upon the very armes and head of him that cryed'. The Sea Owle is no more.

The first travellers noted the tameness of the birds of the Galapagos: it is not complete, for the finches are wary of familiar enemies, and will perch on a person if a hawk flies over. Their evolved confidence in the face of the unknown was a mistake, and they, with many other islanders, have paid the price. The number of seabirds on Ua Huku in the Marquesas dropped from twenty-two to four after man arrived two thousand years ago. In Hawaii the fossils show that at least eighty native finches, ducks, geese and ibises have disappeared. The first New Zealanders destroyed eight kinds of moa, together with parrots, owls and many others. In Polynesia, two thousand unique terrestrial rails were lost, and the last Wake Island rail, first described in 1903 by Lord Rothschild, was devoured by Japanese soldiers in 1945: a career from discovery to extinction of four short decades.

Ancient lakes, too, have paid for their sheltered lives. Ten million people live around Lake Tanganyika, and thousands of tons of fish are taken each year. The Nile perch, introduced as a food fish into several of the Rift lakes, has driven many of the native cichlids to extinction. The perch is now exported to Europe and will no doubt be moved to wherever it can grow, so that many cichlids yet unknown will soon be gone.

At the time of the *Beagle* most islands were as isolated as they had been since they began. Now, everything has changed. The world's land masses – like its economies, with five trillion dollars' worth of goods travelling by sea each year – have become connected into a single body. So have their animals and plants. Isolation is now a rare commodity, and the most distant scrap of land has joined the continents. In the United States, one plant species in ten has come from elsewhere, in Britain almost half – and on Ascension Island in the Atlantic more than 80 per cent of plants are immigrants. The grand simplification is not yet complete, but soon will be. The larger any piece of land, the more forms of life it contains, but the number does not rise in exact relationship to its area. In that fact is a dismal message for the future. If all the world's islands were to be joined together into a supercontinent – a new Pangaea – its magnificent expanse would, given the relationship between the size of islands and the number of species, contain half the kinds of animal that now exist.

As travel brings the earth together, many of the products of evolution will disappear. Darwin, were he to circle the globe today, would find it a less remarkable place than it was when the *Beagle* made its voyage.

Summary of last and present Chapters. In these chapters I have endeavoured to show, that if we make due allowance for our ignorance of the full effects of all the changes of climate and of the level of the land, which have certainly occurred within the recent period, and of other similar changes which may have occurred within the same period; if we remember how profoundly ignorant we are with respect to the many and curious means of occasional transport, – a subject which has hardly ever been properly experimentised on; if we bear in mind how often a species may have ranged continuously over a wide area, and then have become extinct in the intermediate tracts, I think the difficulties in believing that all the individuals of the same species, wherever located, have descended from the same parents, are not insuperable. And we are led to this conclusion, which has been arrived at by many naturalists under the designation of single centres of creation, by some general considerations, more especially from the importance of barriers and from the analogical distribution of sub-genera, genera, and families.

With respect to the distinct species of the same genus, which on my theory must have spread from one parent-source; if we make the same allowances as before for our ignorance, and remember that some forms of life change most slowly, enormous periods of time being thus granted for their migration, I do not think that the difficulties are insuperable; though they often are in this case, and in that of the individuals of the same species, extremely grave.

As exemplifying the effects of climatal changes on distribution, I have attempted to show how important has been the influence of the modern Glacial period, which I am fully convinced simultaneously affected the whole world, or at least great meridional belts. As show-ing how diversified are the means of occasional transport, I have discussed at some little length the means of dispersal of fresh-water productions.

If the difficulties be not insuperable in admitting that in the long course of time the individuals of the same species, and likewise of allied species, have proceeded from some one source; then I think all the grand leading facts of geographical distribution are explicable on the theory of migration (generally of the more dominant forms of life), together with subsequent modification and the multiplication of new forms. We can thus understand the high importance of barriers, whether of land or water, which separate our several zoological and botanical provinces. We can thus understand the localisation of sub-genera, genera, and families; and how it is that under different latitudes, for instance in South America, the inhabitants of the plains and mountains, of the forests, marshes, and deserts, are in so mysteri-ous a manner linked together by affinity, and are likewise linked to the extinct beings which formerly inhabited the same continent. Bearing in mind that the mutual relations of organism to organism are of the highest importance, we can see why two areas having nearly the same physical conditions should often be inhabited by very different forms of life; for according to the length of time which has elapsed since new inhabitants entered one region; according to the nature of the communication which allowed certain forms and not others to enter, either in greater or lesser numbers; according or not, as those which entered happened to come in more or less direct competition with each other and with the aborigines; and according as the immigrants were capable of varying more or less rapidly, there would ensue in different regions, independently of their physical con-ditions, infinitely diversified conditions of life, – there would be an almost endless amount of organic action and reaction, – and we should find, as we do find, some groups of beings greatly, and some only slightly modified, – some developed in great force, some existing in scanty numbers – in the different great geographical provinces of the world.

On these same principles, we can understand, as I have en-deavoured to show, why oceanic islands should have few inhabitants, but of these a great number should be endemic or peculiar; and why, in relation to the means of migration, one group of beings, even within the same class, should have all its species endemic, and another group should have all its species common to other quarters of the world. We can see why whole groups of organisms, as batrachians and terrestrial mammals, should be absent from oceanic islands, whilst the most isolated islands possess their own peculiar species of aerial mammals or bats. We can see why there should be some relation between the presence of mammals, in a more or less modified condition, and the depth of the sea between an island and the mainland. We can clearly see why all the inhabitants of an archipelago, though specifically distinct on the several islets, should be closely related to each other, and likewise be related, but less closely, to those of the nearest continent or other source whence immigrants were probably derived. We can see why in two areas, however distant from each other, there should be a correlation, in the presence of identical species, of varieties, of doubtful species, and of distinct but representative species.

As the late Edward Forbes often insisted, there is a striking parallelism in the laws of life throughout time and space: the laws governing the succession of forms in past times being nearly the same with those governing at the present time the differences in different areas. We see this in many facts. The endurance of each species and group of species is continuous in time; for the exceptions to the rule are so few, that they may fairly be attributed to our not having as yet discovered in an intermediate deposit the forms which are therein absent, but which occur above and below: so in space, it certainly is the general rule that the area inhabited by a single species, or by a group of species, is continuous; and the exceptions, which are not rare, may, as I have attempted to show, be accounted for by migration at some former period under different conditions or by occasional means of transport, and by the species having become extinct in the intermediate tracts. Both in time and space, species and groups of species have their points of maximum development. Groups of species, belonging either to a certain period of time, or to a certain area, are often characterised by trifling characters in common, as of sculpture or colour. In looking to the long succession of ages, as in now looking to distant provinces throughout the world, we find that some organisms differ little, whilst others belonging to a different class, or to a different order, or even only to a different family of the same order, differ greatly. In both time and space the lower members of each class generally change less than the higher; but there are in both cases marked exceptions to the rule. On my theory these several relations throughout time and space are intelligible; for whether we look to the forms of life which have changed during successive ages within the same quarter of the world, or to those which have changed after having migrated into distant quarters, in both cases the forms within each class have been connected by the same bond of ordinary generation; and the more nearly any two forms are related in blood, the nearer they will generally stand to each other in time and space; in both cases the laws of variation have been the same, and modifications have been accumulated by the same power of natural selection.

CHAPTER XIII

MUTUAL AFFINITIES OF ORGANIC BEINGS; MORPHOLOGY; EMBRYOLOGY; RUDIMENTARY ORGANS

CLASSIFICATION, and the hidden order of life — Arbitrary and Natural systems — Common descent the key to classification — Cladistics and the rules of arrangement — The comparative anatomy of the genes and the new tree of life — An identical present may conceal a separate past. MORPHOLOGY, theme and variations in related creatures, and in their several parts — The switches of growth — Monstrous animals and plants. EMBRYOLOGY, laws of, how the embryo reveals a past lost in the adult. RUDIMENTARY ORGANS; the cost of unwanted structures, and their origin explained — Summary

A CELEBRATED CHINESE encyclopaedia of the tenth century classifies plants and animals as follows: (a) those that belong to the Emperor, (b) embalmed ones, (c) those that are trained, (d) suckling pigs, (e) mermaids, (f) fabulous ones, (g) stray dogs, (h) those that are included in this classification, (i) those that tremble as if they were mad, (j) innumerable ones, (k) those drawn with a very fine camel's-hair brush, (l) others, (m) those that have just broken a flower vase, (n) those that resemble flies from a distance.

To the author of *The Celestial Emporium of Benevolent Knowledge* – and perhaps to his readers – that catalogue made sense. It is hard to comprehend today. There may be a certain affinity among beasts that break flower vases or tremble as if mad, but to the modern eye they form a less natural group than do, say, suckling pigs, stray dogs or even those that from a distance resemble flies. The philosopher Michel Foucault remarked of the list that: 'In the wonderment of this taxonomy, the thing that is demonstrated in the exotic charm of another system of thought is the limitations of our own'. Biology has begun to show how limited have been our own ideas about the great emporium of existence.

Man is a classifying animal. His world is so full of objects that he must reduce their number by arranging them in groups. They can be sorted in many ways – red or blue, large or small, safe or dangerous, near or far. The brain can be seen at work as it puts objects into pigeonholes. As people think of tools, or names, or vegetables, they activate specific areas of grey matter. Drill, mallet and sandpaper all involve the same few cells, Tom, Dick and Harry another set. Those who have difficulty with the names of tools have the same problem with those of animals because the system for sorting the two is in the same place.

The notion that a spirit level and a file fall into the same category must be learned, because a group called 'tools' makes sense only in context. Apart from utility, such objects have little in common. Why should the brain not separate things belonging to the emperor from those with other owners? In China, after all, to do so might be a matter of life and death. Even so, to sort by ownership does not say much else about any object, be it palace, pekingese or porcelain teapot. They are united by a single property that does not overlap with others.

A thousand years after the oriental list, life can still be organized in many ways. All but one are useful but biased, because they are artificial. The Catholic Church once classified, for culinary reasons, the capybara as a fish. That makes sense for hungry travellers on a Friday, but is not much use to students of South American mammals. To define animals as edible, or endangered, or cute, is a great help in the context of kitchens, zoos and pet shops, but members of such groups have little else in common.

One arrangement, and only one, reveals the order hidden in nature. This classification is evidently not arbitrary like the grouping of stars in constellations. Something more is included than mere resemblance. It is based on a natural structure, on common descent from a distant past. As a result, it reveals the truth of evolution in a way denied to all others.

The contrast is laid bare in the natural history museums of Paris. The Zoology Museum and the Museum of Comparative Anatomy trace their descent from the Jardin des Plantes, founded by Louis XIII. They were, for much of their history, repositories of Nature, present and past; ordered, rational and deductive, their contents arranged by the rules of biological classification. Each exudes the same grave nineteenth-century air as does *The Origin* (although the image of Darwin is hard to find in either).

The Museum of Zoology sank into decline and was closed in 1965 as its specimens crumbled in their cases. Then, in 1994, its Grand Gallery of Evolution was reborn as part of the French reinvention of their capital. The museum's public face has changed. The stuffed animals that once decayed in organized grandeur are now arranged as a series of spectacles: the African savannah, the ecological crisis, or the plants and animals useful to man. Spotlights abound, with animated exhibits popular with children.

The new museum's logic, splendid as its interior may be, is that of a Chinese encyclopaedia. The members of each display are united by a single property; each lives on the African plains, has been domesticated, or is threatened by extinction. To know that does not say much more about what they are. In one exhibit, an elephant and a locust are next to each other, in another a camel and a goose.

The nearby Gallery of Comparative Anatomy and Paleontology is unrenewed. Around its walls are hundreds of pickled guts and through its dusty halls march a myriad of bones, in progress from fish to man (not a Frenchman, but an Italian). Marbles and bronzes – *An Orang-Utan Strangling a Native of Borneo, Man Triumphant over Two Bears* – press home the message of progress (although the unfortunate victim of the orang has for some reason a snail crawling up his leg). The edifice, with its cast-iron beams and Beaux-Arts style, has much charm but is for most of the time almost empty. The distractions of its modernized neighbour are what attract the crowds.

Forlorn as it is, the Anatomy Museum has stayed true to the spirit of evolution. It sets out its animals by affinity. The Linnaean system leads to a hierarchy of order – the American wolf, for instance, belongs in the kingdom Animalia, the subkingdom Metazoa, the phylum Chordata, the subphylum Vertebrata, the superclass

Tetrapoda, the class Mammalia, the subclass Theria, the infraclass Eutheria, the cohort Ferungulata, the superorder Ferae, the order Carnivora, the suborder Fissipeda, the superfamily Canoidea, the family Canidae, the subfamily Caninae, the genus *Canis*, the species *lupus* and the subspecies *occidentalis* (perhaps fortunately, the Caninae – dogs, dingos, foxes and wolves – are not divided, as are some groups, into separate tribes, and the genus *Canis* – dogs, dingos and wolves – has not been split into subgenera).

Linnaeus invented his system as a bookkeeper might, as a scheme for filing similar objects. God created, but Linnaeus arranged; and how he did his job was criticized from the earliest days. A reviewer of 1759 complained, in anticipation of today's museologists, that dogs should not go with foxes and wolves but with horses, as both are found in the farmyard.

His scheme has lasted well (although it has been hijacked by wits who have sneaked in names such as *Ba humbugi* for a snail, and *Agra vation* and *Agra phobia* for a pair of beetles). It is, like the Zoology Museum, to some degree arbitrary, most of all in its higher ranks. The eight thousand or so kinds of bird are divided into a hundred and sixty families, while sixty thousand species of parasitic wasps have to fit into a single family in their part of the catalogue. Nevertheless, his system contains a great truth: that the world has a pattern. Linnaeus did better in drawing life's big picture than did the tribal experts who are so good at recognizing different species of bird. Once above taxonomy's basic unit, their categories become bizarre. For one set of people, all birds bar one go together, but the cassowary fits, in the rural mind, with rats and mice because it is brown and lacks wings. The Linnaean system, in contrast, is arranged as a universal plan that hints that, from the first dawn of life, all organic beings resemble each other in descending degrees, so that they can be classed in groups under groups.

As a result, a taxonomy based on one property predicts much of what else its members will share. All animals that give milk also have hair. In the same way, all those with hard external skeletons and air-tubes (insects included) have jointed limbs, and no animal both gives milk and has its skeleton on the outside. Plants show the same harmony, as those with flowers all have seeds enclosed in a fruit rather than directly exposed. To put animals and plants in order reveals a long chain of affinity – an unseen connection – between different creatures, distant as they might appear.

Propinquity of descent – the only known cause of similarity of organic beings – is the bond, hidden as it is by various degrees of modification, which is partially revealed to us by our classifications. All true classification is genealogical. Just as in a human family, with its brothers, sisters, cousins and second cousins, it hints at descent from shared ancestors further and further in the past. The patterns seen by biologists as they arrange their world are, unlike those of tenth-century China, evidence for a system outside the mind of the classifier.

The logic of life can be seen without bothering with evolution at all. Linnaeus was no evolutionist, and neither is a New Guinea tribesman. The Swedish taxonomist, his Papuan cousin, and the apparatus of natural history all use shared descent as the key to order, without realizing it. To classifiers, of whatever persuasion, related-ness is always more important than mere appearance. Everyone includes within the same species its males, females, adults and young, however different they appear. The growth stages of certain plants are so distinct as to be recognizable as members of the same kind only by experts. The Killarney fern was written off as extinct in the nine-teenth century. It is in fact still widespread – but most of the survivors remain in another part of the lifecycle as a green mat on moist rocks. As soon as that was realized, its status as an endangered species was removed.

What unites Killarney ferns is not shape or size, or habit, for the different stages are quite unalike, but common parenthood. Once continuity of descent – the bond that draws together all members of a group – is established, mere lists of similar objects can be abandoned. If descent is used to unite individuals of the same species, then their arrangement into higher classes, however distinct they are, must also involve an element of shared ancestry. Thus the grand fact in natural history of the subordination of group under group, which, from its familiarity, does not always sufficiently strike us, is fully explained.

The cicadas have been around for three hundred million years. Their image is as a backdrop to the Italian summer, but their capital has always been in Australia, where the various kinds – the greengrocer, the double drummer, the floury baker and many more – are much collected by children. Cicadas show how group becomes divided into group as evolution proceeds. That family of a thousand or so insects is spread over the tropical and subtropical world, in Old World and New, and has so been for almost three hundred million years. Individual genera (the second lowest level in the Linnaean system) are more confined: thus, three-quarters of the Australian genera are restricted to their native continent. Species within a genus are even more localized, with two hundred and forty-six of the two hundred and fifty Australian forms found only there. The arrangement of cicadas – in a museum case or a child's collection – hence retraces, with no need for conscious effort, their shared descent. Life's logic, its grandest fact, reflects its history.

All classifications need rules that do not depend on the arbitrary interests of a curator. Although it took a long time for biology to realize that, it now has a statistical machine to order the world and to search for the logic of relatedness over time.

The new science of cladistics means that ancient forms are not seen merely as missing links on the way to today, to be inserted at the right point in some great chain of being. Instead, the extinct and the extant are put into the same system. Cladistics maps out affinities as impartially as it can. It depends on a single idea: that groups sharing traits not present in others must descend from a common ancestor. The more characters that are added to the mix, the further the root is pushed into the past. Men and whales are related because they have warm blood, four limbs (reduced to a relic in whales) and hair. Whales, men and fish are joined because all have backbones (although fish are not hairy and are excluded from man's more immediate family). Those three fall into the same group as the oak, as all have a cell nucleus enclosed by a membrane. Bacteria are outside even that capacious assembly because, unlike the others, their DNA floats free inside the cell.

Cladistics, a German invention, has strict rules and a complex vocabulary. It can, if not carefully used, give erratic results and is still filled with argument about just what should be plugged into its analyses. It has, nevertheless, transformed our view of the world.

The central rule of the new science is that only shared characters, each derived from the same ancestor, can be used to decide relatedness. As a result, ancestral characters are of no use in working out who is kin. Many animals – humans, chimps and iguanas – have five fingers, but others manage with fewer; thus, horses have a single toe on each foot. A whole set of shared patterns – hair, milk and so on – show that five fingers came first and that horses lost them later. In spite of their hands, humans are not closer to iguanas than they are to horses and the five-fingered hand is of no use in deciding where they should sit. In contrast, all horses, and no other mammals, have a single toe. For them, the hoof is a shared derived character. It places them in a group evolved from a common ancestor. The less any part of the organization is concerned with special habits, the more important it becomes for classification. As a result, attributes only found in a single species – the single tusk of the narwhal, or the erectile penis of the female hyena – are of no use as they, too, contain no information about descent.

Sometimes an impartial judge must be brought in to sort out a family problem. To sort out the affinities of a group of species it helps to appeal to a creature so detached from the animals whose relationship is disputed that it can act as an outgroup, a reference point against which to compare their similarities and differences. Captain Cook, on his first sight of a kangaroo, saw that it had 'a long tail which it carried like a grey hound, in short I should have taken it for a wild dog, but for its walking or running in which it jumped like a Hare or dear.' The platypus was even more baffling. To the naturalist Thomas Bewick it appeared 'to possess a three fold nature, that of a fish, a bird and a quadruped, and is related to nothing that we have hitherto seen.' Although the story was for a time confused by the alleged discovery of fish with pouches, the Antipodeans have now been put in their place. The kangaroo and the mouse bear live young; the platypus and the birds lay eggs. By appealing to what is clearly a distant relative – a fish, say, or a frog – it becomes clear that to lay eggs is the ancestral habit, and that live-bearing comes later. Kangaroos and mice are hence closer than either is to a bird or fish, and the platypus, a further set of characters shows, is – in spite of its birdlike beak (a derived nose) – in a group whose ancient members took the first step on the road to kangaroos.

The cladist's art is enlivened by the comparative anatomy of the genes. DNA provides millions of links to the past, some of which have revealed unexpected patterns. A fluorescent probe based on the genes of one animal can be used to search for its match in another. Wherever a few thousand bases correspond, the bait is taken up and makes a lurid blob on the chromosomes. Man and pig, or man and cow, share more than fifty long sequences. All are evidence of common descent as persuasive as are live young, milk or hair.

The genes of many creatures have now been read from end to end. They reveal groups within groups beyond the imagination of earlier naturalists. Thirty thousand genes have been located in humans, and almost as many in mice. PIGMAP, the cartography of swine, is up to six hundred or so, followed by the genetic atlas of the cow. Even the cat has had more than a hundred genes placed on its chromosomes. A tiny nematode worm, the only animal to have had all its DNA letters read, has nineteen thousand and ninety-nine genes altogether, all of them tracked down. Several bacteria and single-celled parasites have also been deciphered.

Their landscapes have much in common. Whole sections are the same in mice and men, and two thousand human genes have exact homologues in mice. A trudge along the DNA shows more than half of a certain mouse chromosome to be more or less identical in the arrangement of genes to one of our own; and cows are even more like us. Half of all plant genes have a mouse equivalent. The nematode worm shares a fifth of its own heritage with yeast (from which it split a billion years ago). Unexpected parallels emerge from distant places. A gene that in humans causes an inherited disorder of the nervous system has an exact match in yeast (which has no nerves at all).

The scattered information on bones, leaves and DNA can be put into a common frame. Because cladism considers so many characters and so many forms of life at once, the science is a statistical nightmare. Thirty-five million pedigrees are possible when just eleven different species are compared. Even so, it has changed our view of life's tree, from its familiar branches to its deepest roots.

There was once disagreement about how the mammals fit together. Some were an obvious ragbag. The insectivores included hedgehogs, moles and shrews, all put in the same pigeonhole on the basis of their joint fondness for insects, the shape of the skull and their absence from South America. Other categories, such as the carnivores (dogs, cats and the like) made more sense, while yet other alliances (such as the marriage of elephants with sea-cows) seem at first sight odd but hold up when enough characters are looked at. The deeper roots of the mammalian family were quite unknown.

Bones and molecules, objectively arranged, reveal more of the truth. The insectivores as an entity disappear altogether and their members shuffle off to other places. The hedgehog is on a first and separate branch in the mammalian family, and the elephant shrew, the golden mole and the aardvark (all once included as insectivores) join sea-cows, elephants and hyraxes in a conjunction of mammals that evolved in Africa. Other mammals, too, change their alliances. Not only do whales group with hippopotami, but dogs and cats join on as more distant members of their coalition. Humans and apes are, it transpires, quite close to rabbits and bats.

Before cladistics, four-legged vertebrates as a whole – lizards, kangaroos and mammals – were thought to descend from the ancient lobe-finned fish, most of which disappeared four hundred million years ago. Their fins do have a structure at their axis that might have turned into legs. The discovery of a member of this group, the coelacanth, off the coast of Africa in the 1930s was hailed as a 'missing link' between fish and ourselves. Cladistics showed this to be untrue. The coelacanth is not on the same branch as vertebrates with four legs. The honour belongs instead to another great group, the lungfish, who flourished at the same time. Although most modern forms lack fins altogether (and even those of their fossils are not at all leg-like), an objective look at skeletons puts them closer to ourselves. Now, the molecules agree: the coelacanth is indeed further from today's four-legged animals than is any lungfish.

Cladistics sometimes simplifies the past. It shows that all land plants, diverse as they are, are members of a single stock and (unlike the animals, who did the job dozens of times) moved from sea to land but once, a fact upon which the fossils are silent. Sometimes it makes things more complicated. The reptiles are, it transpires, a group no more natural than are those of the Chinese encyclopaedia. Reptiles include lizards and crocodiles, but a whole host of characters show crocodiles to be closer to birds (not, of course, classified as reptiles) than to their supposed sibs. Crocodiles are related to dinosaurs; and cladistics shows that birds are dinosaurs who grew wings and flew away. Indeed, to ask of such animals, ancient or modern, 'Is it a bird or a dinosaur?' is like asking 'Is it an apple or a fruit?'

The birds still have a tie with their newly revealed kin. Crocodiles (and, no doubt, their dinosaur forebears) lay eggs in nests, can chirp and determine sex in a peculiar way. If an egg is incubated at low temperatures it develops into a male, at higher temperatures into a female. Armed with the new insight into their past, an audacious biologist raised chicken eggs at high temperatures and increased the proportion of females, which, for an animal in which most males are wasted, is of much interest to farmers.

The basic rule of evolution remains, however bizarre its products. On the principle of the multiplication and gradual divergence in character of the species descended from a common parent, together with their retention by inheritance of some characters in common, we can understand the excessively complex and radiating affinities by which all the members of the same family or higher group are connected together. They hint at a shared past older than any fossil.

To track down such distant links and draw a new pedigree of life needs characters that are almost universal; genes at work so deep within the cell as to resist almost all change and to retain clues about their earliest origin. A hunt for similarity through the dozen and more genomes now known in all their detail, and the hundreds for which long segments have been worked out, reveals the very framework of existence. About a thousand genes are shared by every organism, however simple or complicated. Although they split more than a billion years ago, their common structure can still be glimpsed. It shows how the grand plan of life has been much modified through the course of evolution.

Such genes show that animals with backbones are close to starfish and that worms and snails live on a different branch from insects and nematodes (both of whom shed their coats as they grow). Their common roots lie within an ancestor who lived hundreds of millions of years before any life appeared in the rocks.

Biology's greater divisions are at first sight self-evident. Men and chimps are close kin, each is less related to worms, and bananas and bacteria are quite separate. However, the new taxonomy has transformed the tree of life into an exotic plant. Men and chimps are indeed more related than are men and bananas, but primates, insects and plants are, the genes show, all mere twigs on the same branch. Its trunk has suffered some radical changes of shape.

Living beings were once divided into five kingdoms of more or less equal size (animals, plants, fungi, protozoa – such as the familiar amoeba – and bacteria). Bacteria were out on somewhat of a limb, as their genes are not contained in a cell nucleus. They seemed otherwise not much more distinct from other kinds than plants were from animals. Now, a radical new logic has emerged. The genes show that plants and animals lie close together. Mushrooms deserve a branch of their own, closer to animals than to plants. Most of the tree belongs not to the lords of creation, or even to mushrooms, but to the bacteria and their previously unrecognized relatives. They put humankind in its place, near bananas.

The difference between men and mice, or mice and plants is, relative to the gulf separating all three from other organisms such as bacteria, tiny. The familiar branches of existence (on one of which men, mice and plants reside) are in reality unimportant twigs. There is no need to travel to distant places to see extraordinary beings. Any zoology text claims that there are more kinds of insect than of anything else. Squash a fly and thousands of microbes unknown to science will be squeezed from its gut. Gardens are familiar, biology under control, but a handful of suburban soil contains a myriad undescribed bacteria with genes quite distinct from the plants and animals around them.

Such minute beings are a diverse and remarkable group. Species thought to be quite similar (such as the agents of botulism and of anthrax) are, the genes show, as distinct as men and maize. *Escherischia coli*, a bacterium present in billions in our guts (and the cause of outbreaks of food poisoning), has about four and a half million DNA units in its genetic instructions. They code for four thousand or so proteins. Although some are similar to those of other beasts, almost half are quite different and give no sign of what they do.

One set of genes is found everywhere. It translates the information coded in the DNA and allows it to make proteins. The job is so essential that such genes changed little over millions of years. To put them through the cladistic machine shows how biology has failed to notice some fundamental splits among its subjects.

Nature's new pedigree has three great domains. One encompasses organisms whose genes are contained within a nuclear membrane, together with, a little further away, a variety of single-celled gut parasites. The bacteria and – quite new – the archaea, tiny entities once seen as a mere subdivision of bacteria but in fact distinct, are each in a class of their own. Their dominion is divided not into a mere five kingdoms, but into dozens.

Some are found in unexpected places. The water in Octopus Spring, near the Old Faithful Geyser in Yellowstone National Park, is hot, clear and alkaline. It contains 'pink filaments' – a whole new group. Their existence needs no oxygen. The nearby Obsidian Pool, whose waters are filled with iron, hydrogen sulphide and carbon dioxide, has scores of new groups of archaea. Undiscovered empires live within this single pond. They are so various that so far they have just been given numbers. The great naturalist Lionel Walter Rothschild is commemorated in the scientific names of fifty-eight birds, eighteen mammals, three fish, two reptiles, a hundred and fifty-three insects, three arachnids, a millipede and a small worm. These new beings will need a host of Rothschilds and whole departments of classics to give them the titles they deserve.

The world beneath the surface has always seemed an alien place, with a mythology of its own. Who sang the national anthem to Queen Victoria from the sewers beneath Buckingham Palace? What became of the plan to pipe their contents to the suburbs for sale in corner shops for garden use, or of Thelma Ursula Beatrice Eleanor, the first baby born on a tube-train? Like all cities, London is a vast and curious three-dimensional world unknown to those who live on its surface. Its inhabitants are beyond the imagination of any tube passenger.

Underground London is a Victorian city, built when life was lived in the atmosphere – a layer no thicker than the varnish on a school globe – and when even the oceans seemed dead. Objects sank by weight, with a layer of cannons suspended beneath drowned ships, themselves below the bodies of their sailors. The biology of the abyss did not begin until the *Challenger* Expedition of 1872, which revealed a hidden world beneath the waves. A century later, our understanding of the deep oceans – and of the interior of the Earth itself – has undergone a sea-change. A strange and rich universe stretches far beneath our feet.

Around the ocean vents in the mid-ocean ridges is a new world. In 1977, the submarine *Alvin* dived two miles below the Pacific. There, lava pours from the sea bed. 'Black smokers', mineral-rich streams of super-heated water, gush into the sea. Other vents, 'white smokers', are surrounded by mats of bacteria. So rapid is the flow that the entire volume of the oceans circulates through the Earth's crust every five million years. As it does, it carries heat and chemicals from deep within.

Such places are home to many bizarre beings. They include bright red tubeworms twelve feet long. The tentacles of these Vestimentifera (worms in dinner jackets, to translate their Latin name) absorb hydrogen sulphide from the hot water, and forward it to dense masses of bacteria within their own bodies. A smoker may need a dinner jacket, but a mouth and gut is optional.

DNA shows that the worms, like oaks, are mere versions of ourselves. The real discoveries are among their microscopic helpers. They, with their fellow microbes of the vents, are quite distinct from worms, oaks and the whole world of the surface.

One of the vent archaeans lives ten thousand feet down, at a pressure of two hundred atmospheres. Its genes come in three closed circles and code for seventeen hundred proteins, most of which have no known matches elsewhere. It needs only inorganic compounds to stay alive, and makes methane as waste. Some of its genes are more like our own than those of bacteria, but this novel beast has many unconventional properties of its own.

The enzymes of such strange beings are forced to work in hot places. They might be useful to bleach paper, or to make jeans with a 'stone-washed' look. Other archaeans turn up just as deep, but in the icy cold, as mats on the bones of drowned whales (first discovered by a Navy submarine as it searched for a lost missile). The skeleton of a whale can support more kinds of life than does the richest hot vent, and can take ten years to decay. Its inhabitants make enzymes able to work at low temperatures, which is useful for companies interested in cold-water detergents. They now fish for new forms in the ocean trenches, with dead whales attached to buoys as bait. Their catch might include the bugs that ate the *Titanic*, a wreck already covered with 'rusticles'; and, within a century or so, doomed to return to a mound of inorganic sludge.

More oddities are hidden beneath our feet. They made their presence known with the collapse of the Cairo sewers soon after they were built. A new bug had eaten the concrete. It was the first of many. Novel kingdoms have now been found two miles into the earth. As some survive at a temperature of a hundred and thirty degrees Celsius, such forms may exist another mile deeper. Dead bacteria gush from ocean vents, evidence that the land under the deepest seas is alive. Some unique creatures were discovered in a search for oil, ten thousand feet under Virginia. They live in what was a stream, buried for two hundred million years below layers of sediment. They have been isolated from the world since the days of the dinosaurs.

The Underworld – like much of the sea, but hot where that is cold – is an austere place. Its denizens use not oxygen but minerals such as iron ore to burn the carbon used for food. They are a million times less abundant than in the topmost layer of soil, and divide not every twenty minutes but once a century. Even so, the new-found land inside the Earth holds one part in a thousand of the entire mass of life. Remote as it seems, the underground universe impinges on our own. Natural gas is the product of archaea, who cause change and decay in the deepest mines (which should worry those who hope to use them to store nuclear waste). Life beneath the land, as yet little studied, already looks so distinct from even its submarine cousins that the gulf between mankind and the animals and plants around us will, in comparison, shrink still further.

Genes mean that the whole of existence can now be arranged into a single natural system. Evolutionists are obsessed with pedigrees. Their desire to draw family trees based on flimsy information with man on the topmost twig has led them into many blind alleys. Now, at last, it is possible to draw a true map of relatedness; a plan seen from above, with no declaration of where life's journey begins or how high or low each branch might be. Like the London Tube, much of its substance is hidden away, but the map cares not at all for that.

The map of the Tube twists the metropolis out of shape to suit the perceptions of travellers. The distance between stations is exaggerated in well-known parts at the expense of distant suburbs. As a result, a place like Wimbledon – larger than the West End or the City of London – dwindles into apparent insignificance. To a tourist, it seems scarcely to exist. The plan of the Paris Métro, in contrast, is less biased as it shows the real intervals between stops, how far each district is from the centre, and the size of every suburb.

Before DNA, our image of the map of life was as distorted as that of a tourist on the Tube, with too much attention paid to its familiar sectors – animals, plants and mushrooms – and very little to its remote extremities. The genes do for life what the *Plan du Métro* does for Paris: they reveal the real and unexpected contours of the great city of existence.

The accompanying diagram will aid us in understanding this perplexing subject. It shows the plan of life as revealed by one set of highly conservative genes, those involved in the machinery that helps assemble proteins. Crowded and unfamiliar neighbourhoods cover – as do the suburbs of any large town – most of life's territory. Two districts, the animals and the plants, are known to all. Like Westminster and Chelsea they loom large in the eyes of visitors, but represent only a small part of the whole. Even the fungi are on the same branch line as those celebrated places. The rest of town is much further away than it seems to the uninformed traveller, and is little known to anyone apart from a few cartographers (and those who live there).

On the true chart of the genes, the bacteria occupy a great Wimbledon of life: a large neighbourhood of their own, with the archaea in the suburb next door. Three groups known as the diplomonads, trichomonads and microsporidians, are diffused through a nearby district. They sound obscure, but among them are the gut parasite that causes severe cases of travellers' diarrhoea, and a single-

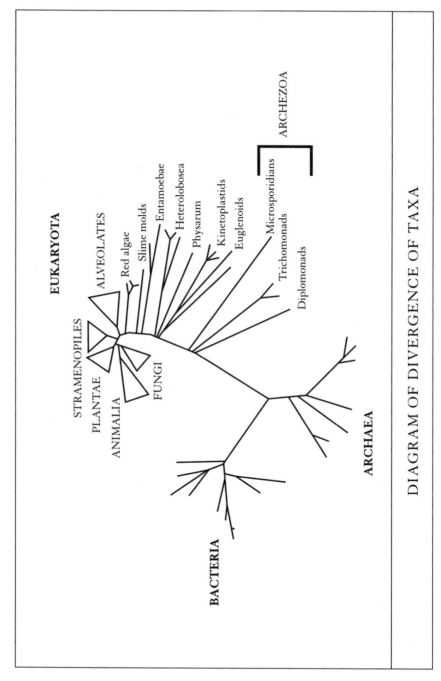

DIAGRAM OF DIVERGENCE OF TAXA

 *Genealogy redrawn from original by M.L. Sogin*

celled creature responsible for inflammation of the vagina. Although they have a cell nucleus, such creatures lack mitochondria.

The rest of town is filled with creatures blessed with both those useful structures. Kinetoplastids – single-celled animals with a tail that lashes them through the water – include agents of disease such as the sleeping sickness found in Africa and the trypanosome that might, perhaps, have infected Darwin in South America and led to his many years of invalidism. Next door is a thriving group of amoeba-like creatures; and two separate groups of slime moulds (much used in the study of development). Alveolates include the ciliates (single-celled animals covered with fine mobile hairs), the agent of malaria, and the dinoflagellates, creatures enclosed within the solid shells that make up much of the chalk that covers southern England.

Some areas traditionally seen as one are, like the East End of London, in fact several, of different character. The algae – seaweeds, waterweeds and the green film found on treetrunks – comprise three separate groups, each with an identity as distinct as is that of the animals from the plants. Red algae, on a branch of their own, include the seaweeds much eaten in Japan, together with others that make reefs. The brown algae contain familiar seaweeds such as kelp, but they belong with diatoms, tiny shelled creatures that abound in the ocean, in a group called the stramenopiles. Green algae, found in freshwater ponds, are different again, and live close to the familiar plants. Plants, animals and fungi are near neighbours in a well-explored but minor part of the metropolis of life.

A Tube map does not show where any journey begins. However, the chart of life suggests that the first genes were those of the simple creatures of the underworld, the archaea, lacking cell nuclei and mitochondria as they do. The idea makes sense: after all, three and a half billion years ago, the surface was bombarded with meteorites big enough to evaporate the oceans, and flooded with ultraviolet strong enough to sterilize an elephant. If so, man – with most of the rest of creation – ascended from Hades rather than stepping down from the Garden of Eden. Genetics has drawn a new map of hell, with humanity firmly placed in the suburbs.

Evolution can sometimes hide a true pedigree. It would be foolish to classify flies, birds and bats as the same because they all have wings. Their similarity comes not from shared descent but from separate answers to the same problem. It is shallow indeed, based on external appearance rather than essential structure, and as a result is not much use in taxonomy. Characters are of real importance for classification only in so far as they reveal descent. Analogical or adaptive characters, although of utmost importance to the welfare of the being, are almost valueless to the systematist.

That can sometimes be very clear. The females of a certain moth and of the African elephant each make the same complex sex pheromone. This attracts males of either kind (which must be riskier for the moth than for the other partner in the relationship). For some reason, the molecule hints at sex, and insect and elephant have each taken it up through different routes.

In evolution, parallel lines often converge. As they do, they conceal real patterns of affinity. Polar bears, Arctic insects and the leaves of northern plants are all covered in fur, but nobody calls them relatives because they share a taste in coats. The fish of those icy waters show how the distinction between true alliance and separate solutions to the same problem can be concealed by the amazing efficiency of natural selection.

The Antarctic perch has an antifreeze, a remarkable protein that allows it to live at two degrees below zero, a temperature at which most fish would turn to ice. The gene responsible resembles another responsible for a digestive enzyme in the fish's own stomach. That job came first. Its role was to prevent the stomach from freezing solid, but when the substance was made in the liver as well, the Antarctic perch came up with an antifreeze for its whole body.

The protein (or another identical in structure) is also found in the Arctic cod, a fish from the other end of the earth – which is a surprise, because the two are but distant relatives. All other evidence – from anatomy, from geology and from geography – agrees that they split forty million years ago, long before the Antarctic froze. They differ in their body structures and in many of their genes.

The antifreeze is, in fact, a separately evolved adaptation in each fish. The Arctic and Antarctic proteins, specialized as each is and with almost the same structure, are not evidence of common descent but – like the furry coats of polar bears and northern plants – of a shared solution to an evolutionary challenge. Natural selection has done its job so well that it has fashioned two identical molecules at opposite ends of the earth. Their convergence shows how animals belonging to two most distinct lines of descent may readily become adapted to similar conditions, and thus assume a close external resemblance; but that such resemblances will not reveal – will rather tend to conceal – their blood-relationship to their proper lines of descent. For the two antifreeze proteins, an identical present conceals a separate past.

Morphology. The members of the same class, independently of their habits of life, resemble each other in the general plan of their organization. This resemblance is often expressed by the term 'unity of type'; or by saying that the several parts and organs in the different species of the class are homologous. The whole subject is included under the general name of Morphology. This is the most interesting department of natural history, and may be said to be its very soul. What can be more curious than that the hand of a man, formed for grasping, that of a mole for digging, the leg of the horse, the paddle of the porpoise, and the wing of the bat, should all be constructed on the same pattern, and should include the same bones, in the same relative positions?

Today's students despise comparative anatomy (as the study of such structures is now called). Many refuse to cut up animals on what they see as moral grounds and the remainder find the whole business tedious and irrelevant. They move with relief to the double helix. Although they may not realize it, they are practising anatomy without a licence. The nineteenth century used bones, but now we have molecules. Much of today's biology is no more than the study of the structure of genes in related organisms. Instead of cutting up guts or skulls, biologists now sequence the DNA that makes them. To do so reveals most of life to be a set of themes and variations.

All complex beings are built from a series of repeated and modified segments. Sometimes that is plain. The earthworm is made up of dozens of sections, many of which look much the same, but others are modified to include eyes, sex organs and the like. It can regenerate its whole body from a fragment of its original. Such an ability raises profound issues. The Swiss theologian Charles Bonnet was much concerned: 'Must we admit that there are as many souls in these Worms as there are portions of these same Worms?' The biological questions are easier to answer. For the earthworm, each segment retains within itself the essence of a complete animal; proof that, different as they appear, each part of its body is constructed on the same plan. Natural selection, during a long-continued course of modification, has seized on a certain number of the primordially similar elements, many times repeated, and has adapted them to the most diverse purposes.

The strange resilience of the mouse hints as to how much repetition there must be even within the mammalian frame. A technical trick makes it possible to knock out particular genes. Sometimes, these incomplete mice are in deep trouble and die young. More often, they live blithely on, although what seem essential parts of their machinery – a gene for collagen, the structural material of much of the body, or for another that passes signals around the cell – have been removed. Duplication is a useful insurance policy against the wiles of geneticists; but how and why these extra copies evolved, nobody knows.

In biology, repetition is everywhere. Worms, insects and leeches are built of a series of modules, and even in higher animals elements of such a plan can plainly be seen. Parts many times repeated are eminently liable to vary in number and structure. A glance at the teeth of any vertebrate shows how a simple and much-duplicated element has been modified for its job of biting, tearing or chewing. The mammal body does, nevertheless, retain a certain individuality in its various parts: a cat's skull has not, it seems, much in common with the tip of its tail.

Vertebrates – the group to which humans, snakes, birds and frogs belong – have some modest relatives. They include the lamprey and the lancelet. The lancelet looks like an anaemic anchovy fillet, but has no eyes, no ears, no jaw, no discernible skull even. The Chinese thought it to be a maggot grown from the corpse of a divine crocodile bearing the god of literature; but many Western scholars denied even that it had a brain. In the days when students could stand the sight of blood, both lamprey and lancelet were much dissected. To anatomize their genome rather than its products discloses their role in history, and hints at how mammals gained the organ of literature itself.

The mammalian skull has scores of bones. The eye and ear are triumphs of complexity and the brain so elaborate as – so far – to be unable to understand itself. The lancelet is less pretentious. It has changed little from the ancient animals that gave rise to humans, snakes and birds. The great German biologist Ernest Haeckel was convinced of its closeness to the earliest vertebrates. So certain was he of its importance that lancelets on caviar were served on his sixtieth birthday (although the menu also makes the enigmatic claim that this was followed by a main course of *Archaeopteryx* with Sauerkraut).

Where did our skull come from and why should the brain be enclosed in a box composed of so numerous and such extraordinarily shaped pieces of bone? Is it a modified piece of an ancient body, or an extra structure bolted on to an insensible frame as a container for a new and mighty organ of thought? Because lancelets lack that noble structure (or even a discernible brain) it was long thought that vertebrates evolved the skull as a new element, a sort of bonnet-ornament attesting to the power of the machine within. The pattern of repeats shows that, instead, vertebrates have an old head on young shoulders.

The genes in charge of the machinery for making skulls are arranged in groups. One set is found in all vertebrates, from lancelet to man. A young fish or a human embryo has a series of regular blocks of tissue that develop into the complicated bones and muscles of the body. In the adult the simplicity of their recurring structure is lost, but the genes show that there are parts in the same individual which become unlike and serve for diverse purposes.

A test of a young lancelet with a genetic probe based on the skull machinery of mammals reveals that, although the animal lacks a head, it has the DNA later used and extended to make it. Brainless though it seems, the genes that in other creatures make a discrete forebrain are also at work. Each is active at the same stage as in mice or chickens. The advance guard of the brain and skull have, it seems, been around since the start. Some of the details have changed. The lancelet has a single light-sensitive organ at the tip of its head – but a gene almost identical to that for the twin eyes of mammals builds its lone eye. Cyclops though they may have been, we do not stand head and shoulders above our ancestors.

The road to mammals called for other changes. It takes more to make a mouse than a lancelet; and mice or humans have extra copies of all lancelet genes, because, somewhere on the road from their common ancestor, their number doubled, twice over. One group, the ray-finned fishes, had yet another doubling after they split from the lineage that led to mammals.

All genes need to know not just what to make, but where, when and how often to make it. Changes in time or place can have great effects on body form as a structure loses its way on the inborn map. Some mutations alter not the shape or function of an organ, but where and when it appears. By so doing, they allow the primordial elements to make their presence felt. In fruit-flies, single genetic changes can cause the middle section of the body to duplicate itself, to give a fly with four rather than two wings. To make a thorax takes thousands of genes, but a simple command can, it seems, set the whole army into motion. The additional organs appear because an order is given in the wrong place and the local cell machinery obeys it.

The switches in control of development are called the homeobox genes. They are arranged in groups of ten or so. Invertebrates have a single copy of each group, stretching over a hundred thousand or so DNA bases. Mammals have four, with many tasks in the embryo (and others in the adult; with mutations in homeoboxes leading to baldness, inability to make milk, and even to leukaemia). Ray-finned fish have seven or eight copies of the crucial genes. They lay claim to twenty-five thousand species, from sturgeon to salmon – as many as all other vertebrates combined. Such fish put mammals to shame, with seahorses, tuna, anglerfish, flatfish, cichlids, eels, lionfish and thousands more. Perhaps their extra homeoboxes, the masters of development, allow them to experiment with new and eccentric sets of body form.

The switches are arranged in series, each in charge of a separate part of the body's battalions. Their sequence, from front to back, is the same as that of the organs for which they are responsible. As a result, homeoboxes lead from the front. Those nearer the head can control structures behind them, while those further back have less influence on parts of the body further forward. Changes in such genes persuade different segments to develop into head, heart or tail. The pattern of central command is common to mice, fruit-flies, snails and every other animal. Ownership of a homeobox may, indeed, be as good a way as any other of defining what an 'animal' might be. They decide what their bearer will look like. Baleen whales have a slight quirk near one of the genes, with four DNA bases missing from part of the mechanism. When transferred to a mouse embryo, the tiny change moves the activity of certain neck segments further back, explaining, perhaps, the whale's great head and enigmatic smile.

Plants, too, from dandelions to redwoods, are built on a multiplied plan. It is familiar to almost everyone that in a flower the relative position of the sepals, petals, stamens and pistils, as well as their intimate structure, are intelligible in the view that they consist of metamorphosed leaves, arranged in a spire. Monstrous plants often give direct evidence of the possibility of one organ being transformed into another. Any gardener understands the power of monstrosity: roses, in their native state a plain and unremarkable flower, have been persuaded to double up their petals from the five or so in wild roses to ten times as many in varieties such as 'Elizabeth Taylor' and 'Blushing Bride'. As in the multi-winged fruit-flies, such freaks arise through changes in genes that control the number of copies of a compound structure.

Flowering plants grow in a simple way, from a group of cells at the tip of each shoot. As these divide, they make small clusters of active cells, each destined to be a leaf or a petal. Different though the flower of a snapdragon or a rose might appear, one made as a mirror image, with a left side and a right, the other arranged like the spokes of a wheel, those organs are at an early stage of growth exactly alike. Genetics also shows that, complicated although the parts of a flower might be, they are built to simple instructions. Various mutations change the number and position of flower parts such as sepals or petals, and some can persuade a two-sided flower to take up a disc shape. The structure of the proteins involved is similar in very different flowers and resembles those made by certain vertebrate genes able to bind (as do the products of homeobox genes) to DNA.

To damage just two crucial genes can persuade a flower to develop as a simple leaf. A mutation in another member of its triumvirate of control can double its entire structure. A garden rose's complexity depends, like that of a worm, on a set of modified and repeated units, honed by natural selection from simple parts. Life has, it seems, developed like opera, from short pastorals with a single motif and a limited range of players to the great music dramas of the farm and the forest.

Biology has also begun to reveal a complex set of themes and variations within the genes themselves. Many proteins are built from a series of modules shared with others that do quite different jobs. Thus, the numerous proteins involved in blood-clotting are made up of four distinct sections, shuffled together in various ways to make a useful device. Not all the members of the clotting chain have all the units, but all share pieces in common. Some elements of the blood-clot apparatus also appear in genes that control the growth of cells, or digest meat. They demonstrate how hidden motifs may appear, in true Wagnerian style, deep within the cellular plot, and how the complex structures of today retain – like the rose, the earthworm, or the song of the Rhine Maidens as Valhalla burns – distant echoes of a simpler past.

Embryology. 'We saw the Emperors standing all together huddled under the Barrier cliff some hundreds of yards away. The little light was going fast: we were much more excited about the approach of complete darkness and the look of wind in the south than we were about our triumph. After indescribable effort and hardship we were witnessing a marvel of the natural world, and we were the first and only men who had ever done so; we had within our grasp material which might prove of the utmost importance to science; we were turning theories into facts with every observation we made, – and we had but a moment to give'.

Thus Apsley Cherry-Garrard in his extraordinary book (its language secretly improved by George Bernard Shaw) *The Worst*

Journey in the World. It is the tale of a trip through the Antarctic winter of 1911 to collect the eggs of the emperor penguin; part of the journey (ill prepared as only a British expedition could be) that ended with the failure to reach the Pole and the death of Captain Scott, weighed down by his precious rock specimens, a few miles from safety.

But why risk death to fight in total darkness across the ice? What was this 'material which might prove of the utmost importance to science'? The answer lies in *The Origin*, in its claim that each animal relives its ancient history as an embryo. To biologists of Cherry-Garrard's generation the story of an individual as it developed was that of the species to which it belonged. To read the narrative that begins when sperm meets egg and ends with death was to retrace the past. The eggs of the emperor penguin, he believed, held the clue to the origin of birds: 'These three embryos from Cape Crozier . . . were striven for in order that the world may have a little more knowledge, that it may build on what it knows rather than on what it thinks'.

Biology, unlike physics, has few laws and those made are soon broken. In 1866 Haeckel came up with his 'biogenetic law': new forms are, he said, the result of extra stages added on to the development of those that went before. He published sketches of the embryos of birds, reptiles and mammals (somewhat doctored to support the idea) that seemed to prove his case. The emperor penguin was the most primitive avian form, pushed to the edge of the world by competition from more recent – and more advanced – relatives to the north. Its embryo would reveal the origin of birds.

After many troubles, and a studied rudeness by the Natural History Museum when Cherry-Garrard tried to deposit his specimens, the eggs turned out to be well on in their growth and of no use as a test of Haeckel's theory. We now know that penguins split off from a winged ancestor forty million years ago and are no more primitive than is the albatross. The science upon which the Antarctic travellers depended was wrong, but life before birth reveals more about evolution than they ever dreamed. The question that seemed so important to naturalists of their day – the development of a complex organism from a simple egg – is on its way to a solution.

The Origin makes no mention of 'evolution', a word whose sense has gone full circle since it began. In its Latin root it referred to the unfolding of a scroll. As Cicero so memorably put it: '*Quid poetarum evolutio voluptatis affert*' – What pleasure does the reading of the poets provide! The word was first used in biology to describe the changes in shape of an embryo as it developed. Not until much later did 'evolution' begin to suggest the gradual transformation of one form into another. Now, its definitions have come together and the study of development is unrolling the scroll of biological history.

The power of the unborn was recognized within five years of *The Origin*. Charles Kingsley's *The Water Babies* is a tale of a young chimney-sweep drowned, transformed into an infant and washed clean of his sins. If evolution culminated in that miracle, the Englishman, why could it not bring forth the greater miracle, the soul – and where better to find it than in an English baby? The drowned Tom 'when he woke . . . found himself swimming about in the stream . . . having round the parotid region of his fauces a set of external gills, just like those of a sucking eft, which he mistook for a lace frill, till he pulled at them, found he hurt himself and made up his mind that they were part of himself'. He relives his innocent past as a young newt or eft, then a sinful child, and – at last – enters into the post-evolutionary state with which the scroll of life should end. Not for the last time, science and theology were in perfect harmony.

The new embryology is based, like that of Charles Kingsley, on a search for resemblances among young animals that may be lost later in development. Now at last the embryo is revealed as a picture, more or less obscured, of the common parent-form of each great class of animals. It uncovers hidden patterns of relatedness because an embryo, sheltered as it is from the need to adapt to the world outside, may retain more of its past than does an adult. An embryonic elephant has kidneys rather like those of a dugong, adapted to life in water. Perhaps long ago an amphibious mammal split into two, one irrevocably taking to the water while the other returned to land (but still fond of a bath and of squirting itself with a trunk that was once a snorkel).

The case is different when an animal during any part of its embryonic career is active, and has to provide for itself. Then, natural selection acts to modify even the youngest, and their history is lost. The period of activity may come on earlier or later; but whenever it comes on, the adaptation of the larva to its conditions of life is just as perfect and as beautiful as in the adult animal. Take, for instance, clams, oysters, mussels and the like. All look much the same as adults, but their larvae are quite distinct. Some float for months as plankton, while others develop within their mothers' bodies; some are hunters, while others burrow into the flesh of fish. Selection can, when called upon, work as hard on the embryo as on any other stage of life.

A few animals have the choice of whether to advance into adult-hood or to stay as juveniles. Tom went backwards into the water with his 'set of external gills' that allowed him to give up air-breathing. The axolotl, a North American eft, goes the other way. In some places, it spends its time in lakes, with gills as impressive as any water baby's, and reproduces in this stage. In others, it grows through its adolescence and clambers on to land where it has habits close to those of any other salamander. Crosses in the laboratory reveal that a single gene controls the difference between them.

Some axolotls can be prompted to abandon their childish habits with a dose of thyroid hormone, which makes them grow legs. Others resist such unwanted maturity and stay in the water-baby stage even after a dose of the chemical. Somewhere in the chain of development is a switch that turns on the biological factory needed to make a land axolotl. It reveals the hidden hand that rules the embryo's fate and shows how a few pieces of DNA can direct a series of underlings that, when set into action, make legs, lungs and every-thing else a terrestrial animal might need.

Such hidden potential is everywhere. The new embryology shows that most tissues in most creatures have the capacity to develop into quite unexpected structures. Most of the time they do not (although plenty of creatures have a change in lifestyle with age at least as great as does the axolotl), but, now and again, they reveal their powers and, by so doing, their evolutionary past.

The descendants of the first Duke of Habsburg found themselves, over seven centuries, in control of Austria, Spain, Portugal, Hungary, Holland and Luxembourg; of parts of Germany, Poland, France and Italy; and even, for a time, of Mexico. Their nations varied from semi-democracies to the revolutionary chaos of Central America.

Each was, nevertheless, ruled by a member of the same family and each, with more or less enthusiasm, followed the instructions of their hereditary monarch. Such hierarchies of ancient control exist throughout the animal kingdom.

The eyes of mice and of fruit-flies are quite unalike. One is a camera with a single lens, the other a whole host of small devices arranged into a complex machine. Fruit-flies and mice each have mutations that lead to the loss or great reduction of the eye. Although their organs are so different, the gene that reduces or removes the eye in flies is almost identical in structure to that which reduces its size in mice.

A certain fly mutation affects a length of DNA that – like the master switch of the axolotl – turns on a whole series of eye genes. When it is persuaded to act in unsuitable parts of an embryo, eyes grow in unlikely places, such as the wing or the leg. Quite remarkably, the normal version of the mouse-eye gene, moved to a fly embryo, induces the growth of eyes, with the typical fly structure, wherever it lands – legs, wings or anywhere else. It steps into its ancient and dictatorial role and the cells of its new host respond. The emperor of the eye is much the same in mammals and insects, although these groups last shared an ancestor a billion years ago. Either an ancient light-sensitive structure – a patch of cells – took two different routes down the lines that led to flies and to mice, or two separate visual organs elected to obey the same ruler. Whatever the truth, to grow fly eyes under the instructions of a mouse gene shows how universal is the history revealed before birth.

Other parallels in the development of flies and backboned animals prove how far the genetic gridlines stretch. The ear, like the eye, has an unexpected ancestry. Fruit-flies cannot hear much, but are easily irritated because they are covered by bristles that sense vibration. The same genes are at work in the inner ears of fish, chickens and mice. In the same way, the gene specifying where the rear half of each fly segment should be does the same job in the young lancelet. Genes switched on at the tip of the young limbs of mice, birds, starfish and flatworms reveal another deep and ancient connection; a foot race that ended up with wings, a starfish's tube feet, horses' hooves and the hollow legs of crabs, all descended from some simple and primitive limb.

One of the more outlandish suggestions of the old anatomy was that humans and their relatives are invertebrates turned upside down. In 1822, the French biologist Geoffroy St Hilaire (often referred to in *The Origin*, but more famous in France as the guardian of the first giraffe to enter the country) saw a dissected lobster lying on its back. He was at once struck by its apparent similarity to a man lying on his stomach. In both, the spinal cord is above the intestine, and the oesophagus above the heart. Perhaps, he thought, life had flipped over on the way to the vertebrates. The idea seemed absurd: French biology again on its own eccentric path, with the rest of the world out of step. A century and a half later, the anglophones had reason to be embarrassed when a gene active in the cells of a frog embryo that make part of its back turned out to be at work on the lower surface of the body in developing fruit-flies. There has indeed been a great somersault on the way to ourselves.

For eyes, ears, limbs, and deciding which way is up, the embryo is left as a sort of picture, preserved by nature, of the ancient and less modified condition of each animal. Its image is proof of the course of evolution from a time older than the earliest fossil.

Rudimentary, atrophied, or aborted organs. The Mass is a Christian eternal. As an exchange with God it has its roots in the earliest Church. Wherever he might be, a believer finds himself at home and – except for the language differences that emerged with the abandonment of Latin – part of a universal ceremony.

Not, however, in Ethiopia. In a minor triumph of comic prose, Evelyn Waugh (who visited the country in 1930) described the responses of an expert to his first experience of a local Mass: '"That was the offertory . . . No, I was wrong; it was the consecration . . . I think it is the secret Gospel . . . the Epistle . . . I have noticed some very curious variations in the Canon of the Mass . . . particularly with regard to the kiss of peace." Then the Mass began.'

His liturgical confusion has a message for biology. The Ethiopian Church was cut off for a thousand years, safe from the reforms that seized the rest of Christendom. As its rite passed down the generations, the errors grew until whole sections made no sense even to those who celebrated it. They are of interest to historians as they reveal a history elsewhere swept away by the march of progress. All characters degenerate as soon as their job is done. Once selection ceases, the chaos of Nature sets in and evolution loses its way. A relic may hence be better evidence of the past than is the most exquisite adaptation.

A glance around the world shows general decay: parasites reduced to sacks of guts and genitals, useless organs such as the human appendix, and great stretches of DNA that make no protein. One worm, after an active life, settles on the sea floor and, like a professor given tenure, absorbs its brain. Whole communities may turn to degeneracy. More than a hundred thousand different kinds of animal – spiders, insects, fish and salamanders – live in caves. Most are pallid imitations of their open-air relatives: slow, blind, calm and solitary when compared to their cousins who live in the glare of day.

Such characters, bearing as they do the stamp of inutility, have long been seen as links with the past. A traveller to Tibet spoke of a 'species of human being with short straight tails, which, according to report, were extremely inconvenient to them, as they had to dig holes in the ground before they could attempt to sit down'. That organ (and a few children are born with tails) was much used in pre-Darwinian times to make a link between humans and apes. A structure that has lost its purpose proved descent from an ancestor to whom it was still useful.

A complicated structure can work in only one way but can go wrong in many. A cross between blind cave-fish from different places gives offspring with larger eyes than either parent. Each cave has lost different parts of the machinery and to put two of them together cobbles together something better than either can manage on its own. Other rudimentary organs, too, sometimes retain their potentiality and are merely not developed. Why is the milk of human kindness made by only half the population? After all, men have nipples and the capacity to use them. Males given certain chemicals lactate with no difficulty. Even a heavy dose of alcohol can do the same as the liver loses its ability to suppress each man's guilty secret, his female hormones. Teenage boys, in a natural desire to see what might be done with their bodies, now and then stimulate their own nipples and (no doubt to their amazement) may eject milk. The

Dyak fruitbat is the sole mammal to reach the logical conclusion, as its males suckle their young; but as the males of many others invest lots of care in children it seems strange that they do not go the whole hog and make food themselves.

A man's pert but useless nipples are his real stamp of inutility. They have no job because of evolution. Many males leave at fertilization and do not hang around for birth. Their energies are invested in more sperm rather than in the results of an earlier fling. There is no market for their milk. Because the female is already committed through an expensive pregnancy to her child, it does not pay her to abandon it in the hope of success with another. Selection has turned her into a milk machine. The male's job, if any, is to protect his child as he keeps a weather-eye open for what else might turn up. His nipples are just a reminder that he is of common stock with his partner.

An organ serving two purposes may become rudimentary or utterly aborted for one, even the more important purpose, and remain perfectly efficient for the other. Tiny bones inside the ear of mammals – the hammer, the anvil and the stirrup, as they are called – act as a series of levers to transmit sound waves from the eardrum to the sense cells. They trace their evolution to organs in ancient fish, useful in quite different ways, but now decayed. The most primitive fishes had not jaws but supports called gill-arches. The first jaws were made of hijacked copies of these structures, although the rest stayed on as struts for gills. As fish evolved towards mammals, the arches were lost; but some of the tiny bones of the middle ear remain as their descendants. The rest of the ear has the same utilitarian history. Fish have a lateral line, a network of receptors that stretches along the body and detects changes in water pressure. The sensory part of our own organ of hearing descends from the ancient pressure-sensor of fish (and, the genes show, in the end from an ancestor shared with the irritable hairs of flies) but all other signs of its presence have gone.

Plant parts, too, can have a change of career. Climbers may use reduced leaves (as in the grape), leaflets (the trumpet vine), paired structures on either side of the leaf base (passion-flowers) or parts of the flower (blackberries) as they clamber up their supports. Each tendril or hook has lost one job to take up another, as evidence of the expedience of evolution when offered a structure that might be induced to help in a new way. Natural selection does not hesitate to pick up and use whatever becomes available. But, for most of the time, decay is just decay and life has, with a sigh of relief, given up what it does not need.

Rudimentary organs, such as teeth in the upper jaws of whales and ruminants, can often be detected in the embryo, but afterwards wholly disappear. Kiwis, while still in the egg, have wings as big as those of embryonic albatrosses; but they never grow. The instructions may be retained even when the structures themselves are lost. A bird can be persuaded to grow teeth in its developing beak if a few appropriate mouse cells are moved nearby to switch on those long-dormant genes. Once again, economics is at work. As the main expense of making wings or teeth comes later in life, natural selection is not much interested in making cuts in places where no real savings are made.

There is a message for evolution hidden in its discarded baggage: it is heavy, and costs a lot to carry. An unencumbered traveller can use the energy in other ways. The biggest burdens are shed first. The speed with which they are lost hints at the cost of keeping them. H. G. Wells, in *The War of the Worlds*, has his Martians reduced to mere gutless heads that suck the blood from men and women. They come from an old, dry planet, short of oxygen. On Earth, they struggle against gravity – a 'cope of lead'. They have abandoned all but the capacity to think.

Take, in contrast, that most artificial of beasts, the chicken. It has taken the opposite route to the Martians. It joined the household in Asia and was used to divine the future (entrails were most reliable) long before the Romans turned it into food. Because their wild relatives, jungle fowl, are still with us, today's chickens show how evolution has cast off its burdens. They have become animated machines to turn food into flesh, with lighter bones, weaker muscles and smaller brains than jungle fowl. Why spend energy on thought when man does it for you? Brains have been sacrificed to make guts. Chickens have intestines three times as long and twice as efficient as those of their ancestors, but much reduced intellects.

Jungle fowl, unlike their descendants, can still fly. Wings are expensive and have been abandoned again and again. Females are more liable to lose them than are males. A simple but brutal experiment shows why. Cut off the wing of a young cricket, and she lays more eggs. Birds, too, have shed the organ gained with such effort by their dinosaur ancestors. Some are unique to coral islands a mere hundred thousand years old and have cast off much of the structure of their wings in that time. The dodo, a turkey-sized flightless pigeon, went further along its evolutionary path than anything achieved by the Philoperisteron. Madagascar was once full of elephant-birds (flightless, as the name suggests), most Pacific islands had their own ground rails, and the Galapagos retains its flightless cormorant. In New Zealand, more than thirty different kinds are confined to the ground – moas, kiwis, ducks, geese, rails, crows, wrens and a parrot. With no enemies to flee from, and nowhere to go, why fly?

Penguins are active birds that use their wings for swimming, but most flightless forms are sluggish in comparison to their winged kin. Small islands are hungry places and a slowdown can be the only way to cope. Island birds lose more than wings. Most are smaller than their mainland relatives – sometimes half the size – and consume a fifth as much energy. The bones of reduced elephants found in Cyprus gave Jonathan Swift the idea for Lilliput, a land of insular dwarfs. Lethargy needs less food, and the torpid survive where those full of energy would starve. Six-ton mammoths were reduced to two-ton midgets on Wrangel Island in the Arctic Ocean and red deer marooned on Jersey after the ice age withered in a mere six thousand years to the size of a St Bernard dog. They were forced to shrink or starve. Sex, too, is a costly pastime. Potatoes kept as clones can still flower, but their sexual advertisements are feeble compared to those of their wild relatives. For potatoes advertising is expensive. Its hoardings soon rust when not burnished by natural selection.

At the molecular level, useless structures may soon disappear. A certain virus uses a specialized piece of protein to copy itself. If placed in a test-tube with its raw material, the gene still works. As it does, it evolves to suit its new home, and within three hundred generations its copying efficiency goes up by fifteen times. The enzyme is ruthless in its demand for speed and dumps most of its material on the way, until it is reduced to a sixth the length of the original. As a rump of what it was, the streamlined virus cannot invade new cells, but in the test-tube it has no need to.

Given all this emphasis on efficiency, it is a surprise to find great tracts of repetition and decay within the DNA. There are not just thousands of repeats of the same message, but hundreds of dilapidated ruins of what were once working genes. Such pseudo-genes, as they are known, are everywhere, in mammals at least. The haemoglobin gene family has half a dozen, each corrupted almost beyond recognition. Long ago, a mutation destroyed the switch that turns the gene on, or inserted an instruction that it should stop doing its job, or damaged its ability to edit its message. Some pseudogenes are the remnants of viral attack and have been read back into the DNA from an edited version of the genetic message, to be scattered where they fall. Evolution at once lost interest: as soon as the gene stopped work it was, in effect, invisible. Such structures sit for millions of years and crumble until their shape can barely be discerned.

Why do rotting genes persist and why is life happy to accept them when it so soon gets rid of unwanted wings or eyes? Perhaps genes are cheaper than the flesh and bones they make. Pseudogenes, like other rudimentary organs, retain more of their family relatedness than do those still scrutinized by natural selection. A pedigree based on decay reveals descent with modification better than one in which most changes have been removed. As a result, true relatedness is often best revealed in characters that may be considered rudimentary or atrophied; both in DNA and in the leg-bones that draw the resemblance between whales and deer. Thus, in their functional parts, the antifreeze proteins of the Arctic cod and Antarctic perch are the same. They say nothing about the affinity of the two fish. However, the genes for antifreeze contain sections of DNA that make no product. Such DNA – a rudimentary or useless character indeed – reveals the real patterns of kinship. The Arctic cod's inclusions are quite different from those of the Antarctic animal. Unlike the protein itself, they tie it to its true relatives, who include the fish pursued without mercy by man.

Such withered lengths of DNA may be compared to the letters in a word, still retained in the spelling, but become useless in the pronunciation they serve as a clue in seeking for its derivation. A message that has become gibberish, and is transmitted – like the Coptic Mass – without regard for meaning, retains more of the past than one visible to natural selection, the great editor who removes all errors and eliminates history as he does so.

Summary. In this chapter I have attempted to show, that the subordination of group to group in all organisms throughout all time; that the nature of the relationship, by which all living and extinct beings are united by complex, radiating, and circuitous lines of affinities into one grand system; the rules followed and the difficulties encountered by naturalists in their classifications; the value set upon characters, if constant and prevalent, whether of high vital importance, or of the most trifling importance, or, as in rudimentary organs, of no importance; the wide opposition in value between analogical or adaptive characters, and characters of true affinity; and other such rules; – all naturally follow on the view of the common parentage of those forms which are considered by naturalists as allied, together with their modification through natural selection, with its contingencies of extinction and divergence of character. In considering this view of classification, it should be borne in mind that the element of descent has been universally used in ranking together the sexes, ages, and acknowledged varieties of the same species, however different they may be in structure. If we extend the use of this element of descent, – the only certainly known cause of similarity in organic beings, – we shall understand what is meant by the natural system: it is genealogical in its attempted arrangement, with the grades of acquired difference marked by the terms varieties, species, genera, families, orders, and classes.

On this same view of descent with modification, all the great facts in Morphology become intelligible, – whether we look to the same pattern displayed in the homologous organs, to whatever purpose applied, of the different species of a class; or to the homologous parts constructed on the same pattern in each individual animal and plant.

On the principle of successive slight variations, not necessarily or generally supervening at a very early period of life, and being inherited at a corresponding period, we can understand the great leading facts in Embryology; namely, the resemblance in an individual embryo of the homologous parts, which when matured will become widely different from each other in structure and function; and the resemblance in different species of a class of the homologous parts or organs, though fitted in the adult members for purposes as different as possible. Larvae are active embryos, which have become specially modified in relation to their habits of life, through the principle of modifications being inherited at corresponding ages. On this same principle – and bearing in mind, that when organs are reduced in size, either from disuse or selection, it will generally be at that period of life when the being has to provide for its own wants, and bearing in mind how strong is the principle of inheritance – the occurrence of rudimentary organs and their final abortion, present to us no inexplicable difficulties; on the contrary, their presence might have been even anticipated. The importance of embryological characters and of rudimentary organs in classification is intelligible, on the view that an arrangement is only so far natural as it is genealogical.

Finally, the several classes of facts which have been considered in this chapter, seem to me to proclaim so plainly, that the innumerable species, genera, and families of organic beings, with which this world is peopled, have all descended, each within its own class or group, from common parents, and have all been modified in the course of descent, that I should without hesitation adopt this view, even if it were unsupported by other facts or arguments.

INTERLUDE

ALMOST LIKE A WHALE?

Difficulties on the theory of humans as apes — Evidence that Man is a product of evolution — Recapitulation of circumstances in which selection works upon ourselves — Causes of the general belief in the immutability of Man — How far the theory of Natural Selection may be extended — Effects of its adoption on the study of Man — Concluding remarks

THE LINES OF doggerel that begin this book are a parody. They mock a work called *The Progress of Civil Society; A Didactic Poem in Six Books*, written in 1796 by Richard Payne Knight (whose botanical brother Andrew is, by coincidence, mentioned in *The Origin*). His instructive poem is a broad – a universal – survey of humankind set in terms of the animal world and annotated with the apparatus of Enlightenment scholarship. Hookham Frere's burlesque ascends from aerial mackerel and aquatic bears into a kind of gothic dementia ('Ah! Who has seen the mailed lobster rise,/ Clap her broad wings, and soaring claim the skies?/ When did the owl, descending from her bower,/ Crop, midst the fleecy flocks, the tender flow'r,/ Or the young heifer plunge, with pliant limb,/ In the salt wave, and fish-like strive to swim?'). On the way it makes a serious point about the limits to biology in human affairs.

The parody has marginal comments and notes to match the dozens that decorate Payne Knight's futile attempt to use science to explain the obvious. That to the first line (in which a feathered race skims, in leaden verse, the air) reads, simply, '[1] Birds fly'.

Humans are a footnote to biology. Evolution is often seen as a progress towards civil society, an overture to *Homo sapiens*. That view ignores the substance of life in favour of its details. The last chapter of *The Origin of Species* is an abstract of the book's 'long argument' in simple, even urgent, terms. It returns with almost poetic intensity to the woodpeckers, wasps and oaks that make Darwin's case; and contains its most famous sentence: 'Light will be thrown on the origin of man and his history'. His final statement appears as an appendix to the present work.

My own last chapter, unlike its original, is not a digest of the whole, but a detail of history, a test of how our own past is illuminated by evolution; of whether man follows the rules that govern the whale and the AIDS virus.

When Wallace asked Darwin whether he planned to include evidence from humans in *The Origin*, he answered, 'I think I shall avoid the whole subject, as so surrounded with prejudices; though I fully admit that it is the highest and most interesting problem for the naturalist'. He was right. In 1859 not a single human fossil was known, but now thousands have been found. Genes reveal the true relatedness of people, peoples and primates. Darwin's entire argument can be made in human terms and *The Origin* restated with evidence taken from our own evolution. Variation, a struggle for existence, natural selection, fossils, embryos, geography, comparative anatomy and instinct; all the elements are there. The evidence that mankind evolved is impossible to resist. We know so much about our own biology that the progress (if such it is) towards the animal that civilized society has a lesson for evolution as a whole. Its logic is that of *The Origin* itself.

Darwin's case begins with life on the farm, but man himself has become the greatest domestic animal of all. A fifth of the earth is covered by crops or cities as proof of how much the world has been forced into the household. The passage was not easy. Although cows gained from a move to the field, for one species farms reduced the quality of life. Before agriculture, the people of the Middle East ate a hundred and fifty kinds of plant. Afterwards, that fell to half a dozen or so. Those who tilled the soil were smaller and less healthy than their ancestors, with worse teeth and signs of anaemia written in their bones. Women's elbows were ruined by constant grinding and men paid the price in social stress, with far more beaten to death or stabbed than before.

The malign effects of farming are almost over. Since the Second World War, Europeans have grown by half an inch a decade. One man in four in Britain is six feet tall. In fifty years that is likely to be the average and we will have regained the physique, if not the psyche, of the hunter-gatherers. People who stayed as hunters or fishermen – such as the inhabitants of Pacific islands – now find it hard to deal with the excesses of the Western diet and suffer from diabetes as a result. Domestication changed man's genes, as it did those of his plants and animals.

Cities did further damage to those who built them and provided natural selection with a new forum. In the Middle East, hunters lived in settlements hundreds strong. Nine thousand years ago – before the great Black Sea flood – came the first metropolis. Catal Hüyük in Turkey had a population of ten thousand, crammed into tiny houses decorated with images of volcanoes and of headless men attacked by vultures, with the remains of ancestors buried beneath their floors. For the first time, pestilence found a pool of people in which to sustain itself and, their descendants show, the genes responded. For much of its history, even London could not maintain its numbers except by movement from the countryside, so pervasive was disease. Illnesses such as cystic fibrosis today might be the remnants of genes that, long ago, protected against the cholera which swept through the new nurseries of pestilence.

However domestic man may be, he is (unlike most of his servants) still filled with diversity. Our variation under nature is such that we have a great ability to tell ourselves apart. Francis Galton, Darwin's cousin, set out to spot a typical English face – the John Bull of those days – in Kensington Gardens. He failed because each person he saw was so unalike. A distinctive face is remembered better than others, with Prince Charles noticed sooner than Tony Blair. Much of the variation is inherited, and to study faces in families shows that rather few genes are involved.

Diversity is in faces, on cells and within the DNA. The sections that code for useful proteins – the eye lens, or the framework of the cell – vary not at all, while others, such as the short segments that make the 'DNA fingerprint', differ to such a degree that no two people who have ever lived (or ever will live) are alike. Molecular genetics once had a Linnaean view of man: that to establish his nature it is enough to study a single specimen (although the chart of human genes was drawn from a sort of chimera in which lengths of DNA were lifted at random from different people). Rather late in the day, evolution has been granted a look-in, with a first step to a world map of molecular variation. Human DNA has about three million variable sites. A scheme to set up a grid of a hundred thousand points is well under way in a search for the parts of the genome that may affect liability to disease.

Such genes are as universal in humans as in their dogs and may be revealed in much the same manner. When people mate with relatives, they bring forth unhealthy children. The children of cousins have a survival rate three or four per cent lower than average. They have been unlucky in the biological lottery and inherit double copies of genes that win no prizes. A simple sum based on the health of inbred children shows that everyone must carry single copies of one or two genes that, if present in double dose, would kill them. As is true for domestic animals and plants, inbreeding reveals a hidden universe of human variation.

Most of the diversity separates one person from the next. Some, though, reflects differences among populations. For most genes the centre of variety is in Africa, with trends of decrease through Europe and Asia to the small islands of the Pacific and the New World. Within Africa live peoples as distinct as the Zairean pygmies and the tall Nilotics. Europe, in contrast, can do no better than Greeks and Swedes. Between people and between places, Africa is different. Two Africans are, on the average, more dissimilar than two Europeans, and two African villages a few miles apart can be as distinct as whole nations elsewhere. By comparison with that continent, the rest of the world is monotonous indeed.

By comparison with other mammals, however, man himself is a tedious beast. Although the world is divided by politics it is united by genes. Chimps have two species – the common chimpanzee and the bonobo – and gorillas are divided into mountain and lowland kinds,

but man and aardvark are unique among mammals as solitary
species, each within a single genus. Our relatives have evolved more
than we have because they have been around for much longer. Man
is a newcomer and has not had time to build up the diversity found
in chimps or gorillas. The most remarkable thing about humankind
is how uniform it is.

As any chimp knows, man is also amazingly common. He has
released himself from the struggle for existence. The abundance of
any creature is related to its size, with fewer whales than mice. The
fit is consistent, and almost all mammals from mouse to whale sit on
the same line (cars follow the same rule, with more Renault Clios
than Lagunas). Man is ten thousand times more plentiful than
expected from his size. Civilization has removed the constraints that
limit all other animals. In spite of many claims that resources are
near exhaustion (in 1865 that Britain's coal was almost finished, in
1951 and 1972 that the world's oil would be gone within a decade),
most minerals are still abundant and prices are falling. In the 1970s
there were apocalyptic forecasts of famine, which did not arrive; and
food production continues to go up – most of the time – the crops get
cheaper. Even so, with man already using almost half the total production of all plants, a limit must be reached. On ecological grounds
Americans are the most evolved of all, for each needs as much energy
as a sperm whale. Their nation is already so advanced that it would
take two complete Planet Earths to feed the world to the level of the
United States.

If humans were on the curve that ties mice to elephants, the global
population would be around half a million (rather than six billion)
and Britain would contain the same number of people as it does seals.
The explosion began with agriculture and speeded up with industry.
Although the rate of increase has begun to slow, our numbers are
almost certain to double before they begin to fall. Today's respite
from the Malthusian struggle will not last.

There never has been a rest from natural selection. Genes are
checked at all times for their ability to cope; and many people fail the
examination. In the Western world only around one baby in a
hundred dies from an overt genetic disease and, given a survival rate

of 98 per cent to the age of thirty, it might seem that natural selection has little raw material left to play with. However, genes were involved in far more deaths a century ago. For most of history, half of all children died from cold, from starvation or from infection. Although squalor may have been the direct cause, DNA was hard at work behind the scenes because some babies failed and some succeeded by virtue of what they inherited.

Cold, hunger and disease are still hard at work. Nineteen out of every twenty deaths of children are in the Third World, compared to half the mortality at the age of seventy. Although smallpox vanished twenty years ago, and polio should be extinct within a decade, not all the news is good: there was no cholera in South America for a century, but now it has a million cases a year. Many societies live on the edge. In West Africa, children born at the time of the rains, the 'hungry season', are ten times more liable to die in their teens than are others, as they never recover from the experience. Because genes influence how well they survive, and because death in childhood counts for much more than one delayed, their demise is fuel for selection.

Many of the trends in human size, shape and appearance that cross the world may reflect past natural selection; but – even for striking differences such as skin colour – we are embarrassingly ignorant about what it was. For other patterns (such as the absence of the B blood group in the New World) we know even less. Sometimes there is a hint of the truth. Some people are unable to deal with Nature's poisons and get cancer as a result. The genes involved are related to those that allow insects to live on toxic plants and have emerged in a new guise in the modern world. A tenth of Europeans lack one of those genes. They find it hard to break down the drugs used to treat heart conditions and a dose helpful to most is lethal to them. Natural selection is, once again, at work; but for humans – as for other creatures – whether, and how hard, it influences the mass of diversity, from fingerprints to DNA, is as yet unknown.

The same is true of sexual selection. Darwin, in a letter twenty years before *The Origin*, weighed the case for and against marriage: 'Children – (if it Please God) – Constant companion, (& friend in old age) who will feel interested in one, – object to be played with. better than a dog anyhow ... but terrible loss of time.– cannot read in the

evenings – fatness & idleness – less money for books &c – if many
children forced to gain one's bread ... banishment & degradation
into indolent, idle fool.' In the end uxoriousness triumphed over
literature and he married his cousin (who had ten children to pass on
the Darwinian legacy). In humans, too, selection works on sexual
success as much as on survival, with a balance of cost and benefit for
each party to the reproductive debate.

In sea-elephants and flowers, males have more variation in sexual
success than do females, with plenty of chance for male competition
and female choice. Men, too, prefer young, attractive and faithful
women; while women look for the same in their own partners, but a
decent bank balance helps. That revelation by the new science of
sociobiology has been claimed to result from sexual selection.
Without evidence of genetic variation in looks or wealth it is hard
to know whether the rules that govern sea-elephants also apply to
humans. Human behaviour is certainly susceptible to a change in the
balance of the sexes. In China, because of selective abortion, a hundred and fourteen boys are born to every hundred girls. This has led
to a generation of frustrated bachelors – and to an increase in the
value of females, who are either forced into marriage or attracted
with large dowries. In time, no doubt, the attraction of sons will
wane as they are forced to stay single.

The days of confusion about human inheritance have been succeeded
by what seems like clarity. It takes seventy to a hundred thousand
genes to make a man (about the same number of parts needed to
build a modest executive jet) and the Human Genome Project – the
scheme to read the order of the DNA bases – should be complete in
2003, a mere fifty years after the discovery of the double helix. A single laboratory can now read off millions of its letters a year and (in
some at least) the information is fed on to the Internet, to protect it
from those who hope to patent genes. Two people out of three in the
Western world die for reasons connected to the genes they carry
(most of them because their inheritance is less able to cope with the
environment – tobacco, alcohol, salt, sugar or stress – which they
face). So important has genetics become in medicine, and so far has
it advanced, that to study humans is often the first (rather than the

last) resort in understanding the science. As a result, our own inborn weaknesses illustrate the apparent difficulties of evolutionary theory better than anything else.

Malaria kills two million people a year (and half a billion more suffer from the disease). Its agent is a single-celled parasite passed on by mosquitoes. All the parties – man, mosquito and parasite – have responded to natural selection. Among the mosquitoes, insecticide resistance is everywhere. In the parasite, resistance to drugs means that death rates from malaria have risen by ten times and that in some places medicine is a single remedy from defeat. In spite of great hopes for vaccines or new treatments in a continent in which some states spend half their wealth on arms, most of Africa cannot even afford to control the insects.

Man, too, has done a lot of evolving in the face of malaria. How he did it rebuts many of the supposed problems of Darwinism.

What once appeared a single illness is in fact four, caused by parasites not much related to each other. Each does its damage as it invades the red cell and spreads to liver and brain, with fevers, anaemia and worse. Some groups are highly susceptible. When the disease moved to the New World, one tribe was reduced in numbers from fifteen thousand to seventy within a century. Within Africa, malaria's native continent, too, some individuals – and some peoples – deal with it far better than do others.

Evolution is to blame. Malaria is a great test of Darwinism. It shows how that theory's apparent weaknesses are in fact its best support. An ever-changing enemy is fought with whatever comes to hand and different peoples achieve the same end through different means. Often, the same gene arises in distant places, and has tasks unrelated to disease. Sometimes, resistance imposes a terrible burden on those involved; but that matters not at all if the cost of the defences is outweighed by the benefits they bring.

The parasite spends much of its time inside red blood cells, where it feeds on haemoglobin. A change in the structure of the molecule can put paid to it. The genes involved protect those who inherit a single copy; but they also cause the commonest inherited diseases in the world, because in double dose they damage their carriers. One person in fifteen, worldwide, has a copy of an altered haemoglobin, and tens of millions suffer as they pay the price for the advantage that such genes give to their kin.

The symptoms of sickle-cell anaemia were described in 1670. They come from a single change in the DNA, a shift in a certain base that alters the stability of the oxygen-carrying molecule. In those with two copies of the gene, the red cells collapse into a crescent shape when oxygen is short. As the cells buckle, the parasite may die; but the sickled cells clog tiny blood vessels and lead to pain, kidney failure, heart problems and more. The gene maintains itself because children who inherit just one copy resist malaria and are otherwise well. Their unhealthy relatives – those condemned by the laws of Mendel to inherit two damaged versions of the haemoglobin gene – pay the price for the survival of their more plentiful relatives. The spread of the sickle-cell gene depends on a balance of kinship, cost and benefit as relentless as that of the beehive, and involves as little conscious effort.

Millions of Africans trace their descent from the single individual within whom the mutation happened long ago. Sickle-cell is also found in India, but there the DNA around the crucial site reveals an origin not in Africa but in an ancient Asian. For the sickle-cell gene (as for the antifreeze of the Arctic and Antarctic fish), an identical present conceals an independent past.

Haemoglobin has hundreds of other changes that, in their several ways, help defeat the parasite. Some involve single shifts in its building blocks, while others cut out great lengths of the protein chain. Changes in quite separate genes alter the red cell's economy to make things hard for the invader. Some render cells more sensitive to wastes so that infected cells, and their unwelcome visitor, die. The identity molecules on white blood cells also protect against malaria's worst effects as some versions allow the immune system to destroy the parasite before it enters the brain. Certain peoples resist infection for reasons quite unknown. In the Sudan, the Fulani live mixed among the Rimaibe (who used to be their slaves). Everyone is bitten at least once a night – but most Fulani stay clear of the disease, while almost all their neighbours suffer from it.

The Duffy blood group has two forms. The agent of malaria uses one variant as a docking site. The other, common in West Africa, gives protection against the disease (a fact discovered when blacks were infected with malaria in an attempt to treat syphilis but failed to contract it). The African version removes the whole structure and leaves the parasite baffled. The Duffy receptor remains on all other tissues, but on the red cells it is switched off. The protein also serves another purpose: in the brain, where it helps to direct cells to where they should be. In West Africans it has become rudimentary or utterly aborted for its docking job, but remains perfectly efficient for the other.

Such diversified adaptations in the same species are just part of the war against infection. Some anti-malaria genes were first recognized as diseases in their own right; as organs of small importance, indeed. People of African origin are susceptible to high blood pressure because they are bad at clearing salt from the body. Salty blood is, it transpires, a defence against the parasite. Many suffer from a high (and often fatal) level of iron in the liver. Iron, too, kills the agent of malaria; and the illness is a side-effect. All this is evidence of the many unknown laws of correlation of growth, which, when one part of the organization is modified and the modifications are accumulated by natural selection, will cause other changes, often of the most unexpected nature.

In a few places – East Anglia, the United States, around the Mediterranean – the parasite has been driven out. The protective genes are still there, and will long remain (as they will, perhaps, after the disease has gone altogether from the world). They will be seen as relict characters, evidence of a battle against a defeated enemy. Some, like sickle-cell, will show themselves as apparent flaws that have slipped through the net of natural selection. Because the history of the illness is known, such genes can, instead, be recognized as evidence of selection's efficiency and its uncanny ability to offset a benefit against what it costs.

Society is much the same. Human behaviour is, without doubt, crafted by evolution, and we have inborn instincts as strong as those of any animal. Our species has long been seen as an ape remade by thought: to one expert, 'the ancestors of the Gorilla and the Chimpanzee gave up the struggle for mental supremacy because they were satisfied with their circumstances.' Humans, it appears, were not. The brain makes us human. It has doubled in size in the past two million years. Its progress was not smooth, as some of our forerunners had brains larger than our own (albeit on a heftier frame). Bodies became smaller, but the brain did not follow. Our intellect might result not, as we flatter ourselves, from the benefit conferred by a great mind, but by a small body.

Most of what makes all primates what they are resides within the skull. Our relatives are pretty smart: but why? Perhaps it is because the fruit trees favoured by many are patchily distributed and a mental map is needed to remember where they are. Social life, too, needs grey matter to tell who is who and how to treat the neighbours. Comparative anatomy hints at the past. The bigger the group, the more complex the society. The size of the brain fits that of the community, with a relationship much better than anything to do with what a particular species eats. Society, not shopping, swelled our heads.

However they arose, its contents are expensive. A newborn baby uses more than half its energies on what is in its skull. The human brain is a kilogram heavier than that of a similar-sized mammal – a hefty pig, say; but we do not use much more energy than pigs. Men have smaller guts than other primates and have, unlike chickens, invested in brains over bellies. The brain's costs come early on, as the organ reaches almost adult size by the time a child goes to school. As Winston Churchill said, 'There is no finer investment for any community than putting milk into babies'. A giant intellect explains why so much is needed.

Far more genes are active there than in any other tissue, and they can mutate and evolve as much as others. Many mental disorders have a simple genetic basis. Others – some forms of schizophrenia, depression and mania included – may also be influenced by inheritance. Genes also affect personality, with claims of inborn variation in how sociable or misanthropic people might be and of their ability to deal with words or with tools. When it comes to intelligence, schools mould what nature provides (which is why Eton stays in business); but without some foundation in reason, the most expensive education is wasted. Variation in a gene controlling the growth of cells (brain cells included) plays a small part in differences in intelligence. More will, no doubt, be found. However many might emerge, man's behaviour – like that of the Cocos finch – will always be defined more by habit than instinct, by what he learns rather than by what he is.

Behaviour apart, man is – of course – not much more than just another primate. No doubt, the laws of hybridism that keep him separate can be blamed to some degree on intellect. There are rumours of test-tube crosses between men and chimps. One such creature made it to the front pages, although on close examination he turned out to be just an unusually rational ape. A cross between the two primates might at least be possible. Humans and chimps share 98.8 per cent of their DNA, humans and gorillas rather less. In chromosomes, too, we are similar. The main difference between the great apes and ourselves is that two chimpanzee chromosomes are fused and three more are reshuffled (which means that any hybrid would be sterile).

Our affinity to chimps should be seen in context. It leaves thousands of genes unaccounted for; and, in any case, the diversity of mammals as a whole hides genetic conservatism, with a mere thirty or so large sections of DNA re-ordered when humans are compared with cats. One cell surface molecule has changed in man alone. It acts as a docking site for diseases such as malaria (to which humans are uniquely susceptible) and as a means of communication between cells – those in the brain included. Perhaps it played a part in man's struggle for mental supremacy and helped keep him distinct from his kin.

Man may differ little in his anatomy from other primates, but his past has a certain interest if only on grounds of familiarity. Darwin was cautious in *The Origin*, but in *The Descent of Man*, published twelve years later, could afford to be direct: 'Man is descended from a hairy quadruped, furnished with a tail and pointed ears, probably arboreal in its habits, and an inhabitant of the Old World'. To console those insulted by such a notion, he noted that if the human pedigree was no longer 'of noble quality', at least it was 'of prodigious length'.

Now, we have fossils as proof of his claim. The human record is, like all others, incomplete, although it has become much less so in the past half-century. Dozens of Neanderthals have been found. Another almost complete set of bones comes from the famous 'Lucy', who lived more than three million years ago. These are exceptions. Primate remains are as liable to upheaval as are those of less noble beings. Hundreds of fossil teeth, and not much more, are found along the Lower Omo River in Ethiopia, because teeth are tough and last when other relics have gone. Most human fossils are small fragments of bone, upon each of which great mountains of theory have been built.

Lucy, for example, was smaller than her fellows. Perhaps, some suggest, males were then much larger than females and life was based on sexual battles. However, it now seems that there were two different-sized species around at the time, so that her supposedly unequal society disappeared as soon as more fossils were found. Other finds have also been much speculated on. On such remains were built a theory of man as a killer ape (in the language of science, an 'osteodontokeratic culture') because the skulls were damaged and bones scored as if by knives. In fact, the cracks were the accidents of time, and the cuts are grooves made as bones tumbled down a riverbed.

Science needs theories, but – with its shortage of facts – human evolution has rather more than it can cope with. Take the art of walking. Lucy and her predecessors could stand erect (although the organ of balance in her inner ear is more like that of apes than humans, so that perhaps they did not do so very well). It is still a delicately poised talent, as anyone who overdoes the drink finds out. But why did we change from horizontal animals who, now and again, took to the vertical, to an upright beast who must struggle against gravity with a body evolved at ninety degrees to where it usually finds itself?

In the days when man was seen as striving to reach a manifest destiny, the answer was obvious: our ancestors stood up to free their hands for tasks more important than walking. Of course, the idea is foolish: it is like saying that the brain evolved in order to watch television. Evolution did not plan ahead to the days of the *Jerry Springer Show*. Man stood up, as he did everything else, for pragmatic and short-term reasons. But what could they have been?

Perhaps a vertical animal found it easier to hunt, or to follow herds of grazers in the hope of plunder. Perhaps our ancestor was himself hunted and, by getting on his hind legs, could see danger in good time. There may have been a change of diet towards fruit high on trees or, as the climate dried, a change in habits as food became more scattered and had to be carried back to a distant camp. Being erect is a sexual display, and a male who could keep upright for longer might do better with the females. Those who lie down on a hot day expose more to the sun (which is why sunbathers rarely do so standing up) and, in the tropics at least, pay the price. All these ideas are reasonable but, given that the event happened five million years ago, are hard to test. With so little from the past, anthropology is one of the few sciences in which it is possible to be famous for having an opinion, and until more facts emerge such speculation is bound to remain.

Incomplete as it is, and over-interpreted as it may be, the record of the past is forceful evidence of the reality of human evolution. The primates began some sixty-five million years ago in a warm and wet Africa. Some six thousand kinds have lived since then (and two hundred or so remain today, from the quarter-pound mouse lemur to the gorilla, a thousand times heavier). Once, the world had many more species of apes than of monkeys, but now just five great apes are left (one of which is us) while monkeys flourish; proof that there was no inevitable progress towards mankind.

About fifteen million years after the emergence of the first primate, the predecessors of apes appear. The molecular clock suggests that the lines to chimp, to gorilla and to humans split some six million years ago. A dozen or so hominines – as the branch upon which we belong is called – have lived since then (although how many are real entities is hard to assess). Over that period, there appeared *Ardipithecus* (four and a half million years old and of intermediate form); several kinds of *Australopithecus* (an African primate from around a million years later which was, roughly speaking, a human below the neck but an ape above); *Homo habilis* (an animal defined as having crossed the cerebral Rubicon of brain size needed to qualify for our own family that may have been the first tool-user nearly two and a half million years ago); and *Homo erectus* (a large-brained ape that looked rather like a man). Each emerged first in Africa and most spread, at one time or another, to Asia and to Europe.

The first members of our own species, *Homo sapiens*, arose about a hundred and fifty thousand years ago as large, thick-skulled, but recognizably human apes. By the time the miseries of the ice ages were over we were, more or less, ourselves, with smaller brains in a thin skull on a slim and elegant body. Why we shrank is not certain. Perhaps, as in the dwarf mammoths on islands, shortage of food did the job, or perhaps a shift to a kinder society cut down the need for a sturdy frame.

It once seemed natural to arrange the fossils in a sequence that clambered up an evolutionary tree in single file to man. Every find had a name and a place in the hierarchy of the almost-human. Two species, almost by definition, could not live together, as the cultural superior would at once drive out its brutish ancestor.

As more bones turned up, the story became less clear. For much of the time, two or more kinds lived together in an uneasy coexistence rather like that of men and chimps today. In East Africa, for instance, two species of large-brained *Homo* lived alongside smaller-brained *Australopithecines* of several types, with perhaps half a dozen forms present at once. Some enigmatic fossils, such as the 'Black Skull' found at Lake Turkana, combine primitive and advanced features and suggest that patterns of change were complex indeed.

In spite of a century's claims of the discovery of 'missing links', it is quite possible that no bone yet found is on the direct genetic line to ourselves. With so many kinds to choose from, so few remains of each, and such havoc among their relics, none of the fossils may have direct descendants today. The proportion of the people alive even a hundred thousand years ago who contributed to modern pedigrees is small; and the chance that any surviving relic, one among lost billions, belonged to that élite is quite minute.

Neanderthals are the most familiar of fossils. Once dismissed as the remains of diseased Napoleonic soldiers, and then hailed as our immediate predecessors, their true history is one of a dead end on the road of human evolution and of extinction at the hands of those who had taken the correct turning. With their stocky bodies, they were adapted to cold; and evolved in Europe or the Middle East, their home until they disappeared thirty thousand years ago. They used stone tools, but in an uninspired way, and stayed apart from their intellectual neighbours.

Although Neanderthals, at first sight so similar to modern humans, were once placed on the last rung before mankind, fossil DNA hints that they may not even be on the same ladder. Their mitochondrial genes are quite distinct from our own. They were not the ancestors of human genes but followed a separate path. For mitochondria, at least, Neanderthals and ourselves split half a million years ago. Today's genes also hint that our immediate ancestors are also lost for ever. The longer the history of any population, the more variation it contains, because there has been more time for mutations to build up. The more abundant it is, the less the chance of an accidental loss of the new variants as they arise. Any large and ancient group of animals hence contains lots of different inherited forms, many of which – given time – become common. A new or sparse population, in contrast, has little diversity and the altered forms are rare.

Surnames are rather like genes (although they reproduce without benefit of sex). The probable date of the shared ancestor of everyone with a certain surname depends on the size and age of the population in which they live. Half a dozen Sidebothams in a village of twenty people are almost certain to stem from the same recent ancestor. Six randomly chosen people with the name scattered among millions of Londoners (scores of whom claim that noble label) will have to go much further back to find the ur-Sidebotham from whom they descend. A city in which most people have the same name is therefore likely to have expanded from a small village, while one with thousands of surnames traces its origin to a huge and variable ancient populace. Genes, like names, can be used to make guesses about the past.

Compared to other primates, humans are not very diverse, most variants are rare, and most people – and most places – are much the same. Chimpanzees are three times more distinct one from the other than are men, with fifty times as much divergence among separate populations. The logic of the genes shows that chimps, now rare (with a world population of only a couple of hundred thousand), were once common and that the human race, abundant as it is today, was scarce. Its average size over hundreds of thousands of years may have been a mere ten to twenty thousand people. If so, our immediate ancestors were a small band who occupied a few hundred square miles and, more than likely, left no fossils at all. To draw the human family tree reveals an explosion of change, a starburst like that of AIDS viruses, with its centre a hundred thousand years ago. It may mark the expansion and spread of modern *Homo sapiens*, but the chances of finding where we came from are small indeed.

Wherever the Garden of Eden may have been, man became a traveller as soon as he escaped it. Although a shortage of bones makes it hard to be certain, there were several journeys out of (and perhaps even back to) Africa.

The great arena of evolution was in Asia, Europe and Africa. The habitable world was not filled until a thousand years ago, with the settlement of New Zealand. Many barriers stood in the way. Some, narrow though they are, proved hard to cross. The New World, most agree, was not reached until about fourteen thousand years ago, across the Bering Land Bridge between what is now Siberia and Alaska. Within a thousand years people had reached southern Chile (where hints have been found of an occupation twice as old). By contrast, humans have been in Australia for fifty thousand years, and crossed the Sunda Strait – recognized by Wallace as a break between the continents – to do so.

The simplest of all barriers is distance. Although that has been defeated by the wheel, it is easy to forget how isolated all of us once were. Numbers shot up after farming began, but even then, most people stayed at home. Although the conventional view of history is of rape and pillage – men, from Attila the Hun onwards, forcing their genes on to women – DNA reveals another past. The genes passed through females, on the mitochondria, are less localized than those of the male chromosome, the Y. Women, it seems, have travelled more than men (perhaps because they move to find a husband in the next village), with a rate of gene migration eight times higher through females than through males.

Like Red Sea fish in the Mediterranean, men destroy what they meet. Alfred Russel Wallace noted that 'we live in a zoologically impoverished world, from which all the hugest, and fiercest, and strangest forms have recently disappeared.' The culprit is plain. Humans reached Australia fifty thousand years ago. They came across tortoises as big as a Volkswagen Beetle, carnivorous kangaroos, and flightless birds twice the size of an emu. Within ten thousand years, all were gone. The first Americans were even more efficient. Throwing sticks, atlatls, increased the leverage on a spear and gave it the power of a Magnum rifle. Five hundred years after the hunters reached the Great Plains, mammoths, camels and horses were extinct or almost so and the survivors – wild sheep and bison – were much reduced in size. Their demise is a reminder that, for most species, extinction at the hands of a successor is inevitable.

Travel as he might, man is a lowland animal. Eleven of the world's fifteen largest cities are on the coast. Mountain – and northern – peoples have been pushed around by ice ages as much as have dung beetles. In China, the tribes of the hills are distinct while the masses of the coastal plains are more uniform. Even in Italy, hill villages are, because of their isolation, more different from each other in their genes (and in their surnames) than are the cities of the plain.

DNA reconstructs the history of human migration. The trends across the world reflect bonds of shared descent, as modified by natural selection. Peoples from northern places are taller and broader than those from the tropics. What counts is the relation between the mass of the body and its surface area. To stay warm it pays to be spherical. Eskimos (and Neanderthals) have thick bodies with short limbs, while most Africans are slimmer with longer arms and legs. Patterns of body shape are consistent to north and south, and those in skin colour (little though we understand them) have parallels in the New World and the Old.

On islands, too, people have changed, more by the accidents of migration than by natural selection, as most islands have not been occupied for long. All over the world, isolated by water, by mountains or by bigotry, they have evolved. By chance, certain genes that are rare at home took a trip and at once became relatively common. The inhabitants of Tristan da Cunha and several Pacific islands suffer from inherited blindness, and religious isolates such as the Amish of North America also have much genetic disease.

Human evolution can be studied in the same objective way as that of any animal. Genetics, geology, and geography illuminate our past. Morphology – comparative anatomy – is the most powerful tool of all. Queen Victoria noticed as much on her first visit to London Zoo, in 1842, and was not amused: 'The Orang Outang is too wonderful ... He is frightful and painfully and disagreeably human.' She was right. Anti AIDS medicines are tested on chimps because their bodies are so like ours. The cladistic rules put humans into the same family as chimps and gorillas, as all derive from a common ancestor not shared by anything else. The laws that put birds with crocodiles cannot be broken simply to satisfy our wish to be in a class of our own.

Man is a primate, and in some ways not a very special one. He can do more than any other creature, but has not changed much to do so. The strangest thing about human evolution is how little there has been. Nothing else is so widespread and nobody fills so many gaps in the economy of nature. Many animals carry out tasks almost as wonderful as those achieved by men, but through biology rather than intellect. For them, success at one task means failure at all others. In the past hundred thousand – in the past hundred – years, human lives have been transformed, but bodies have not. We did not evolve, because our machines did it for us. As Darwin put it in *The Descent of Man*: 'The highest possible stage in moral culture is when we recognize that we ought to control our thoughts'. Human progress has made a simple but crucial move, from body to mind. That mind is built from genes but what it can do has long transcended DNA.

Many sociologists (and a few biologists) hope for a comparative anatomy of the mind; but that can never succeed. When it comes to what makes us different from other creatures, science can answer all the questions except the interesting ones. The human intellect stands alone. As there is nothing else like it, the rules of classification come into play. If an object is one of a kind, it is impossible to know where to put it. The problem with the mind, or any uniquely human attribute, is simple: it is, like the narwhal's tusk or the female hyena's penis, unique.

To understand how far a body part has come, we must know

where it started from. That can be done for cell membranes, for gills, breasts, kidneys or opposable thumbs, because some animals have them and some do not. Their pattern – present in groups who descend from a shared ancestor and not in others – is a map through the past. It is the problem of language: if there were but one world tongue (as, with the spread of English, there may one day be) it would be impossible without further evidence to tell from whence it came. Shared derived characters – words common to Welsh, English and French, or even to Chinese – are needed to untangle the hierarchy of evolution. The conscious mind is different: we all have one, but, as far as we can tell, nothing else does. As a result, to speculate about its evolution is largely futile.

This book began with mankind's latest challenge, an attack by a virus that came from an ape. We have evolved in response to AIDS through the most familiar kind of selection, but the disease also shows how humans are different; how we transcend the biological rules that apply everywhere else.

Some HIV-positive people survive for decades, for good evolutionary reasons. Their luck results from inherited diversity; not in viruses, but in men. The survivors tend to carry certain forms of the receptor lock into which the HIV key fits so that the virus finds it hard to break in to their cells. Access is denied in many ways, with a whole series of bolts and bars, each of which comes in several versions. One person in six inherits an altered form of part of the defence mechanism. For them, the onset of symptoms is delayed for years. About one in twenty is lucky enough to have two copies of this altered gene. This shuts out the virus altogether. A variant of a second receptor is present in about a quarter of all Asians, and rather fewer Africans and Europeans. Although its bearers may pick up the virus, they last much better than average. If – as it will – the disease continues to kill, only those with a heritage that allows them to survive for long enough to have children will succeed. Their genes will spread and perhaps, in time, mankind might evolve resistance to AIDS.

There has been another response to the illness. It far surpasses what evolution can do. People's minds have changed much more than their DNA. For the first few years after the outbreak, the gay community was in denial. AIDS tests were criticized as a step towards concentration camps and the New York Health Commissioner banned them. However, as news of the disaster spread homosexuals changed their habits in the quickest response to a health threat ever seen. Bath-houses closed, the numbers of partners dropped, and unprotected anal sex went out of fashion. At the same time, those with the disease died or became so ill as to lose all allure. The rate of transmission among gay men plummeted by nine-tenths and, in the developed world, continues to do so. In Europe the number of new cases went down by half between 1993 and 1998. In parts of Africa, too, there has been a tenfold rise in condom use within a decade and a matching fall in the number of infections. A change in behaviour means that, in time, evolution will be beside the point.

That is true for most of what makes us human. The many attempts to explain the oddities of human life – sex life included – in terms of that of bees or chimpanzees show how easy it is to fall unthinking into the capacious arms of Darwin. Many universals link the human race. Most, like our need for vitamin C or for company, descend in modified form from our ancestors. There are also differences among people and among societies that seem to cry out for an evolutionary explanation. The task is impossible, because – like the mind itself – there is no outgroup; nothing apart from ourselves with which to compare them.

That elementary fact is often forgotten. For some, to explain any pattern of society all that is needed is to stir in a Darwinian nostrum. If the anthropological soufflé fails to rise – reach for another bottle. As in the kitchen, the ingredients can be varied to taste. Mix them with enough enthusiasm and, with a single bound, life is explained. Its infinite varieties are justified with adaptive stories to fit. In fact, most are an incidental to the healing power of lust, greed or political expedience. Sexual fidelity or promiscuity seems, according to taste, the natural thing to do; families are often a joy, and patriotism arouses some, but we do not need natural selection to tell us why. So much light has now been thrown on the origin of man and his history that the limits of what biology can say about his present condition are starkly exposed.

Such vulgar Darwinism is of interest only in its diversity. Evolution is a political sofa that moulds itself to the buttocks of the last to sit upon it. Alfred Russel Wallace, in old age, turned to socialism. He saw biology's real message: 'All shall contribute their share either of physical or mental labour, and . . . shall obtain the full and equal reward for their work. The future progress of the race will be rendered certain . . . by a special form of selection which will then come into play'. Marxists in their thousands agreed. In China they named their children 'Natural Selection' or 'Struggle for Existence' in homage to the Law of Nature that was to transform society.

Marx's monument in Highgate is opposite the grave of Herbert Spencer, who coined that unfortunate phrase 'the survival of the fittest' and the notion of Social Darwinism, an idea much used to explain the excesses of capitalism. One devotee saw millionaires as 'naturally selected in the crucible of competition'. The steel magnate Andrew Carnegie agreed. 'Before Spencer,' he said, 'all for me had been darkness, after him, all had become light – and right'. Sidney Webb, to justify his own political Third Way, spoke in overtly Darwinian terms of 'the inevitability of gradualism'.

Spencer and Wallace – and even Webb – at least knew their biology. F. R. Leavis once derided what he called the culture of the Sunday papers: a civilization based on colour supplements. The new culture of the science pages uses a nodding acquaintance with evolution to promote an ethical agenda. For some social theorists, science is no more than fuel for parable, its evidence selected in neat accord with political bias. As for Carnegie, biology makes everything light – and right. It is the universal excuse.

Evolution deserves better than that; and this book has, I hope, shown as much. Its contents are not a simple homage to Darwin, much as he deserves it. He had a wonderful phrase to describe those unaware of his theory: they 'look at an organic being as a savage looks at a ship, as at something wholly beyond his comprehension'. Nowadays, nobody can use the defence of ignorance when faced with the need to understand life's diversity: evolution forces all of us to look at the world in a new way.

Some of Darwin's ideas survive while others have been proved wrong. *The Origin of Species* endures as a work of art as much as of science. Its message remains. Man, the highest of animals, and the most exalted object which we are capable of conceiving, emerged from the war of Nature, from famine and death as much as did all others. Humans, alone, have gone further. As a result, much of what makes us what we are does not need a Darwinian explanation. The birth of Adam, whether real or metaphorical, marked the insertion into an animal body of a post-biological soul that leaves no fossils and needs no genes. To use the past to excuse the present is to embrace Payne Knight's pathetic fallacy, that society can be explained in terms of the animal world. However, the new insight that biology gives into our own history releases us from the narcissism of a creature that is one of a kind. It shows that humans are part of creation, because we evolved.

CHAPTER XIV

RECAPITULATION AND CONCLUSION

Recapitulation of the difficulties on the theory of Natural Selection — Recapitulation of the general and special circumstances in its favour — Causes of the general belief in the immutability of species — How far the theory of natural selection may be extended — Effects of its adoption on the study of Natural history — Concluding remarks

As THIS WHOLE volume is one long argument, it may be convenient to the reader to have the leading facts and inferences briefly recapitulated.

That many and grave objections may be advanced against the theory of descent with modification through natural selection, I do not deny. I have endeavoured to give to them their full force. Nothing at first can appear more difficult to believe than that the more complex organs and instincts should have been perfected not by means superior to, though analogous with, human reason, but by the accumulation of innumerable slight variations, each good for the individual possessor. Nevertheless, this difficulty, though appearing to our imagination insuperably great, cannot be considered real if we admit the following propositions, namely, – that gradations in the perfection of any organ or instinct, which we may consider, either do now exist or could have existed, each good of its kind, – that all organs and instincts are, in ever so slight a degree, variable, – and, lastly, that there is a struggle for existence leading to the preservation of each profitable deviation of structure or instinct. The truth of these propositions cannot, I think, be disputed.

It is, no doubt, extremely difficult even to conjecture by what gradations many structures have been perfected, more especially amongst broken and failing groups of organic beings; but we see so many strange gradations in nature, as is proclaimed by the canon, 'Natura non facit saltum,' that we ought to be extremely cautious in saying that any organ or instinct, or any whole being, could not have arrived at its present state by many graduated steps. There are, it must be admitted, cases of special difficulty on the theory of natural selection; and one of the most curious of these is the existence of two or three defined castes of workers or sterile females in the same community of ants but I have attempted to show how this difficulty can be mastered. With respect to the almost universal sterility of species when first crossed, which forms so remarkable a contrast with the almost universal fertility of varieties when crossed, I must refer the reader to the recapitulation of the facts given at the end of the eighth chapter, which seem to me conclusively to show that this sterility is no more a special endowment than is the incapacity of two trees to be grafted together, but that it is incidental on constitutional differences in the reproductive systems of the intercrossed species. We see the truth of this conclusion in the vast difference in the result, when the same two species are crossed reciprocally; that is, when one species is first used as the father and then as the mother.

The fertility of varieties when intercrossed and of their mongrel offspring cannot be considered as universal; nor is their very general fertility surprising when we remember that it is not likely that either their constitutions or their reproductive systems should have been profoundly modified. Moreover, most of the varieties which have been experimentised on have been produced under domestication; and as domestication apparently tends to eliminate sterility, we ought not to expect it also to produce sterility.

The sterility of hybrids is a very different case from that of first crosses, for their reproductive organs are more or less functionally impotent; whereas in first crosses the organs on both sides are in a perfect condition. As we continually see that organisms of all kinds are rendered in some degree sterile from their constitutions having been disturbed by slightly different and new conditions of life, we need not feel surprise at hybrids being in some degree sterile, for their constitutions can hardly fail to have been disturbed from being compounded of two distinct organisations. This parallelism is supported by another parallel, but directly opposite, class of facts; namely, that the vigour and fertility of all organic beings are increased by slight changes in their conditions of life, and that the offspring of slightly modified forms or varieties acquire from being crossed increased vigour and fertility. So that, on the one hand, considerable changes in the conditions of life and crosses between greatly modified forms, lessen fertility; and on the other hand, lesser changes in the conditions of life and crosses between less modified forms, increase fertility.

Turning to geographical distribution, the difficulties encountered on the theory of descent with modification are grave enough. All the individuals of the same species, and all the species of the same genus, or even higher group, must have descended from common parents; and therefore, in however distant and isolated parts of the world they are now found, they must in the course of successive generations have passed from some one part to the others. We are often wholly unable even to conjecture how this could have been effected. Yet, as we have reason to believe that some species have retained the same specific form for very long periods, enormously long as measured by years, too much stress ought not to be laid on the occasional wide diffusion of the same species; for during very long periods of time there will always be a good chance for wide migration by many means. A broken or interrupted range may often be accounted for by the extinction of the species in the intermediate regions. It cannot be denied that we are as yet very ignorant of the full extent of the various climatal and geographical changes which have affected the earth during modern periods; and such changes will obviously have greatly facilitated migration. As an example, I have attempted to show how potent has been the influence of the Glacial period on the distribution both of the same and of representative species throughout the world. We are as yet profoundly ignorant of the many occasional means of transport. With respect to distinct species of the same genus inhabiting very distant and isolated regions, as the process of modification has necessarily been slow, all the means of migration will have been possible during a very long period; and consequently the difficulty of the wide diffusion of species of the same genus is in some degree lessened.

As on the theory of natural selection an interminable number of intermediate forms must have existed, linking together all the species in each group by gradations as fine as our present varieties, it may be asked, Why do we not see these linking forms all around us? Why are not all organic beings blended together in an inextricable chaos? With respect to existing forms, we should remember that we have no right to expect (excepting in rare cases) to discover *directly* connecting links between them, but only between each and some extinct and supplanted form. Even on a wide area, which has during a long period remained continuous, and of which the climate and other conditions of life change insensibly in going from a district occupied by one species into another district occupied by a closely allied species, we have no just right to expect often to find intermediate varieties in the intermediate zone. For we have reason to believe that only a few species are undergoing change at any one period; and all changes are slowly effected. I have also shown that the intermediate varieties which will at first probably exist in the intermediate zones, will be liable to be supplanted by the allied forms on either hand; and the latter, from existing in greater numbers, will generally be modified and improved at a quicker rate than the intermediate varieties, which exist in lesser numbers; so that the intermediate varieties will, in the long run, be supplanted and exterminated.

On this doctrine of the extermination of an infinitude of connecting links, between the living and extinct inhabitants of the world, and at each successive period between the extinct and still older species, why is not every geological formation charged with such links? Why does not every collection of fossil remains afford plain evidence of the gradation and mutation of the forms of life? We meet with no such evidence, and this is the most obvious and forcible of the many objections which may be urged against my theory. Why, again, do whole groups of allied species appear, though certainly they often falsely appear, to have come in suddenly on the several geological stages? Why do we not find great piles of strata beneath the Silurian system, stored with the remains of the progenitors of the Silurian groups of fossils? For certainly on my theory such strata must somewhere have been deposited at these ancient and utterly unknown epochs in the world's history.

I can answer these questions and grave objections only on the supposition that the geological record is far more imperfect than most geologists believe. It cannot be objected that there has not been time sufficient for any amount of organic change; for the lapse of time has been so great as to be utterly inappreciable by the human intellect. The number of specimens in all our museums is absolutely as nothing compared with the countless generations of countless species which certainly have existed. We should not be able to recognise a species as the parent of any one or more species if we were to examine them ever so closely, unless we likewise possessed many of the intermediate links between their past or parent and present states; and these many links we could hardly ever expect to discover, owing to the imperfection of the geological record. Numerous existing doubtful forms could be named which are probably varieties; but who will pretend that in future ages so many fossil links will be discovered, that naturalists will be able to decide, on the common view, whether or not these doubtful forms are varieties? As long as most of the links between any two species are unknown, if any one link or intermediate variety be discovered, it will simply be classed as another and distinct species. Only a small portion of the world has been geologically explored. Only organic beings of certain classes can be preserved in a fossil condition, at least in any great number. Widely ranging species vary most, and varieties are often at first local, – both causes rendering the discovery of intermediate links less likely. Local varieties will not spread into other and distant regions until they are considerably modified and improved; and when they do spread, if discovered in a geological formation, they will appear as if suddenly created there, and will be simply classed as new species. Most formations have been intermittent in their accumulation; and their duration, I am inclined to believe, has been shorter than the average duration of specific forms. Successive formations are separated from each other by enormous blank intervals of time; for fossiliferous formations, thick enough to resist future degradation, can be accumulated only where much sediment is deposited on the subsiding bed of the sea. During the alternate periods of elevation and of stationary level the record will be blank. During these latter periods there will proably be more variability in the forms of life;

during periods of subsidence, more extinction.

With respect to the absence of fossiliferous formations beneath the lowest Silurian strata, I can only recur to the hypothesis given in the ninth chapter. That the geological record is imperfect all will admit; but that it is imperfect to the degree which I require, few will be inclined to admit. If we look to long enough intervals of time, geology plainly declares that all species have changed; and they have changed in the manner which my theory requires, for they have changed slowly and in a graduated manner. We clearly see this in the fossil remains from consecutive formations invariably being much more closely related to each other, than are the fossils from formations distant from each other in time.

Such is the sum of the several chief objections and difficulties which may justly be urged against my theory; and I have now briefly recapitulated the answers and explanations which can be given to them. I have felt these difficulties far too heavily during many years to doubt their weight. But it deserves especial notice that the more important objections relate to questions on which we are confessedly ignorant; nor do we know how ignorant we are. We do not know all the possible transitional gradations between the simplest and the most perfect organs; it cannot be pretended that we know all the varied means of Distribution during the long lapse of years, or that we know how imperfect the Geological Record is. Grave as these several difficulties are, in my judgement they do not overthrow the theory of descent with modification.

Now let us turn to the other side of the argument. Under domestication we see much variability. This seems to be mainly due to the reproductive system being eminently susceptible to changes in the conditions of life so that this system, when not rendered impotent, fails to reproduce offspring exactly like the parent-form. Variability is governed by many complex laws, – by correlation of growth, by use and disuse, and by the direct action of the physical conditions of life. There is much difficulty in ascertaining how much modification our domestic productions have undergone; but we may safely infer that the amount has been large, and that modifications can be inherited for long periods. As long as the conditions of life remain the same, we have reason to believe that a modification, which has already been inherited for many generations, may continue to be inherited for an almost infinite number of generations. On the other hand we have evidence that variability, when it has once come into play, does not wholly cease; for new varieties are still occasionally produced by our most anciently domesticated productions.

Man does not actually produce variability; he only unintentionally exposes organic beings to new conditions of life, and then nature acts on the organisation, and causes variability. But man can and does select the variations given to him by nature, and thus accumulate them in any desired manner. He thus adapts animals and plants for his own benefit or pleasure. He may do this methodically, or he may do it unconsciously by preserving the individuals most useful to him at the time, without any thought of altering the breed. It is certain that he can largely influence the character of a breed by selecting, in each successive generation, individual differences so slight as to be quite inappreciable by an uneducated eye. This process of selection has been the great agency in the production of the most distinct and useful domestic breeds. That many of the breeds produced by man have to a large extent the character of natural species, is shown by the inextricable doubts whether very many of them are varieties or aboriginal species.

There is no obvious reason why the principles which have acted so efficiently under domestication should not have acted under nature. In the preservation of favoured individuals and races, during the constantly-recurrent Struggle for Existence, we see the most powerful and ever-acting means of selection. The struggle for existence inevitably follows from the high geometrical ratio of increase which is common to all organic beings. This high rate of increase is proved by calculation, by the effects of a succession of peculiar seasons, and by the results of naturalisation, as explained in the third chapter. More individuals are born than can possibly survive. A grain in the balance will determine which individual shall live and which shall die, – which variety or species shall increase in number, and which shall decrease, or finally become extinct. As the individuals of the same species come in all respects into the closest competition with each other, the struggle will generally be most severe between them; it will be almost equally severe between the varieties of the same species, and next in severity between the species of the same genus. But the struggle will often be very severe between beings most remote in the scale of nature. The slightest advantage in one being, at any age or during any season, over those with which it comes into competition, or better adaptation in however slight a degree to the surrounding physical conditions, will turn the balance.

With animals having separated sexes there will in most cases be a struggle between the males for possession of the females. The most vigorous individuals, or those which have most successfully struggled with their conditions of life, will generally leave most progeny. But success will often depend on having special weapons or means of defence, or on the charms of the males; and the slightest advantage will lead to victory.

As geology plainly proclaims that each land has undergone great physical changes, we might have expected that organic beings would have varied under nature, in the same way as they generally have varied under the changed conditions of domestication. And if there be any variability under nature, it would be an unaccountable fact if natural selection had not come into play. It has often been asserted, but the assertion is quite incapable of proof, that the amount of variation under nature is a strictly limited quantity. Man, though acting on external characters alone and often capriciously, can produce within a short period a great result by adding up mere individual differences in his domestic productions; and every one admits that there are at least individual differences in species under nature. But, besides such differences, all naturalists have admitted the existence of varieties, which they think sufficiently distinct to be worthy of record in systematic works. No one can draw any clear distinction between individual differences and slight varieties; or between more plainly marked varieties and subspecies, and species. Let it be observed how naturalists differ in the rank which they assign to the many representative forms in Europe and North America.

If then we have under nature variability and a powerful agent always ready to act and select, why should we doubt that variations in any way useful to beings, under their excessively complex relations of life, would be preserved, accumulated, and inherited? Why, if man can by patience select variations most useful to himself, should nature fail in selecting variations useful, under changing conditions of life, to her living products? What limit can be put to this power, acting during long ages and rigidly scrutinising the whole constitution, structure, and habits of each creature, – favouring the good and rejecting the bad? I can see no limit to this power, in slowly and beautifully adapting each form to the most complex relations of life. The theory of natural selection, even if we looked no further than this, seems to me to be in itself probable. I have already recapitulated, as fairly as I could, the opposed difficulties and objections: now let us turn to the special facts and arguments in favour of the theory.

On the view that species are only strongly marked and permanent varieties, and that each species first existed as a variety, we can see why it is that no line of demarcation can be drawn between species, commonly supposed to have been produced by special acts of creation, and varieties which are acknowledged to have been produced by secondary laws. On this same view we can understand how it is that in each region where many species of a genus have been produced, and where they now flourish, these same species should present many varieties; for where the manufactory of species has been active, we might expect, as a general rule, to find it still in action; and this is the case if varieties be incipient species. Moreover, the species of the large genera, which afford the greater number of varieties or incipient species, retain to a certain degree the character of varieties; for they differ from each other by a less amount of difference than do the species of smaller genera. The closely allied species also of the larger genera apparently have restricted ranges, and they are clustered in little groups round other species – in which respects they resemble varieties. These are strange relations on the view of each species having been independently created, but are intelligible if all species first existed as varieties.

As each species tends by its geometrical ratio of reproduction to increase inordinately in number; and as the modified descendants of each species will be enabled to increase by so much the more as they become more diversified in habits and structure, so as to be enabled to seize on many and widely different places in the economy of nature, there will be a constant tendency in natural selection to preserve the most divergent offspring of any one species. Hence during a long-continued course of modification, the slight differences, characteristic of varieties of the same species, tend to be augmented into the greater differences characteristic of species of the same genus. New and improved varieties will inevitably supplant and exterminate the older, less improved and intermediate varieties; and thus species are rendered to a large extent defined and distinct objects. Dominant species belonging to the larger groups tend to give birth to new and dominant forms; so that each large group tends to become still larger, and at the same time more divergent in character. But as all groups cannot thus succeed in increasing in size, for the world would not hold them, the more dominant groups beat the less dominant. This tendency in the large groups to go on increasing in size and diverging in character, together with the almost inevitable contingency of much extinction, explains the arrangement of all the forms of life, in groups subordinate to groups, all within a few great classes, which we now see everywhere around us, and which has prevailed throughout all time. This grand fact of the grouping of all organic beings seems to me utterly inexplicable on the theory of creation.

As natural selection acts solely by accumulating slight, successive, favourable variations, it can produce no great or sudden modification; it can act only by very short and slow steps. Hence the canon of 'Natura non facit saltum,' which every fresh addition to our knowledge tends to make more strictly correct, is on this theory simply intelligible. We can plainly see why nature is prodigal in variety, though niggard in innovation. But why this should be a law of nature if each species has been independently created, no man can explain.

Many other facts are, as it seems to me, explicable on this theory. How strange it is that a bird, under the form of woodpecker, should have been created to prey on insects on the ground; that upland geese, which never or rarely swim, should have been created with webbed feet; that a thrush should have been created to dive and feed on sub-aquatic insects; and that a petrel should have been created with habits and structure fitting it for the life of an auk or grebe! and so on in endless other cases. But on the view of each species constantly trying to increase in number, with natural selection always ready to adapt the slowly varying descendants of each to any unoccupied or ill-occupied place in nature, these facts cease to be strange, or perhaps might even have been anticipated.

As natural selection acts by competition, it adapts the inhabitants of each country only in relation to the degree of perfection of their associates; so that we need feel no surprise at the inhabitants of any one country, although on the ordinary view supposed to have been specially created and adapted for that country, being beaten and sup-planted by the naturalised productions from another land. Nor ought we to marvel if all the contrivances in nature be not, as far as we can judge, absolutely perfect; and if some of them be abhorrent to our ideas of fitness. We need not marvel at the sting of the bee causing the bee's own death; at drones being produced in such vast numbers for one single act, and being then slaughtered by their sterile sisters; at the astonishing waste of pollen by our fir-trees; at the instinctive hatred of the queen bee for her own fertile daughters; at ichneumonidae feeding within the live bodies of caterpillars; and at other such cases. The wonder indeed is, on the theory of natural selection, that more cases of the want of absolute perfection have not been observed.

The complex and little known laws governing variation are the same, as far as we can see, with the laws which have governed the production of so-called specific forms. In both cases physical conditions seem to have produced but little direct effect; yet when varieties enter any zone, they occasionally assume some of the characters of the species proper to that zone. In both varieties and species, use and disuse seem to have produced some effect; for it is difficult to resist this conclusion when we look, for instance, at the logger-headed duck, which has wings incapable of flight, in nearly the same condition as in the domestic duck; or when we look at the burrowing tucutucu, which is occasionally blind, and then at certain moles, which are habitually blind and have their eyes covered with skin; or when we look at the blind animals inhabiting the dark caves of America and Europe. In both varieties and species correction of growth seems to have played a most important part, so that when one part has been modified other parts are necessarily modified. In both varieties and species reversions to long-lost characters occur. How inexplicable on the theory of creation is the occasional appearance of stripes on the shoulder and legs of the several species of the horse-genus and in their hybrids! How simply is this fact explained if we believe that these species have descended from a striped progenitor, in the same manner as the several domestic breeds of pigeon have descended from the blue and barred rock-pigeon!

On the ordinary view of each species having been independently created, why should the specific characters, or those by which the species of the same genus differ from each other, be more variable than the generic characters in which they all agree? Why, for instance, should the colour of a flower be more likely to vary in any one species of a genus, if the other species, supposed to have been created independently, have differently coloured flowers, than if all the species of the genus have the same coloured flowers? If species are only well-marked varieties, of which the characters have become in a high degree permanent, we can understand this fact; for they have already varied since they branched off from a common progenitor in certain characters, by which they have come to be specifically distinct from each other; and therefore these same characters would be more likely still to be variable than the generic characters which have been inherited without change for an enormous period. It is inexplicable on the theory of creation why a part developed in a very unusual manner in any one species of a genus, and therefore, as we may naturally infer, of great importance to the species, should be eminently liable to variation; but, on my view, this part has undergone, since the several species branched off from a common progenitor, an unusual amount of variability and modification, and therefore we might expect this part generally to be still variable. But a part may be developed in the most unusual manner, like the wing of a bat, and yet not be more variable than any other structure, if the part be common to many subordinate forms, that is, if it has been inherited for a very long period; for in this case it will have been rendered constant by long-continued natural selection.

Glancing at instincts, marvellous as some are, they offer no greater difficulty than does corporeal structure on the theory of the natural selection of successive, slight, but profitable modifications. We can thus understand why nature moves by graduated steps in endowing different animals of the same class with their several instincts. I have attempted to show how much light the principle of gradation throws on the admirable architectural powers of the hive-bee. Habit no doubt sometimes comes into play in modifying instincts; but it certainly is not indispensable, as we see, in the case of neuter insects, which leave no progeny to inherit the effects of long-continued habit. On the view of all the species of the same genus having descended from a common parent, and having inherited much in common, we can understand how it is that allied species, when placed under considerably different conditions of life, yet should follow nearly the same instincts; why the thrush of South America, for instance, lines her nest with mud like our British species. On the view of instincts having been slowly acquired through natural selection we need not marvel at some instincts being apparently not perfect and liable to mistakes, and at many instincts causing other animals to suffer.

If species be only well-marked and permanent varieties, we can at once see why their crossed offspring should follow the same complex laws in their degrees and kinds of resemblance to their parents, – in being absorbed into each other by successive crosses, and in other such points, – as do the crossed offspring of acknowledged varieties. On the other hand, these would be strange facts if species have been independently created, and varieties have been produced by secondary laws.

If we admit that the geological record is imperfect in an extreme degree, then such facts as the record gives, support the theory of descent with modification. New species have come on the stage slowly and at successive intervals; and the amount of change, after equal intervals of time, is widely different in different groups. The extinction of species and of whole groups of species, which has played so conspicuous a part in the history of the organic world, almost inevitably follows on the principle of natural selection; for old forms will be supplanted by new and improved forms. Neither single species nor groups of species reappear when the chain of ordinary generation has once been broken. The gradual diffusion of dominant forms, with the slow modification of their descendants, causes the forms of life, after long intervals of time, to appear as if they had changed simultaneously throughout the world. The fact of the fossil remains of each formation being in some degree intermediate in character between the fossils in the formations above and below, is simply explained by their intermediate position in the chain of descent. The grand fact that all extinct organic beings belong to the same system with recent beings, falling either into the same or into intermediate groups, follows from the living and the extinct being the offspring of common parents. As the groups which have descended from an ancient progenitor have generally diverged in character, the progenitor with its early descendants will often be intermediate in character in comparison with its later descendants; and thus we can see why the more ancient a fossil is, the oftener it stands in some degree intermediate between existing and allied groups. Recent forms are generally looked at as being, in some vague sense, higher than ancient and extinct forms; and they are in so far higher as the later and more improved forms have conquered the older and less improved organic beings in the struggle for life. Lastly, the law of the long endurance of allied forms on the same continent; of marsupials in Australia, of edentata in America, and other such cases, – is intelligible, for within a confined country, the recent and the extinct will naturally be allied by descent.

Looking to geographical distribution, if we admit that there has been during the long course of ages much migration from one part of the world to another, owing to former climatal and geographical changes and to the many occasional and unknown means of dispersal, then we can understand, on the theory of descent with modification, most of the great leading facts in Distribution. We can see why there should be so striking a parallelism in the distribution of organic beings throughout space, and in their geological succession throughout time; for in both cases the beings have been connected by the bond of ordinary generation, and the means of modification have been the same. We see the full meaning of the wonderful fact, which must have struck every traveller, namely, that on the same continent, under the most diverse conditions, under heat and cold, on mountain and lowland, on deserts and marshes, most of the inhabitants within each great class are plainly related; for they will generally be descendants of the same progenitors and early colonists. On this same principle of former migration, combined in most cases with modification, we can understand, by the aid of the Glacial period, the identity of some few plants, and the close alliance of many others, on the most distant mountains, under the most different climates; and likewise the close alliance of some of the inhabitants of the sea in the northern and southern temperate zones, though separated by the whole inter-tropical ocean. Although two areas may present the same physical conditions of life, we need feel no surprise at their inhabitants being widely different, if they have been for a long period completely separated from each other; for as the relation of organism to organism is the most important of all relations, and as the two areas will have received colonists from some third source or from each other, at various periods and in different proportions, the course of modification in the two areas will inevitably be different.

On this view of migration, with subsequent modification, we can see why oceanic islands should be inhabited by few species, but of these, that many should be peculiar. We can clearly see why those animals which cannot cross wide spaces of ocean, as frogs and terrestrial mammals, should not inhabit oceanic islands; and why, on the other hand, new and peculiar species of bats, which can traverse the ocean, should so often be found on islands far distant from any continent. Such facts as the presence of peculiar species of bats, and the absence of all other mammals, on oceanic islands, are utterly inexplicable on the theory of independent acts of creation.

The existence of closely allied or representative species in any two areas, implies, on the theory of descent with modification, that the same parents formerly inhabited both areas; and we almost invariably find that wherever many closely allied species inhabit two areas, some identical species common to both still exist. Wherever many closely allied yet distinct species occur, many doubtful forms and varieties of the same species likewise occur. It is a rule of high generality that the inhabitants of each area are related to the inhabitants of the nearest source whence immigrants might have been derived. We see this in nearly all the plants and animals of the Galapagos archipelago, of Juan Fernandez, and of the other American islands being related in the most striking manner to the plants and animals of the neighbouring American mainland; and those of the Cape de Verde archipelago and other African islands to the African mainland. It must be admitted that these facts receive no explanation on the theory of creation.

The fact, as we have seen, that all past and present organic beings constitute one grand natural system, with group subordinate to group, and with extinct groups often falling in between recent groups, is intelligible on the theory of natural selection with its contingencies of extinction and divergence of character. On these same principles we see how it is, that the mutual affinities of the species and genera within each class are so complex and circuitous. We see why certain characters are far more serviceable than others for classification; – why adaptive characters, though of paramount importance to the being, are of hardly any importance in classification; why characters derived from rudimentary parts, though of no service to the being, are often of high classificatory value; and why embryological characters are the most valuable of all. The real affinities of all organic beings are due to inheritance or community of descent. The natural system is a genealogical arrangement, in which we have to discover the lines of descent by the most permanent characters, however slight their vital importance may be.

The framework of bones being the same in the hand of a man, wing of a bat, fin of the porpoise, and leg of the horse, – the same number of vertebrae forming the neck of the giraffe and of the elephant, – and innumerable other such facts, at once explain themselves on the theory of descent with slow and slight successive modifications. The similarity of pattern in the wing and leg of a bat, though used for such different purposes, – in the jaws and legs of a crab, – in the petals, stamens, and pistils of a flower, is likewise intelligible on the view of the gradual modification of parts or organs, which were alike in the early progenitor of each class. On the principle of successive variations not always supervening at an early age, and being inherited at a corresponding not early period of life, we can clearly see why the embryos of mammals, birds, reptiles, and fishes should be so closely alike, and should be so unlike the adult forms. We may cease marvelling at the embryo of an air-breathing mammal or bird having branchial slits and arteries running in loops, like those in a fish which has to breathe the air dissolved in water, by the aid of well-developed branchiae.

Disuse, aided sometimes by natural selection, will often tend to reduce an organ, when it has become useless by changed habits or under changed conditions of life; and we can clearly understand on this view the meaning of rudimentary organs. But disuse and selection will generally act on each creature, when it has come to maturity and has to play its full part in the struggle for existence, and will thus have little power of acting on an organ during early life; hence the organ will not be much reduced or rendered rudimentary at this early age. The calf, for instance, has inherited teeth, which never cut through the gums of the upper jaw, from an early progenitor having well-developed teeth; and we may believe, that the teeth in the mature animal were reduced, during successive generations, by disuse or by the tongue and palate having been fitted by natural selection to browse without their aid; whereas in the calf, the teeth have been left untouched by selection or disuse, and on the principle of inheritance at corresponding ages have been inherited from a remote period to the present day. On the view of each organic being and each separate organ having been specially created, how utterly inexplicable it is that parts, like the teeth in the embryonic calf or like the shrivelled wings under the soldered wing-covers of some beetles, should thus so frequently bear the plain stamp of inutility! Nature may be said to have taken pains to reveal, by rudimentary organs and by homologous structures, her scheme of modification, which it seems that we wilfully will not understand.

I have now recapitulated the chief facts and considerations which have thoroughly convinced me that species have changed, and are still slowly changing by the preservation and accumulation of successive slight favourable variations. Why, it may be asked, have all the most eminent living naturalists and geologists rejected this view of the mutability of species? It cannot be asserted that organic beings in a state of nature are subject to no variation; it cannot be proved that the amount of variation in the course of long ages is a limited quantity; no clear distinction has been, or can be, drawn between species and well-marked varieties. It cannot be maintained that species when intercrossed are invariably sterile, and varieties invariably fertile; or that sterility is a special endowment and sign of creation. The belief that species were immutable productions was almost unavoidable as long as the history of the world was thought to be of short duration;

and now that we have acquired some idea of the lapse of time, we are too apt to assume, without proof, that the geological record is so perfect that it would have afforded us plain evidence of the mutation of species, if they had undergone mutation.

But the chief cause of our natural unwillingness to admit that one species has given birth to other and distinct species, is that we are always slow in admitting any great change of which we do not see the intermediate steps. The difficulty is the same as that felt by so many geologists, when Lyell first insisted that long lines of inland cliffs had been formed, and great valleys excavated, by the slow action of the coast-waves. The mind cannot possibly grasp the full meaning of the term of a hundred million years; it cannot add up and perceive the full effects of many slight variations, accumulated during an almost infinite number of generations.

Although I am fully convinced of the truth of the views given in this volume under the form of an abstract, I by no means expect to convince experienced naturalists whose minds are stocked with a multitude of facts all viewed, during a long course of years, from a point of view directly opposite to mine. It is so easy to hide our ignorance under such expressions as the 'plan of creation,' 'unity of design,' and to think that we give an explanation when we only restate a fact. Any one whose disposition leads him to attach more weight to unexplained difficulties than to the explanation of a certain number of facts will certainly reject my theory. A few naturalists, endowed with much flexibility of mind, and who have already begun to doubt on the immutability of species, may be influenced by this volume; but I look with confidence to the future, to young and rising naturalists, who will be able to view both sides of the question with impartiality. Whoever is led to believe that species are mutable will do good service by conscientiously expressing his conviction; for only thus can the load of prejudice by which this subject is overwhelmed be removed.

Several eminent naturalists have of late published their belief that a multitude of reputed species in each genus are not real species; but that other species are real, that is, have been independently created. This seems to me a strange conclusion to arrive at. They admit that a multitude of forms, which till lately they themselves thought were special creations, and which are still thus looked at by the majority of naturalists, and which consequently have every external characteristic feature of true species, – they admit that these have been produced by variation, but they refuse to extend the same view to other and very slightly different forms. Nevertheless they do not pretend that they can define, or even conjecture, which are the created forms of life, and which are those produced by secondary laws. They admit variation as a *vera causa* in one case, they arbitrarily reject it in another, without assigning any distinction in the two cases. The day will come when this will be given as a curious illustration of the blindness of preconceived opinion. These authors seem no more startled at a miraculous act of creation than at an ordinary birth. But do they really believe that at innumerable periods in the earth's history certain elemental atoms have been commanded suddenly to flash into living tissues? Do they believe that at each supposed act of creation one individual or many were produced? Were all the infinitely numerous kinds of animals and plants created as eggs or seed, or as full grown? and in the case of mammals, were they created bearing the false marks of nourishment from the mother's womb? Although naturalists very properly demand a full explanation of every difficulty from those who believe in the mutability of species, on their own side they ignore the whole subject of the first appearance of species in what they consider reverent silence.

It may be asked how far I extend the doctrine of the modification of species. The question is difficult to answer, because the more distinct the forms are which we may consider, by so much the arguments fall away in force. But some arguments of the greatest weight extend very far. All the members of whole classes can be connected together by chains of affinities, and all can be classified on the same principle, in groups subordinate to groups. Fossil remains sometimes tend to fill up very wide intervals between existing orders. Organs in a rudimentary condition plainly show that an early progenitor had the organ in a fully developed state; and this in some instances necessarily implies an enormous amount of modification in the descendants. Throughout whole classes various structures are formed on the same pattern, and at an embryonic age the species closely resemble each other. Therefore I cannot doubt that the theory of descent with modification embraces all the members of the same class. I believe that animals have descended from at most only four or five progenitors, and plants from an equal or lesser number.

Analogy would lead me one step further, namely, to the belief that all animals and plants have descended from some one prototype. But analogy may be a deceitful guide. Nevertheless all living things have much in common, in their chemical composition, their germinal vesicles, their cellular structure, and their laws of growth and reproduction. We see this even in so trifling a circumstance as that the same poison often similarly affects plants and animals; or that the poison secreted by the gall-fly produces monstrous growths on the wild rose or oak-tree. Therefore I should infer from analogy that probably all the organic beings which have ever lived on this earth have descended from some one primordial form, into which life was first breathed.

When the views entertained in this volume on the origin of species, or when analogous views are generally admitted, we can dimly foresee that there will be a considerable revolution in natural history. Systematists will be able to pursue their labours as at present; but they will not be incessantly haunted by the shadowy doubt whether this or that form be in essence a species. This I feel sure, and I speak after experience, will be no slight relief. The endless disputes whether or not some fifty species of British brambles are true species will cease. Systematists will have only to decide (not that this will be easy) whether any form be sufficiently constant and distinct from other forms, to be capable of definition; and if definable, whether the differences be sufficiently important to deserve a specific name. This latter point will become a far more essential consideration than it is at present; for differences, however slight, between any two forms, if not blended by intermediate gradations, are looked at by most naturalists as sufficient to raise both forms to the rank of species. Hereafter we shall be compelled to acknowledge that the only distinction between species and well-marked varieties is, that the latter are known, or believed, to be connected at the present day by intermediate gradations, whereas species were formerly thus connected. Hence, without quite rejecting the consideration of the present existence of intermediate gradations between any two forms, we shall be led to weigh more carefully and to value higher the actual amount of difference between them. It is quite possible that forms now generally acknowledged to be merely varieties may hereafter be thought worthy of specific names, as with the primrose and cowslip; and in this case scientific and common language will come into accordance. In short, we shall have to treat species in the same manner as those naturalists treat genera, who admit that genera are merely artificial combinations made for convenience. This may not be a cheering prospect; but we shall at least be freed from the vain search for the undiscovered and undiscoverable essence of the term species.

The other and more general departments of natural history will rise greatly in interest. The terms used by naturalists of affinity, relationship, community of type, paternity, morphology, adaptive characters, rudimentary and aborted organs, &c., will cease to be metaphorical, and will have a plain signification. When we no longer look at an organic being as a savage looks at a ship, as at something wholly beyond his comprehension; when we regard every production of nature as one which has had a history; when we contemplate every complex structure and instinct as the summing up of many con-trivances, each useful to the possessor, nearly in the same way as when we look at any great mechanical invention as the summing up of the labour, the experience, the reason, and even the blunders of numerous workmen; when we thus view each organic being, how far more interesting, I speak from experience, will the study of natural history become!

A grand and almost untrodden field of inquiry will be opened, on the causes and laws of variation, on correlation of growth, on the effects of use and disuse, on the direct action of external conditions, and so forth. The study of domestic productions will rise immensely in value. A new variety raised by man will be a far more important and interesting subject for study than one more species added to the infinitude of already recorded species. Our classifications will come to be, as far as they can be so made, genealogies; and will then truly give what may be called the plan of creation. The rules for classifying will no doubt become simpler when we have a definite object in view. We possess no pedigrees or armorial bearings; and we have to discover and trace the many diverging lines of descent in our natural genealogies, by characters of any kind which have long been inherited. Rudimentary organs will speak infallibly with respect to the nature of long-lost structures. Species and groups of species, which are called aberrant, and which may fancifully be called living fossils, will aid us in forming a picture of the ancient forms of life. Embryology will reveal to us the structure, in some degree obscured, of the prototypes of each great class.

When we can feel assured that all the individuals of the same species, and all the closely allied species of most genera, have within a not very remote period descended from one parent, and have migrated from some one birthplace; and when we better know the many means of migration, then, by the light which geology now throws, and will continue to throw, on former changes of climate and of the level of the land, we shall surely be enabled to trace in an admirable manner the former migrations of the inhabitants of the whole world. Even at present, by comparing the differences of the inhabitants of the sea on the opposite sides of a continent, and the nature of the various inhabitants of that continent in relation to their apparent means of immigration, some light can be thrown on ancient geography.

The noble science of Geology loses glory from the extreme imperfection of the record. The crust of the earth with its embedded remains must not be looked at as a well-filled museum, but as a poor collection made at hazard and at rare intervals. The accumulation of each great fossiliferous formation will be recognised as having depended on an unusual concurrence of circumstances, and the blank intervals between the successive stages as having been of vast duration. But we shall be able to gauge with some security the duration of these intervals by a comparison of the preceding and succeeding organic forms. We must be cautious in attempting to correlate as strictly contemporaneous two formations, which include few identical species, by the general succession of their forms of life. As species are produced and exterminated by slowly acting and still existing causes, and not by miraculous acts of creation and by catastrophes; and as the most important of all causes of organic change is one which is almost independent of altered and perhaps suddenly altered physical conditions, namely, the mutual relation of organism to organism, – the improvement of one being entailing the improvement or the extermination of others; it follows, that the amount of organic change in the fossils of consecutive formations probably serves as a fair measure of the lapse of actual time. A number of species, however, keeping in a body might remain for a long period unchanged, whilst within this same period, several of these species, by migrating into new countries and coming into competition with foreign associates, might become modified; so that we must not overrate the accuracy of organic change as a measure of time. During early periods of the earth's history, when the forms of life were probably fewer and simpler, the rate of change was probably slower; and at the first dawn of life, when very few forms of the simplest structure existed, the rate of change may have been slow in an extreme degree. The whole history of the world, as at present known, although of a length quite incomprehensible by us, will here-after be recognised as a mere fragment of time, compared with the ages which have elapsed since the first creature, the progenitor of innumerable extinct and living descendants, was created.

In the distant future I see open fields for far more important researches. Psychology will be based on a new foundation, that of the necessary acquirement of each mental power and capacity by gradation. Light will be thrown on the origin of man and his history.

Authors of the highest eminence seem to be fully satisfied with the view that each species has been independently created. To my mind it accords better with what we know of the laws impressed on matter by the Creator, that the production and extinction of the past and present inhabitants of the world should have been due to secondary causes, like those determining the birth and death of the individual. When I view all beings not as special creations, but as the lineal descendants of some few beings which lived long before the first bed of the Silurian system was deposited, they seem to me to become ennobled. Judging from the past, we may safely infer that not one living species will transmit its unaltered likeness to a distant futurity. And of the species now living very few will transmit progeny of any kind to a far distant futurity; for the manner in which all organic beings are grouped, shows that the greater number of species of each genus, and all the species of many genera, have left no descendants, but have become utterly extinct. We can so far take a prophetic glance into futurity as to foretell that it will be the common and widely-spread species, belonging to the larger and dominant groups, which will ultimately prevail and procreate new and dominant species. As all the living forms of life are the lineal descendants of those which lived long before the Silurian epoch, we may feel certain that the ordinary succession by generation has never once been broken, and that no cataclysm has desolated the whole world. Hence we may look with some confidence to a secure future of equally inappreciable length. And as natural selection works solely by and for the good of each being, all corporeal and mental endowments will tend to progress towards perfection.

It is interesting to contemplate an entangled bank, clothed with many plants of many kinds, with birds singing on the bushes, with various insects flitting about, and with worms crawling through the damp earth, and to reflect that these elaborately constructed forms, so different from each other, and dependent on each other in so complex a manner, have all been produced by laws acting around us. These laws, taken in the largest sense, being Growth with Reproduction; inheritance which is almost implied by reproduction; Variability from the indirect and direct action of the external conditions of life, and from use and disuse; a Ratio of Increase so high as to lead to a Struggle for Life, and as a consequence to Natural Selection, entailing Divergence of Character and the Extinction of less-improved forms. Thus, from the war of nature, from famine and death, the most exalted object which we are capable of conceiving, namely, the production of the higher animals, directly follows. There is grandeur in this view of life, with its several powers, having been originally breathed into a few forms or into one; and that, whilst this planet has gone cycling on according to the fixed law of gravity, from so simple a beginning endless forms most beautiful and most wonderful have been, and are being, evolved.

FURTHER READING

THERE IS NO obvious reason why the theory of evolution should attract the finest scientific writers: but it is so. From Darwin to the great popularizers of today, natural selection – and what it implies – has inspired the best of all scientific literature.

Darwin stands alone. Not many people with today's literary tastes would open *The Origin* for pleasure (although it contains much to enjoy). *The Voyage of the Beagle*, though, is a classic of travel-writing that reads as well now as it did in 1839. When it comes to Patagonia, Darwin was the Bruce Chatwin of his day. His *Autobiographical Sketch*, brief though it is, is also a simple and often moving account of a long life well spent. There is a multitude of biographies. Some emphasize the social context of Darwin's life and (inevitably) the coincidence between the theory of evolution by natural selection and the affluent Victorian world in which he was brought up, while others are more concerned with his own story and his scientific views, detached from political speculation. Janet Browne's extensive biography, *Charles Darwin*, with its first volume, *Voyaging* (Pimlico Books, 1966), is comprehensive, clear and dispassionate. Adrian Desmond and James Moore's *Darwin* (Penguin Books, 1992) puts his ideas firmly into the context of his own society.

Darwin has been well served by his successors. Some of science's most gifted authors – Richard Dawkins, Jared Diamond and Stephen Jay Gould – have produced a series of books that illuminate his work, bring it up to date and put it into historical, literary and social context. My own favourites, among many, are Dawkins's *Climbing Mount Improbable* (Penguin Books, 1997), Diamond's *The Rise and Fall of the Third Chimpanzee* (Vintage Books, 1991) and Gould's *Dinosaur in a Haystack* (Penguin Books, 1997).

Evolution suffers from a surfeit of both anecdotes and mathematics. The former have had a fairer run of it (and many are displayed in Richard Milner's *Encyclopaedia of Evolution*, published by Facts on file, New York and Oxford, 1990). For a real insight into the subject it is hard to escape from at least a nod at mathematical reality. The standard texts on the modern theory of evolution are

Douglas Futuyma's *Evolutionary Biology* (Sinaver Associates, Sunderland, Ma., third edn, 1998), John Maynard Smith's *Evolutionary Genetics* (Oxford University Press, second edn, 1998) and Mark Ridley's *Evolution* (Blackwell Science, Oxford, second edn, 1996). All give a solid introduction to the subject (and to the many problems that have been so lightly skipped over in my own book). In most modern texts (unlike *The Origin* itself) plants do not get a fair look-in. That balance is corrected by K. J. Niklas's *The Evolutionary Biology of Plants* (University of Chicago Press, 1997).

As the theory of evolution encompasses the whole of the biological sciences, there are dozens of popular works that illuminate part of the Darwinian story. Richard Fortey in his *Life: an Unauthorised Biography* (HarperCollins, London, 1997) gives a vivid account of evolution's raw material, the advance of existence through time. Matt Ridley's *The Red Queen* (Penguin Books, 1994) discusses an aspect of the subject not enough considered in my own book: natural selection as a race in which not all shall have prizes.

Any science writer has to mine the scientific literature, technical and tedious as it often is, in the hope of extracting a few gems. This book has been no exception: and at times through the thousand and more books and papers consulted here it appeared that the seam was almost exhausted. I list just a few of the most recent (and more accessible) publications as an entrée to the literature; all appeared in the 1990s and through them it is possible to refer back to earlier work. Each week, the journals *Science* and *Nature* keep up to date with the almost limitless world of evolution. *Trends in Research in Ecology and Evolution* produces reviews of advances in the field. The main technical journal is *Evolution*, complemented by *The American Naturalist*, *The Proceedings of the National Academy of Sciences* and *The Philosophical Transactions of the Royal Society of London*. More specialist journals include *The Journal of Molecular Evolution*, *The Journal of Animal Ecology, Behavioural Ecology and Sociobiology*, *Systematic Biology*, *Paleobiology*, and many more. *The Cottage Gardener and Country Gentleman's Companion* (much consulted by Darwin) is, alas, defunct.

In science, the Internet is overtaking the printed word, and there are several excellent sites that keep bang up to date with modern biology. Among them are The WWW Virtual Library: Evolution (http://golgi.harvard.edu/biopages/evolution.html), MendelWeb (http://www.netspace.org/MendelWeb), the Society for the Study of Evolution (http://lsvl.la.asu.edu/evolution), the Tree of Life (http://phylogeny.arizona.edu/tree/phylogeny.html) and the Natural History Museum in London (http://nhm.ac.uk). There is also the regularly updated and highly informative *HMS Beagle* at http://biomednet.com/hmsbeagle. No doubt others will come and go.

Introduction

Gatesy, J., 'More DNA support for a Cetacea/Hippopotamidae clade: the blood-clotting protein gene gamma-fibrinogen', *Molecular Biology and Evolution* 14: 537–43, 1997.

Goudsmit, J., *Viral Sex: the Nature of AIDS*, Oxford University Press, 1997.

Hoelzel, R., 'Genetics and ecology of whales and dolphins', *Annual Review of Ecology and Systematics* 25: 377–99, 1994.

Hooper, E., 'Sailors and star-bursts, and the arrival of HIV', *British Medical Journal* 315: 1689–91, 1997.

Korber, B., Theiler, J., and Wolinsky, S., 'Limitations of a molecular clock applied to considerations of the origins of HIV-1', *Science* 280: 1868–71, 1998.

Molla, A., *et al.*, 'Ordered accumulation of mutations in HIV protease confers resistance to ritonavir', *Nature Medicine* 2: 760–66, 1996.

Montgelard, C., Catzeflis, F. M., and Douzery, E., 'Phylogenetic relationship of artiodactyls and cetaceans as deduced from the comparison of cytochrome B and 12S rRNA mitochondrial sequence', *Molecular Biology and Evolution* 14: 550–59, 1997.

Thewissen, J. G. M., and fish, F. E., 'Locomotor evolution in the earliest cetaceans: functional model, modern analogues, and paleontological evidence', *Paleobiology* 23: 482–90, 1997.

Chapter I: Variation under Domestication

Bradley, D. G., *et al.*, 'Mitochondrial diversity and the origins of African and European cattle', *Proceedings of the National Academy of Sciences* 93: 5131–5, 1996.

Croke, V., *The Modern Ark: The Story of Zoos, Past Present and Future*, Charles Scribner's Sons, New York, 1997.

Frankham, R., 'Conservation genetics', *Annual Review of Genetics* 29: 305–27, 1993.

Kendrick, K. M., *et al.*, 'Mothers determine sexual preference', *Nature* 395: 229–30, 1998.

Ostrander, E. A., and Giniger, E., 'Insights from model systems: what man's best friend can teach us about human biology and disease', *American Journal of Human Genetics* 61: 475–80, 1997.

Serpell, J., *The Domestic Dog; its evolution, behaviour and interactions with people*, Cambridge University Press, 1995.

Tudge, C., *Last Animals at the Zoo*, Oxford University Press, 1992.

Vila, C., *et al.*, 'Multiple and ancient origins of the domestic dog', *Science* 276: 1687–9, 1997.

Chapter II: Variation under Nature

Barrate, E. M., *et al.*, 'DNA answers the call of pipistrelle bat species', *Nature* 387: 138–9, 1997.

Berry, R. J., and Bronson, F. H., 'Life history and bioeconomy of the house mouse', *Biological Reviews* 67: 519–50, 1992.

Din, W., *et al.*, 'Origin and radiation of the house mouse – clues from nuclear genes', *Journal of Evolutionary Biology* 9: 519–39, 1996.

Garner, M., 'Identification of Yellow-legged Gulls in Britain', *British Birds* 90: 25–62, 1997.
Mallet, J., 'A species definition for the Modern Synthesis', *Trends in Research in Ecology and Evolution* 10: 294–9, 1995.
Meffe, G., and Carroll, C. R., *Principles of Conservation Biology*, second edn, Sinauer Associates, Sunderland, Ma., 1997.
Rhymer, J. M., and Simberloff, D., 'Extinction by hybridisation and introgression', *Annual Review of Ecology and Systematics* 27: 83–109, 1996.
Shuker, K., *The Lost Ark: New and Rediscovered Animals of the Twentieth Century*, HarperCollins, London, 1993.

Chapter III: Struggle for Existence

Begon, M., Harper, J. L., and Townsend, C. R., *Ecology: Individuals, Populations, Communities*, third edn, Blackwell Science, Oxford, 1996.
Behrenfeld, M. J., and Kolber, Z. S., 'Widespread iron limitation of phytoplankton in the South Pacific', *Science* 283: 840–43, 1999.
Boonstra, R., *et al.*, 'The impact of predator-induced stress on the snowshoe hare cycle', *Ecological Monographs* 79: 471–94, 1998.
Budiansky, S., *Nature's Keepers: The New Science of Nature Management*, Phoenix, London, 1996.
Green, R. E., 'The influence of numbers released on the outcome of attempts to introduce exotic bird species to New Zealand', *Journal of Animal Ecology* 66: 25–35, 1997.
Greenwood, J. J. D., *et al.*, 'Relations between abundance, body size and species number in British birds and mammals', *Philosophical Transactions of the Royal Society* B 351: 265–78, 1996.
Kurlansky, M., *Cod: A Biography of the fish that changed the World*, Jonathan Cape, London, 1997.
Newton, I., 'The contribution of some recent research on birds to ecological understanding', *Journal of Animal Ecology* 64: 675–96, 1995.
Rebertus, A. J., *et al.*, 'Blowdown history and landscape patterns in the Andes of Tierra del Fuego, Argentina', *Ecology* 78: 678–92, 1997.
Rhymer, J. M., and Simberloff, D., 'Extinction by hybridisation and introgression', *Annual Review of Ecology and Systematics* 27: 83–109, 1996.

Chapter IV: Natural Selection

(various authors), 'Antimicrobial resistance', *British Medical Journal* 317: 609–90, 1998.
Bell, G., *The Basics of Selection*, Chapman and Hall, New York, 1997.
Frank, L. G., 'Evolution of genital masculinization: why do female hyaenas have such a large "penis"?', *Trends in Research in Ecology and Evolution*, 12: 58–62, 1997.
Henson, S. A., and Warner, R. R., 'Male and female alternative reproductive behaviors in fishes', *Annual Review of Ecology and Systematics* 28: 571–92, 1997.
Jennions, M. D., and Petrie, M., 'Variation in mate choice and mating preferences: a review of causes and consequences', *Biological Reviews* 72: 283–327, 1997.

Kearns, K. A., Inouye, D. W., and Waser, N. M., 'Endangered mutualisms: the conservation of plant-pollinator interactions', *Annual Review of Ecology and Systematics* 29: 83–112, 1998.

Lawton, J. H., and May, R. M., *Extinction Rates*, Oxford University Press, 1995.

Linhart, Y., and Grant, M. C., 'Evolutionary significance of local genetic differentiation in plants', *Annual Review of Ecology and Systematics* 27: 237–77, 1996.

Majerus, M. E. N., *Melanism: Evolution in Action*, Oxford University Press, 1998.

Reznick, D. N., *et al.*, 'Evaluation of the rate of evolution in natural populations of guppies', *Science* 275: 1934–7, 1998.

Schluter, D., 'Ecological speciation in post-glacial fishes', in Grant, P. R. (ed.), *Evolution on Islands*, Oxford University Press, 1998.

Welch, A. M., Semlitsch, R. D., and Gerhardt, H. C., 'Call duration as an indicator of genetic quality in tree frogs', *Science* 280: 1928–30, 1998.

Williams, G. C., *Natural Selection: Domains, Levels and Challenges*, Oxford University Press, 1992.

Chapter V: Laws of Variation

Ellegren, H., and Fridolffson, A.-K., 'Male-driven evolution of DNA sequences in birds', *Nature Genetics* 17: 182–4, 1997.

Ritvo, H., *The Platypus and the Mermaid and other figments of the Classifying Imagination*, Harvard University Press, Cambridge, Ma., 1997.

Rossiter, M., 'Incidence and consequences of inherited environmental effects', *Annual Review of Ecology and Systematics* 27: 451–76, 1996.

Chapter VI: Difficulties on Theory

Averof, M., and Cohen, S. M., 'Evolutionary origin of insect wings from ancestral gills', *Nature* 385: 627–30, 1997.

Berenbaum, M. R., Favret, C., and Schuler, M. A., 'On defining "key innovations" in an adaptive radiation: cytochrome P450s and Papilionidae', *American Naturalist* 148: S139–55, 1996.

Farrell, B. D., '"Inordinate fondness" explained: why are there so many beetles?', *Science* 281: 555–9, 1998.

Feder, J. L., *et al.*, 'The effects of winter length on the genetics of apple and hawthorn races of *Rhagoletis pomonella*', *Evolution* 51: 1862–76, 1997.

Golding, G. B., and Dean, A. M., 'The structural basis of molecular adaptation', *Molecular Biology and Evolution* 15: 355–69, 1998.

Lawrence, J. G., and Ochman, H., 'Molecular archeology of the *Escherischia coli* genome', *Proceedings of the National Academy of Sciences* 95: 9413–17, 1998.

Herre, E. A., *et al.*, 'The evolution of mutualisms: exploring the paths between conflict and cooperation', *Trends in Research in Ecology and Evolution* 14: 49–53, 1999.

Kazazian, H. H., and Moran, J. V., 'The impact of L1 retrotransposons on the human genome', *Nature Genetics* 19: 19–24, 1998.

Land, M. F., 'Visual acuity in insects', *Annual Review of Entomology* 42: 147–77, 1997.

Lewin, B., *Genes* VI. Oxford University Press, 1997.

Rivera, M. C., *et al.*, 'Genomic evidence for two functionally distinct gene classes', *Proceedings of the National Academy of Sciences* 95: 6239–44, 1998.

Chapter VII: Instinct

Dreller, C., 'Division of labor between scouts and recruits: genetic influence and mechanisms', *Behavioral Ecology and Sociobiology* 43: 191–6, 1998.

Holldobler, B., and Wilson, E. O., *Journey to the Ants: A Voyage of Scientific Exploration*, Belknap Press of Harvard University Press, Cambridge, Ma., 1994.

Klein, N. K., and Payne, R. B., 'Evolutionary associations of brood parasitic finches (Vidua) and their host species: analyses of mitochondrial DNA restriction sites', *Evolution* 52: 566–82, 1998.

Mock, D. W., and Parker, G. A., 'Siblicide, family conflict and the evolutionary limits of selfishness', *Animal Behaviour* 56: 1–10, 1998.

Rothstein, S. I., 'A model system for coevolution: avian brood parasitism', *Annual Review of Ecology and Systematics* 21: 481–508, 1990.

Seeley, T. D., 'Honey bee colonies are group-level adaptive units', *American Naturalist* 150, supp.: S22–41, 1997.

Stern, D. L., and Foster, W. A., 'The evolution of soldiers in aphids', *Biological Reviews* 71: 27–79, 1996.

Chapter VIII: Hybridism

Grant, P. R., and Grant, E. R., 'Genetics and the origin of bird species', *Proceedings of the National Academy of Sciences* 94: 7768–75, 1997.

Howard, D. J., and Berlocher, S. H., *Endless Forms: Species and Speciation*, Oxford University Press, 1998.

Marchetti, K., Nakamura, H., and Gibbs, H. L., 'Host-race formation in the common cuckoo', *Science* 282: 471–2, 1998.

Orr, H. A., 'Haldane's Rule', *Annual Review of Ecology and Systematics* 28: 195–218, 1997.

Parker, G. A., and Partridge, L., 'Sexual conflict and speciation', *Philosophical Transactions of the Royal Society* B 353: 261–74, 1998.

Saetre, G. P., *et al.*, 'A sexually selected character displacement in flycatchers reinforces premating isolation', *Nature* 387: 589–92, 1997.

Seehausen, O., Van Alphen, J. J. M., and Witte, F., 'Eutrophication that curbs sexual selection', *Science* 277: 1808–11, 1997.

Swanson, W. J., and Vacquier, V. D., 'Concerted evolution in an egg receptor for a rapidly evolving abalone sperm protein', *Science* 281: 710–12, 1998.

Turelli, M., 'The causes of Haldane's Rule', *Science* 282: 889–91, 1998.

Ungerer, M. C., *et al.*, 'Rapid hybrid speciation in wild sunflowers', *Proceedings of the National Academy of Sciences* 95: 11757–68, 1998.

Chapter IX: On the Imperfection of the Geological Record

Aslan, A., and Behernsmeyer, A. K., 'Taphonomy and time resolution of bone assemblages in a contemporary fluvial system – the East Fork River, Wyoming', *Palaios* 11: 411–21, 1996.

Cooper, A., and Penny, D., 'Mass survival of birds across the Cretaceous-Tertiary boundary: molecular evidence', *Science* 275: 1109–13, 1997.

Fortey, R. A., Briggs, D. E. G., and Wills, M. A., 'The Cambrian evolutionary "explosion" recalibrated', *BioEssays* 19: 429–34, 1997.

Gallois, R. W., *British Regional Geology: the Wealden District*, fourth edn, HMSO, London, 1992.

Kidwell, S. M., and flessa, K. W., 'The quality of the fossil record: populations, species and communities', *Annual Review of Ecology and Systematics* 26: 269–99, 1995.

Kumar, S., and Hedges, S. B., 'A molecular timescale for vertebrate evolution', *Nature* 392: 917–18, 1998.

Padian, K., and Chiappe, L. M., 'The origin and early evolution of birds', *Biological Reviews* 73: 1–42, 1998.

Xiao, S., Zhang, Y., and Knoll, A. H., 'Three-dimensional preservation of algae and animal embryos in a Neoproterozoic phosphorite', *Nature* 391: 553–5, 1998.

Chapter X: On the Geological Succession of Organic Beings

Aldridge, R. J., and Purnell, M. A., 'The conodont controversies', *Trends in Research in Ecology and Evolution* 11: 463–7, 1996.

Briggs, D. E. G., and Crowther, P. R., (eds), *Paleobiology: a Synthesis*, Blackwell Science, Oxford, 1990.

Conway Morris, S., 'Ecology in deep time', *Trends in Research in Ecology and Evolution* 10: 290–94, 1995.

Gokasan, E., *et al.*, 'On the origin of the Bosphorus', *Marine Geology* 140: 183–99, 1997.

Jackson, J. B. C., and Cheetham, A. H., 'Tempo and mode of speciation in the sea', *Trends in Research in Ecology and Evolution* 14: 72–7, 1999.

McShea, D. W., 'Metazoan complexity and evolution: is there a trend?', *Evolution* 50: 477–92, 1996.

Maynard Smith, J., and Szathmary, E., *The Major Transitions in Evolution*, W. H. Freeman, Oxford and New York, 1995.

Rosenzweig, M. L., 'Tempo and mode of speciation', *Science* 277: 1622–32, 1997.

Chapter XI: Geographical Distribution

Brown, J. H., Stevens, G. C., and Kaufman, D. M., 'The geographic range: size, shape, boundaries and internal structure', *Annual Review of Ecology and Systematics* 27: 597–623, 1996.

Cocks, L. R. M., McKerrow, W. S., and van Staal, C. R., 'The margins of Avalonia', *Geological Magazine* 134: 627–636, 1997.

Davis, G. W., and Richardson, D. M., (eds), *Mediterranean-Type Ecosystems: The Function of Biodiversity*, Springer Verlag, Berlin, 1995.

Cox, C. B., and Moore, P. D., *Biogeography: An Ecological and Evolutionary Approach*, fifth edn, Blackwell Science, Oxford, 1993.

Hamblin, W. K., and Christiansen, E. H., *Earth's Dynamic Systems*, seventh edn, Prentice-Hall, NJ, 1995.

Spanier, E., and Galil, B. S., 'Lessepsian migration: a continuous biological process', *Endeavour* 15: 102–6, 1991.

Van Oosterzee, P., *Where Worlds Collide: The Wallace Line*, Cornell University Press, Ithaca, NY, 1997.

Chapter XII: Geographical Distribution – Continued

Frankam, R., 'Do island populations have less genetic variation than mainland populations?', *Heredity* 78: 311–27, 1997.

Garrison, T., *Oceanography: an Invitation to Marine Science*, Wadsworth Publishing, Washington, 1996.

Grant, P. R., *Evolution on Islands*, Oxford University Press, 1998.

Johnson, L. E., and Padilla, D. K., 'Ecological Lessons and Opportunities from the Invasion of the Zebra Mussel *Dreissena polymorpha*', *Biological Conservation* 78: 23–33, 1996.

Rossiter, A., 'The Cichlid fish assemblages of Lake Tanganyika: Ecology, behaviour and evolution of its species flocks', *Advances in Ecological Research* 26: 187–252, 1995.

Steadman, D., 'Prehistoric extinctions of Pacific island birds: biodiversity meets zooarcheology', *Science* 267: 1123–31, 1995.

Thornton, I., *Krakatau: The Destruction and re-Assembly of an Island Ecosystem*, Harvard University Press, 1996.

Whittaker, R. J., *Island Biogeography: Ecology, Evolution and Conservation*, Oxford University Press, 1998.

Chapter XIII: Mutual Affinities of Organic Beings; Morphology; Embryology; Rudimentary Organs

Fong, D. W., Kane, T. C., and Culver, D., 'Vestigialization and loss of nonfunctional characters', *Annual Review of Ecology and Systematics* 26: 249–68, 1995.

Gale, M. D., and Devos, K. M., 'Plant comparative genomics after 10 years', *Science* 282: 656–9, 1998.

Jackson, S., and Diamond, J., 'Metabolic and digestive responses to artificial selection in chickens', *Evolution* 50: 1638–50, 1996.

Katz, L. A., 'Changing perspectives on the origin of eukaryotes'. *Trends in Research in Ecology and Evolution* 13: 493–7, 1998.

Logsdon, J. M., and Doolittle, W. F., 'Origin of antifreeze proteins: a cool tale in molecular evolution', *Proceedings of the National Academy of Sciences* US 94: 3485–7, 1997.

McNab, B. K., 'Energy conservation and the evolution of flightlessness in birds', *American Naturalist* 144: 628–42, 1994.

Pace, N. R., 'A molecular view of microbial diversity and the biosphere', *Science* 276: 734–40, 1997.

Shubin, N., Tabin, C., and Carroll, S., 'Fossils, genes and the evolution of animal limbs', *Nature* 388: 639–48, 1997.

Tatusov, T., Koonin, E. V., and Lipman, D. J., 'A genomic perspective on protein families', *Science* 278: 631–7, 1997.

Voss, S. R., and Shaffer, H. B., 'Adaptive evolution via a major gene effect: Paedomorphosis in the Mexican axolotl', *Proceedings of the National Academy of Sciences* 94: 14185–9, 1997.

Williams, R. W., 'Plant homeobox genes', *BioEssays* 20: 280–82, 1998.

Interlude: Almost like a Whale?

Boyd, R., and Silk, J. B., *How Humans Evolved*, W. W. Norton and Company, New York, 1997.

Harpending, H. C., *et al.*, 'Genetic traces of ancient demography', *Proceedings of the National Academy of Sciences* 95: 1961–7, 1998.

Lewin, R., *Principles of Human Evolution: A Core Textbook*, Blackwell Science, Oxford, 1998.

Modiano, D., *et al.*, 'Different responses to *Plasmodium falciparum* malaria in West African sympatric ethnic groups', *Proceedings of the National Academy of Sciences* 93: 13206–11, 1996.

Ruff, C. B., 'Morphological adaptation to climate in modern and fossil hominids', *Yearbook of Physical Anthropology* 37: 65–107, 1994.

Pringle, H., 'The slow birth of agriculture', *Science* 282: 1446–50, 1998.

Smith, M. W., *et al.*, 'Contrasting genetic influence of CCR2 and CCR5 variants on HIV-1 infection and disease progression', *Science* 277: 959–65, 1997.

Stoneking, M., 'Women on the move', *Nature Genetics* 20: 219–20, 1998.

INDEX

aardvark, 305, 337

abalones, 190

Abu Simbel, 258

acclimatization, 114–15

acquired characteristics, 111–14

Adam, 183, 246

aerial reptiles, fossil, 132–3

Africa: AIDS, 6–7, 13; copper, 105; dogs, 29; genetic diversity, 336; metal extraction, 106; parasitic birds, 163–4, 192–3; use of antibiotics, 96

Agassiz, Louis, 264

agouti, 259

agriculture, 25–6, 36–7, 334–5

AIDS, 2–16; appearance, 4; drugs, 7–9; evolution, 2–4, 13–15, 37; HIV mutation, 7, 8–9; origins, 9–13, 15–16; progress of illness, 5–6, 9, 352–3; responses to, 4–5, 353; spread, 6–7

Ailsa Craig, 264

Akbar, Emperor, 38

Alabama, legislation, 1

Alaska: Bering Land Bridge, 349; Glacier Bay, 267; plant life, 264

albatrosses, 222

alcohol, 114

Aleuts, 272

algae, 146, 160, 162, 313

alleles, 116–17

alligators, 83

alveolates, 313

Alvin (submarine), 309

Ambulocetus, 18

America: East and West, 257–8; immigrants, 59–60; New World, 254–5, 349

American Kennel Club, 33, 34, 156

amino acids, 75, 243, 244

Amish people, 351

ammonites, 208, 238

ampicillin, 96

Anatolian Shepherd dog, 31

Andean birds, 144

angler fish, 88

anoles, 79

Anomalocaris, 219, 223

Antarctic perch, 314, 330

Antarctica: coal, 255; ice sheet, 238; plant life, 263; separation from Australia, 273

antelope, 50

antibiotics, 96, 98

antifreeze protein, 203, 314–15, 330

ants: African driver, 169; Brazilian rain-forest, 166; Caribbean, 62; Galapagos, 102; honeypot, 161, 179; parasites, 160–2; relationship with aphids, 161–2, 182; relationship with fungi, 162–3; sterility, 179–80

aphids, 148, 161–2, 173–4, 179

Appalachian Mountains, 215, 270

apple maggot fly, 135

apples, 134–5, 183–4

Archaea, 308, 309, 310, 313

Archaeopteryx, 207, 218, 219, 220–1

Arctic: cod, 314, 330; wolf, 27

Ardipithecus, 346

Argentina, cattle, 63

Aristotle, 38, 85, 199

armadillo, nine-banded, 172

asses, 35

Aswan Dam, 216

atavism, 115

Atlantic: expansion, 271; ridge, 271–2

Atlantis, 270, 276–7

atrophied organs, 325–30, 371–2

aurochs, 247

Australia: age of rocks, 211, 216; ecosystem, 256, 257; extinct lifeforms, 350; flightless birds, 350; human presence, 349, 350; imports, 60–1; jar-rah, 70; kangaroos, 255, 350; kwongan, 256, 257; pre-Cambrian fossils, 223; separation from Antarctica, 273

Australopithecus, 346

Avalonia, 275–6

axolotl, 323

baboons, 16, 86

bacteria: behaviour, 157; classification, 307, 311; metal extraction, 106; DNA, 152, 245, 307; evolution, 95–9, 248; in aphids, 162; in guts, 96, 151, 162, 278–9, 307; resistance to antibiotics, 96–9, 151

Bagehot, Walter, xxviii

Bahamas, lizards, 79

Baikal, Lake, 290

baleen whales, 319

Baltica, 273

bananas, 307

Barbados: ants, 62; natural history, 138

bark-beetle, 173

barnacles, 64, 88, 220

Basilosaurus, 18

basset hounds, 35

bats, 50, 281, 305, 326

bears: black, xvii–xviii; diet, 249; young, 172

beaver, 259

Beccari, Odoardo, 111

beeches, 66–7, 269

bees: cell making, 166–7; dance, 167–8; DNA, 168; hive society, 165–71, 366; island species, 287; pollination, 85, 190; sex and sterility, 176–7, 178–9; sight, 139, 140; sting, 366

beetles: distribution, 267; evolution, 136; island species, 287; kinds of, 51, 101; life with rela-tives, 172, 173

Bell, Thomas, xxix–xxx

Bering Land Bridge, 349

Berkeley, George, Bishop, xxviii

Berlin Wall, 195

Bermuda, birds, 291–2

Bewick, Thomas, 303

Bible, 1, 209

bighorn sheep, 101

Biosphere Two, 254

birds: behaviour, 158; chromosomes, 121, 196; classification, 306; evolution, 129, 181, 220–1;

first appearance, 220–2; Mediterranean, 256–7; parasitic, 163–4, 192–3

Birnam Wood, 267

bison, European, 37

bizcacha, 259

Black Sea, 236

Black Skull, 347

black smokers, 309

blackberries, 327

blindness, 366; inherited, 351

blood: clots, 143; groups, 338; pressure, 342

bloodhounds, 35

blow-flies, 206

blue whale, 17, 258, 268

blueberry maggot fly, 135

bluetits, 86, 89, 159

bone cells, 245

Bonnet, Charles, 315

bonobo, 337

Borametz, 202–3

Bosphorus, 236

Bounty, HMS, 261

bower-birds, 89, 91

Boxgrove Man, 28

Bradford, William, 271

brains, 157, 248, 317, 343–4

Brazil: ants, 166; coffee, 68

Bryan, William Jennings, 138

Buccleuch, Duke of, 83

Buffon, Comte de, 246

bulldogs, 33

Bulo Burti boubou shrike, 46

Burchell, William, 23, 24

Burgess Shale, 218, 219, 223

Burgon, John, 209

Burton, Richard, 289

burying-beetle, 172

Buss, Island of, 276–7

butterflies: camouflage, 130; chromosomes, 196; classification, 49, 51; flight, 133; Hawaii, 287; swallowtail, 137

C57Bl mice, 53, 54

Cabot, John, 72, 265

cacti, 102, 257

Cairo, sewers, 310

Calgene company, 202

California: chaparral, 256, 257; condor, 60; Death Valley, 268; ecosystem, 256–7; fossils, 213, 219; Gold Rush, 213; plants, 101

Cambrian era, 222–3, 239
Cambrian Explosion, 223–5
camouflage, 78, 90, 130
Camus, Albert, 4
Canada: fishing, 64–5; fish, 103
Canary Current, 260
Canary Isles, 284
cancer, 198, 203, 338
Canis familiaris, 27
Canis lupus, 42
cannibalism, 177–8
Cape Verde Islands, 285
capybara, 259, 298
carbon 14, 212
carbon dioxide, 250, 254, 265
Caribbean: birds, 163; lizards, 79, 286
Carnegie, Andrew, 354
carnivores, 249, 305
carpenter bee, 169, 170
Catahoula leopard cowhog dogs, 32
Catal Hüyük, 335
cats: city, 285; classification, 305; deafness, 123,
    124; edible, 23; Galapagos, 102; genes, 304;
    island populations, 285–6; London, 121; Manx,
    118; Persian, 184; tortoiseshell, 123, 124–5;
    viral diseases, 16; wild, 52
cattle: defences against flies, 141; DNA, 35, 150;
    domestication, 26, 35; face recognition, 36;
    genetic map, 304; in Americas, 63; selection,
    36; teeth, 372; types of, 35
cave animals, 142, 326, 366
cells, 151–2; division, 198
chalk, 226–9
Challenger Expedition (1872), 309
change: gradual, 231–2; rate of, 242–3;
    simultaneous, 239–41; sudden, 239–41
Channel Tunnel, 226
characteristics: acquired, 111–14; reverting to,
    115
Chardonnay grape, 256
Charlemagne, Emperor, 38
Charles II, King, 33, 165
Checkpoint Charlie, 194
Chediak-Higashi Syndrome, 19
cheese-skippers, 206
cheetahs, 38
Cherry-Garrard, Apsley, 320–1
Chesapeake Bay, 238
chestnut trees, 267
chevrotains, 51

Chiang Kai-shek, 240–1
chickens, 79–80, 306, 328
child mortality, 337–8
Chile: beeches, 66; ecosystem, 256, 257; human
    presence, 349; mattoral, 256, 257
chimpanzees: AIDS virus, 15; ancestors, 343;
    diversity, 349; DNA, 344; human kinship, 15,
    307, 351; penis size, 93; species, 336–7
China: balance of sexes, 339; classification, 297,
    298; forensic entomology, 206; fossil record,
    224–5; hill tribes, 350; mice, 52; pandas, 46
Chinese crested dog, 124
chitin, 243
cholera, 335, 338
chows, 35
Christmas, 173, 187
Christmas Island, 261
chromosomes, 48, 118–21, 196–7
Churchill, Winston, 343
cicadas, 189, 301–2
Cicero, 322
cichlids, 191–2, 290–1, 292
ciliates, 313
cities, 104–5, 335
cladistics, 302–6, 351
clams, 322
Clark, William, 99
classification: Chinese, 297–8; cladistics, 302–6;
    comparative anatomy of genes, 306–11;
    Darwin on, 375–6; difficulties, 45–52;
    Linnaean system, 299–301; map of genes,
    311–13
Cleopatra's Needle, 214
climate, 67–8, 265–6, 293
climbing plants, 327
clones, 104, 158
coal, 255, 276, 337
coalmines, 76, 243
coconuts, 286
Cocos Island finches, 158
cod, 72, 80, 103; Arctic, 314, 330
coelacanths, 101, 305
coffee, 67–8
collections, palaeontological, 212–13
collies, 31–2
Colombia, coffee, 68
Colorado River, 216
Columbus, Christopher, 63
comets, 237–9, 273
competition, 366

condor, California, 60
conodonts, 233–4
Constable, John, 186
continents, 271–7
Cook, Captain James, 280, 281, 303
Copernicus, 209
copper, 105, 106, 196
copper flower, 105, 194
Copts, 243–4
corals, 240, 248, 276
corgis, 32
cormorants, 221, 329
Cornbrash, 210
corpses, decay of, 206–7
correlation of growth, 123, 376
Costa del Sol, 263
cougars, Texas, 56
courtship, 89–90
cowbirds, 163
coyotes, 29
coypu, 259
crabs, 142, 223, 261, 324
creationist movement, 1–3
crickets, 189, 190
crocodiles, 264, 306
crows, 130–1
Cruft's, 27, 33
cuckoo-ducks, 163
cuckoos, 163–4, 180, 182, 192
Curse of India, 102
Cuvier, Georges, 155, 209, 250
Cyprus, elephants, 329
cystic fibrosis, 335

Daguerre, Louis, 130
daisies, 287
dandelions, 189
Darwin, Bernard, xxii
Darwin, Charles: correspondence, xxii; *The Descent of Man*, 344, 351; family, xxi–xxii; on classification, 375–6; on common parentage, 331–2, 377; on difficulties of his theory, 153–5; on embryology, 371, 377; on first crosses, 204–5, 357–8; on geographical distribution, 293–6, 358, 369–70; on geological record, 252–2, 359–61, 368–9, 377–8; on heredity, 110–11; on hybrids, 204–5, 357–8; on instincts, 181–2, 367–8; on intermediate varieties, 359; on marriage, 338–9; on morphology, 376; on natural selection, 106–9, 153–5, 356–7, 364–8,

379; on rudimentary organs, 371–2, 376, 377; *The Origin of Species*, xvii–xix, 232; prose style, xix–xx; reading of Malthus, 62
Darwin, Robert, xxii
Darwin, Erasmus, xxi–xxii, 202–3
dating, 210–12
David, Père Armand, 46
DDT, 122
de la Bèche, Henry, 210
de Tocqueville, Alexis, 76
deer, 50, 175
Defoe, Daniel, 76
Devil's Hole pupfish, 267–8
diatoms, 313
dingo, 29
dinoflagellates, 313
dinosaurs: bird ancestry, 129, 181, 220–1; classification, 306; digestion, 227; distribution, 270, 283; DNA, 243, 244; end of, 227, 238–9; fossil record, 207; lifestyles, 181; size, 248; varieties, 274
disease, 335, 338; genetic, 351
dispersal: during Glacial Period, 263–77; means of, 259–63
distribution, geographical, 293–6, 358, 369–70
divergence of character, 103–9, 364–5, 379
diversity, 143, 145, 335–6
DNA: amount of, 245; ancient, 243–4; bases, 145–6; changes, 142–3; decay, 330; discovery, 119; dog faeces, 34; fingerprint, 336; fossil, 243–5, 348; HIV, 5, 7; identification by, 46–8; instability, 148; maps, 119; mitochondrial, 29, 42, 151, 350; movable, 98–9, 146–50; natural history of, 145; preservation, 243; repair, 121, 149–50; repetition, 330; sequencing machines, 47; vertebrate, 150; whales, 19–21, 145
Dobermann pinschers, 35
dodo, 103, 329
dogs: bipedalism, 158; Chinese crested, 124; classification, 27, 41–2, 305; DNA, 29, 42; copulation, 84; domestication, 25, 28–31, 35; evolution, 241–2; feral, 30, 41; Galapagos, 102; genetic disorders, 35; herding, 31–2; naked Turkish, 123, 124; acquired characteristics, 111; sacred, 24; selection, 25; wild nature, 40–1
dogwood fly, 135
dolphins, 16, 50
domestication, 24–35, 361–2
Douglas fir, 80

doves, 191
dragonflies, 133, 140
du Chaillu, Paul, 16
ducks: flightless, 366; Hawaiian, 288; imprinting, 191; sexual selection, 192; species crosses, 185, 193
Duffy blood group, 341–2
Dumas, Alexandre, 11, 72
dung-beetle, 267
dunnock, 164

eagles, 69–70
ears: development, 324, 327; inner, 53, 345
Earth: dating, 211–12; ice, 264; magnetic field, 272–3; rotation, 250
earthquakes, 234–5
earthworms, 315–16
East Fork River, 217
Eden, Garden of, 183, 268, 313, 349
Ediacara fossils, 223, 224, 225
Edison, Thomas, 130
Ein Mallah, burial, 28
electric fishes, 269, 274
elephant-birds, 329
elephants: bones, 220, 245–6; classification, 305; dwarf, 329; embryos, 322; mitochondria, 151; reproduction, 70; selection, 87; sex pheromone, 314; swimming, 261; transitional links, 208; tusks, 87; types of, 50–1; zoo, 40
Ellis Island, 59, 73
embryology, 320–5, 331–2, 371, 377
embryos, fossil, 224–5
emperor penguins, 320, 321
emus, 269, 274
Engels, Friedrich, 62
Escherischia coli, 307
Eskdale, sparrowhawks, 82–3
Eskimos (Inuit), 32, 350
evolution: use of term, 321–2
extinction, 99–103, 235–9
eyes, 138–41, 154, 323 4

Fang people, 16
farming, see agriculture
fathers, age of, 120–1
ferns, 264, 270, 281, 301
fertilizer, 69
figs, 163, 281
finches: Cocos Island, 158–9; egg-dumping,

192–3; Galapagos, 102, 158, 187, 188, 282, 292; zebra, 91
Finland, 283
fire, 67
first crosses, 185–94, 204–5, 357–8
First World War, 228–9, 262
fish: ear development, 327; freshwater, 288–91; genes, 318; hybrids, 203; ray-finned, 318; sex, 88; sperm, 247
flatworms, 324
flax: 102, 113–14
flies, 115, 140, 141, 189
flood legends, 236
flowers: colours, 367; island, 287, 288; structure, 319–20
flycatchers, 194
flying, 131–4
flying-fish, xv
food chains, 69–70
Forbes, Edward, 295
fossil record: confusion, 217; evidence of extinction, 100; incompleteness, 207–9, 360–1, 377–8; slow or rapid change, 207, 240–1, 359–61, 368–9
Foucault, Michel, 297
foxes: diet, 249; rabies, 279; selective breeding, 30–1
Franco-Prussian War, 262
Franklin, Benjamin, 61
Franklin, John, 265
Freke, Dr, xix
Frere, John Hookham, xvii, 333
friendship, 162
Frobisher, Martin, 276
frogs: classification, 50; oceanic islands, 281; sexual selection, 89, 90; varieties, 274
fruit-flies: acclimatization, 114; chromosomes, 118, 196; diversity, 47; eyes, 323–4; hearing, 324; laboratory breeding, 195, 242; mobile DNA, 146–7, 150; mutation rate, 122, 242; natural selection, 195; sex and death, 80; sexual selection, 89
fruitbat, Dyak, 326
fugu, 146, 147
Fulani people, 341
fundamentalists, 99
fungi, 49, 151, 162–3

Galapagos: American species, 284–5; birds, 282, 292, 329; Darwin's description, xx; extinctions, 102, 235; finches, 102, 158, 187, 188, 282,

Galapagos cont.
   292; frogs, 281, 284; ocean currents, 188, 260;
   plankton, 69
gales, 67
Galileo, xviii, 22, 209
Galton, Francis, 335
geese: Andean, 144; bar-headed, 144; first
   appearance, 221; Hawaiian, 288, 291; hybrid,
   185
genes: as particles, 115, 117; behaviour, 157–8;
   brains, 157, 343–4; classification, 304, 306–8,
   311–13; crosses and hybrids, 196–9; development and, 124–5; disease, 337–8; diversity, 336;
   fossil, 243–5; homeobox, 318–19; human,
   348–9; map, 311–13; speciation, 196
genetic engineering, 201, 202–3
genetically modified plants, 202–3
genetics: laws and principles, 111, 115, 117;
   medical, 339–40
Geological Survey of Great Britain, 210
geology, 209–12
George I, King, xxiv
Germany, 45
Ghanaian Times, 13
Giant's Causeway, 275
Gibraltar, 258, 260, 282–3
Gilgamesh, Epic of, 236
gingko, 264
giraffes, 84, 141, 199, 324
Glacial Period, dispersal during, 263–77, 358
Glacier Bay, Alaska, 267
glaciers, 261, 263–4
Glen Canyon Dam, 216–17
goats: bipedalism, 123–4; DNA, 35, 150; domestication, 26; face recognition, 36; Galapagos,
   102
Goethe, Johann Wolfgang von, 123
Gondwanaland, 222, 273, 283
gonorrhoeae, 97, 99
gophers, 143
gorillas: affinity to humans, 15, 351; ancestors,
   342–3; as food, 16; DNA, 344; lowland, 337;
   mountain, 50, 337; penis size, 92; zoo diet, 38
Grand Canyon, 216–17
grape: Chardonnay, 256; climbing, 327
grasses, 194, 281
grasshoppers, 267
greenhouse effect, 250, 265
Greenland: age of rocks, 211; colonists, 60; fossil
   record, 225; temperature, 266

groundsel, 186
grouse: black, 78; red, 78, 83; snow, 64–5;
   testosterone, 83, 85
guava, 102
Gulf Stream, 260, 266
gulls, 43, 44–5, 58, 163, 191
guppies, 80, 242

Habsburgs, 323
Haeckel, Ernest, 317, 321
haemoglobin, 144, 145, 148, 150, 330, 340–1
hagfish, 233
Haldane's Rule, 196
halibut, 71–2
Hallucigenia, 223
hares: European, 259; snowshoe, 65; varying, 266
Harvey, William, xxviii
Hastings Rarities, 44
Hawaii: bird extinction, 288, 291, 292; frogs,
   281; humpback whales, 19–20, 37; insects, 287;
   origins, 283–4; plants, 287–8
hawkweed, 117
hawthorn fly, 135–6
Hearne, Samuel, xvii
heathers, 257
Hebrides, Outer, 69–70
hedgehogs, 102, 305
Henry, Patrick, 61
Henry VI, King, 38
Heraclitus, 231
herbicide resistance, 204
hermaphrodites, 85–6, 94
Herodotus, 214
herons, 288–9
Himalayacetus, 18, 21
Himalayas, 215, 216
hippopotami, 21, 305
Hispaniola, ants, 62
HIV (Human Immunodeficiency Virus):
   defences against, 352–3; HIV-1, 11–16; HIV-2,
   11, 14–15; genes, 5; origin of, 12–13; subtypes,
   11, 14; types, 11, 15; *see also* AIDS
hive-bee society, 165–70
homeobox genes, 318–19
Homer, 245
hominines, 346–7
Homo erectus, 346
Homo habilis, 346
Homo sapiens, 347, 349
honeycreepers, 288

honeypot ants, 161, 179
Hoover Dam, 264
hormones, 174, *see also* testosterone
horses: domestication, 35; edible, 23; evolution, 242–3; hooves, 324; in America, 63; striped, 366–7
horsetails, 248
howler monkeys, 172
Hudson, Rock, 4
Hudson's Bay Company, 65
Hughes, Griffith, 137–8
Human Genome Project, 339
humans: brain, 343–4; diversity, 335–6; effects of farming, 334–5; evolution, 333–4, 346–7, 349; male nipples, 326–7; migration, 349–51; size, 335, 336; tails, 326; walking, 345–6
humpback whales, 19, 37, 64, 159
Huntington's disease, 147
Hutton, James, 209–10
Huxley, Thomas Henry, 78, 157
hybrid(s): ape-human, 344; fertility, 194–9, 201; fish cancers, 203; legal status, 56; mongrel comparison, 198–203; position of, 130–1; sterility, 185–6, 188–94, 357, 372; survival, 186–8; zones, 55
hyenas: diet, 249; spotted, 85
hypothalamus, 142
hyrax, 305

Iapetus, ocean of, 275, 276
ice, 264–6
Iceland: lava, 272; origins, 273
ichneumonidae, 182, 366
iguanas, 102
inbreeding, 174, 336
incest, 89–90
India, fossil record, 225
Indonesian archipelago, 283
Industrial Revolution, 76–8, 104–5
insecticides, 85; resistance to, 122–3, 340
insects: classification, 306; island, 287; sexual selection, 86–7; sight, 140
instinct, 128, 157, 367–8; cell-making, 165–7; slave-making, 160–4, 182
insurance, 80–1
intelligence, 344, 352
intercrossing, 94–5
intermediate forms, 153, 207, 241, 359, 368–9
International Commission on Zoological Nomenclature, 41
Inuit (Eskimos), 32, 350
investment, 79–80

Ireland: ancient forest, 67; continental boundary, 275–6; snakes, 255, 281
iridium, 237
Irish wolfhounds, 34
iron, liver levels, 342
islands, oceanic, 281–93, 295, 370
Israel: jellyfish, 258; mole-rat, 142
Italy: feral dogs, 30; hill villages, 350

jackals, 249
Jackson Laboratory, 53
Japan: moths, 77; mouse-fanciers, 52
jarrah, 70
Jefferson, Thomas, 99, 200, 246
jellyfish, 258; freshwater, 290
Jews, 112
Johnson, Samuel, 275
jungle fowl, 328
Jurassic Park (film), 220, 243

kangaroos: carnivorous, 350; classification, 303, 305; distribution, 255; giant, 70; tree 106, 190
Kaposi's sarcoma, 6
kelp, 313
Kennel Club of England, 34
Kent, chalk, 226
Kenya, antibiotics, 96
Kerguelen, cats, 285
Ketengban people, 93
Kew, Millennium Seed Bank, 262
Keynes, John Maynard, 248
Killarney fern, 301
kinetoplastids, 313
King Charles spaniels, 33–4
Kingsley, Charles, 131, 322
kinship, 171–80, 330
kiwis, 328
Klagenfurt fountain, 219
Knapp, Michelle, 236
Knight, Richard Payne, 333, 355
Knight, Thomas Andrew, 333
Kolreuter, Josef Gottlieb, 200
Koobi Fora beds, 241
Krakatau: eruption, 279–81, 286; inhabitants, 281, 282
krill, 219

La Brea Tar Pits, 27, 218, 219, 249
La Plata, animals, 259
laboratory mice, 53

Labrador dog, 31
Lamb of Tartary, 202–3
lamprey, 316
lancelet, 233, 316–17
language, 10–11
Lathrop, Abbie, 52–3
Laurasia, 273
lead, 105, 194
leaf-cutter ants, 180
learning, 159
Leavis, F.R., 354
leeches, 316
Leigh, Isaac, 46–7
Lemuria, 283
lemurs, 175, 190, 283
Leonardo da Vinci, 271, 273
leopard frog, 50
Lewis, Meriwether, 99
Li Ch'un Feng, 237
lice, 4, 143, 151
lichens, 162
Lilienthal, Otto, 106
Linnaeus, Carolus, 27, 41, 54, 300–1
Linnean Society, xxix
lions, 38, 40, 172, 249
Live Stock Journal and Fancier's Gazette, 23
liverworts, 263
Lizard Peninsula, 105, 210
lizards: Caribbean, 79, 286; classification, 305;
    clones, 172; DNA, 150; kinship, 172; leg
    length, 79; monitor, 281
lobster, 324–5
locusts, 263
logger-headed duck, 366
London: black wildlife, 78; cats, 121; East End,
    313; Great Plague, 6; growth, 76–7; industry,
    104–5; moths, 77; population, 335; Tower, 38;
    Tube, 308, 310–11; Zoo, 37
London, Jack, 40
Longfellow, Henry Wadsworth, 67
lotus, 190
Louis XIII, King, 299
Lucy (primate), 345
lungfish, 100–1, 244–5, 305
lungs, 154
Lyell, Charles, 99, 235, 373
lynx, 65

macaques: AIDS-like virus, 15; food washing,
    159; society, 113

Madagascar: birds, 329; plants and animals, 283;
    resistance to antibiotics, 98
Madeira: frogs, 281; land-shells, 282; seashells,
    282
magnetic field, 272–3
magnolia, 266
Maida Island, 277
maize, 26, 36, 50
malaria, 340, 341–2, 344
Malawi, Lake, 289, 290
mallard ducks, 193
Malthus, Thomas, 61–2, 64, 70–1, 73, 82, 160
mammals: classification, 304–5; early, 238,
    239–40; genes, 318; size, 248
mammoths, 218, 244, 329
Manchester: growth, 76; industry, 76, 77, 104–5;
    moths, 77, 97
mangabey, sooty, 15
Manx cats, 118
Marie Antoinette, 188
marijuana pollen, 263
Marion Island, cats, 102
market, 68
marriage: age at, 82; Darwin on, 338–9
Marsham, Robert, 265
marsupial mice, 175
Marx, Karl, 62, 166, 354
Mass, Catholic, 325, 330
mastodons, 99, 246, 248
Matteson, Edward E., 213
Mauna Loa, 284
Mayflower, 271
Meander, River, 231–2
Mediterranean: ecosystems, 256–9; history,
    258
megapodes, 181
melanin, 124
melanomas, 198
Melville, Herman, xx, xxvii
memory, 111–12, 114
Mendel, Gregor, xxiv–xxv, 52, 111, 115–18
meningitis, 96
mental disorders, 343
mesonychids, 17–18
Meteor Crater, Arizona, 237
meteorites, 236–7, 313
Mexico: cattle, 63; Yucatán crater, 237, 238
mice: breeding, 52–3, 113, 123, 197–8; chromo-
    somes, 196, 304; diet, 113; diversity, 52–5; DNA,
    149, 304; eyes, 323–4; foetus reabsorption, 178;

mice cont.
  genes, 114, 316; identification, 47; inbred,
  174; laboratory, 53; marsupial, 175; neophobia,
  157; sexual strategy, 84; size, 123; tabby, 124;
  terrible, 50; varieties, 52–5; white-footed, 39;
  wild races, 53–4
midge larvae, 150
midwife toad, 149
migration, human, 349–51
Milton, John, xxii
mimosa, 262
mines, 194
missing links, 129, 250, 347
Mississippi, River, 231–2, 234–5
mitochondria, 151, 160, 311, 348, 350
moa, 222, 329
Moby Dick, 19, 22
moles, 366; golden, 305
mole-rats, 142, 174
mongrels, 198–203, 357
monkey-flowers, 194
monkeys: howler, 172; number of species, 346;
  training, 158; viruses, 15–16
Moon: dating, 211; effect on rotation of earth,
  250; ice, 264
morphology, 315–20, 331, 351, 376
mosquitoes, 122, 123, 340
mosses, 248, 263
moths: island species, 287; melanic, 157; peppered, 77–8, 90, 97; sex pheromone, 314
Moustache Gang, 90
Mozart, Wolfgang Amadeus, 159–60
mules, 185
mummy DNA, 243
Murray Collection, 96, 98
Mus domesticus, 54–5, 197
Mus musculus, 54–5, 197
mushrooms, 307
music, 159
musk-rat, 259
mussels: distribution, 269; freshwater, 46–7, 289;
  larvae, 322; zebra, 289
mutation, 119–23, 185
myoglobin, whale, 75

Nagubago, Lake, 291
Napoleon Bonaparte, 209
national identity, 45
Natural History Museum, London, 212–13, 246,
  262

natural selection: action of, 75–6; action on the
  descendants from a common parent, 104;
  circumstances favourable to, 95–9; compared
  to man's selection, 106, 363–4; divergence of
  character, 103–4, 364–5; extinction caused by,
  99–103; illustrations of its action, 76–81; organs
  of extreme perfection, 137–41; organs of little
  apparent importance, 141–4; origin and transition of organic beings with peculiar habits and
  structure, 134–7; sexual selection, 81–94,
  338–9, 363; speed of action, 79
Neanderthals, 345, 347–8, 350
nematode worms, 49–50, 304, 306
New Bedford, whale-port, 73
New Guinea, dogs, 29
New York, tuberculosis, 98
New York Coffee, Sugar, and Cocoa Exchange,
  68
New Zealand: arrival of man, 291, 349; flowers,
  287; flightless birds, 329; imported birds, 60–1,
  285; snails, 104
Newfoundland: fish stocks, 72, 73; Maida Island,
  277; Stone of Scone, 276
Newton, Isaac, xxviii, 148
Nicholas II, Tsar, 120
nickel, 106
nicotine, 114, 137
Nile, River, 216
Nile perch, 292
nimravids, 249
Niño, El, 187–8
nipples, male, 326–7
Noah, 51, 236, 263, 271
North Downs, 226
North Pole, 272–3
Nova Scotia duck-tolling retriever, 34
nucleus, 151–2, 311, 313
nutcracker, 158

oaks, 186, 257, 267
Obsidian Pool, 308
ocean: currents, 260; floor, 271–2; vents, 309
Octopus Spring, 308
oil, 243, 337
oilseed rape, 204
Okinotorishima, 214
Opabinia, 223
Oppian, 199
orang utans, 15, 351
orchids, 70, 85, 189, 190, 281

oryx, Arabian, 37, 40
ostriches: distribution of flightless birds, 269,
    273–4; evolution, 111–12; fossil record, 222;
    hybrid theory, 199; nesting habits, 164
Owen, Richard, 246
owls, 221
Oxford Botanic Garden, 186
oxygen, 136, 144, 148, 250, 254
oystercatchers, 91, 221
oysters, 322

Pakicetus, 18
Paleozoic era, 239
Paley, William, 93, 141
palms, 50, 67
Panama, Isthmus of, 258
panda, giant, 46
Pangaea, 271, 273, 274
panthers, Florida, 56
parasites, 160–1, 248, 340
Paris: Anatomy Museum, 299–300; Basin, 250;
    Métro, 311; Zoology Museum, 299, 300
parrots, 158, 175, 221, 222
partridges, 269
Passchendaele, battle (1917), 228–9
passenger pigeon, 217–18
passionflower, 327
Paul, apostle, 232
peacocks, 86, 90, 93, 94
peat bogs, 218
Pekinese, 32
penguins: distribution, 258, 273; earliest appear-
    ance, 221; emperor, 320, 321; fossil record, 222,
    261
penicillin, 96, 99, 122
penis size, 92–3
peppered moths, 77–8, 90, 97
perch: Antarctic, 314, 330; Nile, 292
perfection, 137–41, 379
pesticides, 122–3, 148, 340
pets, 16, 31
phallocarp, 93
pharaoh hound, 32
pheasants, 185, 269
Philadelphia Zoo, 38, 39
Philip, Prince (Duke of Edinburgh), 93, 120, 271
Phillips, John, 239
phosphorites, 224
pigeons, 24–5, 78, 92, 117, 222, 367
pigs: breeding, 171; domestication, 26;

Galapagos, 102; gene map, 304; sexual
    misconduct, 110; trials of, 23–4
Pilgrim Fathers, 72, 271, 286
Pillars of Hercules, 260
pineal gland, 142
pine trees, 247, 267
pintail ducks, 193
pipistrelle bat, 50
Pitcairn Island, 260–1
plague bacillus, 98
plankton, 69
plants, structure, 319–20
plasmids, 98–9
Plato, 202
platyfish, 198
platypus, duck-billed, 303, 353
Pleistocene era, 264
Pliny, 52, 54
pneumonia, 4, 99
pocket-gophers, 143
poisons, 104–5, 136–7
polio, 338
pollen, dispersal, 263
pollination: Biosphere Two, 254; co-operation
    and parasitism, 160, 190; islands, 287, 288; male
    genes, 84–5
Polynesia, rails, 292
Polyphemus, 245
Pope, Alexander, 246
poppies, 186, 262
population size, 61–5, 337
potato-lice, 138
potatoes, 26, 200, 329
prawns, 259
Priestley, Joseph, xxii
primates: brains, 343; DNA, 344; evolution, 143;
    guts, 343; penis size, 92–3
Przewalski's horses, 39–40
pseudogenes, 330
pterodactyls, 99
puffer fish, 146
punctuated equilibrium, 242
Pyramids, 214–15, 230
Pyrenean Mountain dog, 31
Pytheas of Massalia, 260

quagga, 243
quails, 269
Quayle, T. Danforth, 237
Quetzalcoatlus, 250

rabbits, 60, 259, 305
rabies, 279
radioactive materials, half-life, 211–12
ragwort, 186
rails, terrestrial, 292, 329
Raleigh, Sir Walter, 60
Rameses II, 258
rats, 102
ray-finned fishes, 318
red deer, 175, 329
Red Sea, 258–9
redstart, 192
reed warblers, 163
reptiles, aerial, 132–3; classification, 305–6
retrovirus, 5, 20
Rewa, Maharaja of, 39
rhea, 222, 269, 273–4
rhinoceros, 24; woolly, 218
rice, 25, 26, 199
Ritonavir, 8–9
RNA, 5, 9–10, 12, 14, 152
Roanoke, colony, 60
Romanes, George, 180
Roosevelt, Kermit, 46
Roosevelt, President Theodore, 38
Roosevelt, Theodore
Rothschild, Lionel Walter, Lord, 292, 308
Royal Society for the Protection of Birds, 193
rudimentary organs, 325–30, 371–2, 376, 377
Ruskin, John, xxiii
Russian Royal Family, 120
Rutherford, Lord, 57
rye, 25

St Helena, 286
St Hilaire, Geoffroy, 325
St Pierre-Miquelon, cats, 285–6
salamanders, 145, 146, 186, 245
salmon, 83, 103–4, 318
Salvinia, 63, 66
San Francisco, AIDS, 4, 9, 15
saola, 50
scallops, 269, 243
Scopes, John, 138
scorpions, 250
Scotland: continental movement, 274–6; rocks, 263–4
Scott, Captain Robert Falcon, 270, 320
sea-anemones, 158
sea-cows, 305

sea-elephants, 339
Sea Owle, 292
sea salt, 211
sea-scorpions, 276
sea-urchins, 190
seals, 258
seashells, 282
seaslugs, 86
seaweeds, 225, 261, 313
seed: banks, 36, 262; dispersal, 261–3, 286
Selborne, swifts, 63, 82
septicaemia, 96
sex cells, 87–8, 190
sexual selection, 81–94, 338–9, 363
Seychelles warblers, 175–6
Shakespeare, William, 165, 292
sharks, 143
Shaw, George Bernard, 75–6, 320
sheep: bighorn, 101; DNA, 150; domestication, 25, 26, 35; evolution, 130; face recognition, 36; selection, 36; sheepdogs, 31; types of, 35
sheepdogs, 31–2
shipworm, 106
shrew, elephant, 305
shrimps, 133–4, 240
Shulgia, King, 38
Siberia: Bering Land Bridge, 349; craters, 238
sickle-cell anaemia, 341
Sierra Club, 217
Signy Island, 263
Silbury Hill, 243
Silurian era, 251–2, 359, 361, 378–9
size: evolution, 245–6; island creatures, 329; sexual selection, 88
skin colour, 115
skulls, 248, 249, 316–17
slave-making instinct, 160–4, 182
sled dogs, 32
slime moulds, 313
sloths: giant, 216, 244, 248; ground, 218
slugs, 86, 94–5, 178
smallpox, 338
Smith, Adam, 166, 268
Smithsonian Institute, 42
snails: classification, 306; colour, 157; evolution, 241–2; fossil record, 213, 223, 240; freshwater, 104, 241; Krakatau, 281; marine, 213, 240; sexual and asexual forms, 104; shells, 243; struggle for existence, 64

snakes: colouring, 90; DNA, 150; Ireland, 255, 281; swimming, 261
Snider-Pellegrini, Antonio, 271
snow grouse, 64–5
Social Darwinism, 40, 354
Society for the Propagation of Horse Flesh as an Article of Food, 23
Somerville, Lord, 25
Somme, Second Battle (1918), 228, 229
South Africa: ecosystem, 256, 257; fynbos, 256, 257
South Downs, 226
South Pole, 272–3
Soviet Union: AIDS, 6; whaling, 73
soybeans, 36
sparrow: dusky seaside, 56; evolution, 242; sexual dominance, 90
sparrowhawks, 82–3, 89
species: breed and, 41; classification, 45–6; crosses, 128; definition, 57; dominant, 365; extinct, 232–3, 243–5; frontiers, 184, 195–6; geographical distribution, 370; modification of, 374–5; numbers of, 49; reproduction, 364–5; reversion to earlier characteristics, 115; sexual, 189; subspecies, 51; sudden appearance of groups of allied, 220–30; variations, 115; varieties or, 52, 55–7, 360, 364, 373–4; *see also* hybrid
Speke, John Hanning, 289
Spencer, George, 110
Spencer, Herbert, 354
sperm, 84, 87–8, 190, 247
sperm whale, 17
spiders, 223, 250
sponges, 224, 248
spores, dispersal, 263, 288
squash, cultivated, 25
squirrels, 78, 89
stag-beetles, 83–4
starfish, 223, 306, 324
Stephens Island wren, 102
sterility: hybrids, 128, 185–6, 188–94, 357, 372; insect societies, 176, 178–80; selection, 184
sticklebacks, 103
Stone of Scone, 274–5
Strachey, William, 291–2
stramenopiles, 313
stromatolites, 222–3
struggle for existence, 64–73, 356, 362, 379
sturgeon, 318

subspecies, 51
Suez Canal, 258
Sunda Strait, 349
sunfish, bluegill, 91–2
sunflower, 198
sunlight, 114, 115
swans, 192
Swift, Jonathan, 329
swifts, 63, 82
swim-bladder, 154
swordtail, 198
syphilis, 122, 151

tails, human, 326
talapoin monkeys, 15
tamarins, golden-lion, 159
Tanganyika, Lake, 289–91, 292
tapeworms, 148
Taung Child, 217, 345
Taxol, 36
taxonomy, 45–52
Teilhard de Chardin, Pierre, xxviii
temperature changes, 264–5
Tennyson, Alfred, Lord, 71, 280
termites, 166, 173, 177
Tertullian, 62
testosterone, 83, 85, 172
Tethys Sea, 18, 273
tetracycline, 96, 98, 99
Thames, River, 228, 264
thrushes, 182, 368
thymus, 142
Tibet: dung-beetle, 267
Tierra del Fuego, diet, 28
tigers, white, 39
Time magazine, 35
toads: midwife, 149; varieties, 274
tobacco, 114–15, 137, 200
tobacco budworm, 137
tobacco hornworm, 114–15
Toft, Mary, xxiv
tomatoes, 183–4, 199–202, 203, 204
tortoises: giant, 102, 350; swimming, 261
transitional varieties, 129–34, 153, 207, 241, 359, 368–9
transplants, 185
tree-frogs, 90, 269
tree-kangaroos, 106, 190
trees, 152, 286–7
trilobites, 224, 238, 274, 276

Trinidad, guppies, 80
Tristan da Cunha, blindness, 350–1
trumpet vine, 327
tsetse-flies, 122, 141
tuberculosis, 98
tubeworms, 309
Turing, Alan, 130
Turkana, 241, 347
turtles: Kemp's Ridley, 60; Thames, 264
Tutankhamen, 4, 32
Twain, Mark, 71
Type, Unity of, 155
types, succession of same, 249–50
typhus, 151

Ua Huku, 292
Uganda, elephants, 87
UK400 Club, 58
umbrella tree, 266
underground life, 310
Unenlagia, 221
ungulates, 20–1
United States: AIDS, 4, 6, 8; apples, 134–5;
    Army, 120; Center for Disease Control, 4;
    Corps of Army Engineers, 232; creationism,
    1–2; Department of Agriculture, 135, 200; dog
    breeding, 26–7, 33; Endangered Species Act,
    56; Fish and Wildlife Service, 56; Food and
    Drug Administration, 202; fruit-flies, 146–7;
    New World, 254–5
Unity of Type, 155
use and disuse, effects of, 111–14
Ussher, Archbishop James, 209

varieties: fertility, 128; mice, 52–3; or species, 52,
    55–7, 360, 364, 373–4; transitional, 129–34
Vavilov, Nikolai, 37
Velociraptor, 220
velvet worms, 223
vents, ocean, 309
vertebrates, 316–17
Vian, Boris, xxi
Victoria, Lake, 191, 289, 291
Victoria, Queen, 26, 308, 351
Virgil, 245
virgin birth, 247
Virginia: colony, 60, 61; fishery, 72
virus, copying, 329, *see also* HIV
volcanoes, 212
voles, red-backed, 65

Voltaire, 215, 237, 238
vultures: Andean, 144; types, 249

Wake Island rail, 292
wall-eye fish, 178
Wallace, Alfred Russel, xxiii, xxix, 283, 334, 349,
    350, 354
Wallace's Line, 283
warblers, 163, 175–6, 192
wasps: colouring, 90; family gender, 175, 176;
    relationship with fig trees, 163; sterility, 176
water buffalo, 35
water-skaters, 140–1
waterweeds, 313
wattles, 92, 257
Waugh, Evelyn, 325
weaver-bird, 163–4
Webb, Sidney, 354
Wedgwood, Josiah, xxii
weevils: black long-snouted, 66; in amber, 244
Wells, H.G., 328
Wen Wang, Emperor, 38
whales: baleen, 319; beluga, 19; blue, 17, 258, 268;
    classification, 305; discoveries, 50; DNA,
    19–21, 145; drowned, 309–10; evolution, 17–19;
    extinction, 103; genes, 319; humpback, 19, 37,
    64, 159; kinship, 21; Mediterranean, 258;
    origin, 2–3; retroviruses, 20; sperm, 17; teeth,
    328; whaling, 72–3; white, 19
wheat, 26, 70, 199
whelks, distribution, 269
Whitby, ammonites, 208
White, Gilbert, 63
white smokers, 309
Whittington, Dick, 76
Wilde, Oscar, 202
wings, 131–4, 328–9
wolves: Arctic, 27; carnivore role, 249; dog
    kinship, 25, 29, 30, 40–2; relationship with
    humans, 28; survival, 25
woodpeckers, 137
worms: brain absorption, 326; classification,
    306–7; fossil record, 225, 247–8; nematode,
    49–50; structure, 315–16; velvet, 223
Wrangel Island, mammoths, 329
wrens, 102, 182, 329

Yangtze fossil beds, 224–5
yeast, 304
yellowhammers, 61

Yellowstone National Park, 249, 308
yew trees, 267
Ypres, 228
Yucatán crater, 237, 238

zebra-finches, 91

zebra mussel, 288–9
zebras, 23
zebu, 35
zinc, 105, 106
Zoological Society of London, 37
zoos, 37–40